AF182354

Design of Pulse Oximeters

Other books in the series

The Physics and Radiobiology of Fast Neutron Beams
D K Bewley

Biomedical Magnetic Resonance Technology
C-N Chen and D I Hoult

Rehabilitation Engineering Applied to Mobility and Manipulation
R A Cooper

Linear Accelerators for Radiation Therapy, second edition
D Greene and P C Williams

Health Effects of Exposure to Low-Level Ionizing Radiation
W R Hendee and F M Edwards

Introductory Medical Statistics, third edition
R F Mould

Radiation Protection in Hospitals
R F Mould

RPL Dosimetry—Radiophotoluminescence in Health Physics
J A Perry

Physics of Heart and Circulation
J Strackee and N Westerhof

The Physics of Medical Imaging
S Webb

The Physics of Three-Dimensional Radiation Therapy: Conformal Radiotherapy, Radiosurgery and Treatment Planning
S Webb

The Physics of Conformal Radiotherapy: Advances in Technology
S Webb

Other titles of interest

Prevention of Pressure Sores: Engineering and Clinical Aspects
J G Webster

Medical Science Series

Design of Pulse Oximeters

Edited by

J G Webster

Department of Electrical and Computer Engineering
University of Wisconsin-Madison

Institute of Physics Publishing
Bristol and Philadelphia

© IOP Publishing Ltd 1997

All rights reserved. No part of this publication may be reproduced, stored in a retrieval system or transmitted in any form or by any means, electronic, mechanical, photocopying, recording or otherwise, without the prior permission of the publisher. Multiple copying is permitted in accordance with the terms of licences issued by the Copyright Licensing Agency under the terms of its agreement with the Committee of Vice-Chancellors and Principals.

British Library Cataloguing-in-Publication Data

A catalogue record for this book is available from the British Library.

ISBN 0 7503 0467 7

Library of Congress Cataloging-in-Publication Data are available

The Editor has attempted to trace the copyright holder of all the figures and tables reproduced in this publication and apologizes to copyright holders if permission to publish in this form has not been obtained.

Series Editors:

R F Mould, Croydon, UK

C G Orton, Karamanos Cancer Institute, Detroit, USA

J A E Spaan, University of Amsterdam, The Netherlands

J G Webster, University of Wisconsin-Madison, USA

Published by Institute of Physics Publishing, wholly owned by The Institute of Physics, London

Institute of Physics Publishing, Dirac House, Temple Back, Bristol BS1 6BE, UK

US Editorial Office: Institute of Physics Publishing, The Public Ledger Building, Suite 1035, 150 South Independence Mall West, Philadelphia, PA 19106, USA

Prepared by the Editor using Microsoft Word 6

Printed in Great Britain by J W Arrowsmith Ltd, Bristol

The Medical Science Series is the official book series of the International Federation for Medical and Biological Engineering (IFMBE) and the International Organization for Medical Physics (IOMP).

IFMBE

The IFMBE was established in 1959 to provide medical and biological engineering with an international presence. The Federation has a long history of encouraging and promoting international cooperation and collaboration in the use of technology for improving the health and life quality of man.

The IFMBE is an organization that is mostly an affiliation of national societies. Transnational organizations can also obtain membership. At present there are 42 national members, and one transnational member with a total membership in excess of 15 000. An observer category is provided to give personal status to groups or organizations considering formal affiliation.

Objectives

- To reflect the interests and initiatives of the affiliated organizations.
- To generate and disseminate information of interest to the medical and biological engineering community and international organizations.
- To provide an international forum for the exchange of ideas and concepts.
- To encourage and foster research and application of medical and biological engineering knowledge and techniques in support of life quality and cost-effective health care.
- To stimulate international cooperation and collaboration on medical and biological engineering matters.
- To encourage educational programmes which develop scientific and technical expertise in medical and biological engineering.

Activities

The IFMBE has published the journal *Medical and Biological Engineering and Computing* for over 34 years. A new journal *Cellular Engineering* was established in 1996 in order to stimulate this emerging field in biomedical engineering. In *IFMBE News* members are kept informed of the developments in the Federation. *Clinical Engineering Update* is a publication of our division of Clinical Engineering. The Federation also has a division for Technology Assessment in Health Care.

Every three years, the IFMBE holds a World Congress on Medical Physics and Biomedical Engineering, organized in cooperation with the IOMP and the IUPESM. In addition, annual, milestone, regional conferences are organized in different regions of the world, such as the Asia Pacific, Baltic, Mediterranean, African and South American regions.

The administrative council of the IFMBE meets once or twice a year and is the steering body for the IFMBE. The council is subject to the rulings of the General Assembly which meets every three years.

For further information on the activities of the IFMBE, please contact Jos A E Spaan, Professor of Medical Physics, Academic Medical Centre, University of Amsterdam, PO Box 22660, Meibergdreef 9, 1105 AZ, Amsterdam, The Netherlands. Tel: 31 (0) 20 566 5200. Fax: 31 (0) 20 6917233. Email: IFMBE@amc.uva.nl. WWW: http://vub.vub.ac.be/~ifmbe.

IOMP

The IOMP was founded in 1963. The membership includes 64 national societies, two international organizations and 12 000 individuals. Membership of IOMP consists of individual members of the Adhering National Organizations. Two other forms of membership are available, namely Affiliated Regional Organization and Corporate Members. The IOMP is administered by a Council, which consists of delegates from each of the Adhering National Organizations; regular meetings of Council are held every three years at the International Conference on Medical Physics (ICMP). The Officers of the Council are the President, the Vice-President and the Secretary-General. IOMP committees include: developing countries, education and training; nominating; and publications.

Objectives

- To organize international cooperation in medical physics in all its aspects, especially in developing countries.
- To encourage and advise on the formation of national organizations of medical physics in those countries which lack such organizations.

Activities

Official publications of the IOMP are *Physiological Measurement, Physics in Medicine and Biology* and the *Medical Science Series*, all published by Institute of Physics Publishing . The IOMP publishes a bulletin *Medical Physics World* twice a year.

Two Council meetings and one General Assembly are held every three years at the ICMP. The most recent ICMPs were held in Kyoto, Japan (1991) and Rio de Janeiro, Brazil (1994). Future conferences are scheduled for Nice, France (1997) and Chicago, USA (2000). These conferences are normally held in collaboration with the IFMBE to form the World Congress on Medical Physics and Biomedical Engineering. The IOMP also sponsors occasional international conferences., workshops and courses.

For further information contact: Hans Svensson, PhD, DSc, Professor, Radiation Physics Department, University Hospital, 90185 Umeå, Sweden, Tel: (46) 90 785 3891, Fax: (46) 90 785 1588, email: Hans.Svensson@radfys.umu.se.

CONTENTS

PREFACE

Pulse oximetry was introduced in 1983 as a noninvasive method for monitoring the arterial oxygen saturation of a patient's blood. Recognized worldwide as the standard of care in anesthesiology, it is widely used in intensive care, operating rooms, emergency, patient transport, general wards, birth and delivery, neonatal care, sleep laboratories, home care and in veterinary medicine. It provides early information on problems in the delivery of oxygen to the tissue. Those problems may arise because of improper gas mixtures, blocked hoses or airways, inadequate ventilation, diffusion, or circulation, etc. More than 35 companies manufacture and distribute the more than 300 000 pulse oximeters presently in use in the USA.

This book emphasizes the design of pulse oximeters. It details both the hardware and software required to fabricate a pulse oximeter as well as the equations, methods, and software required for effective functioning. Additionally, it details the testing methods and the resulting accuracy. The book should be of interest to biomedical engineers, medical physicists, and health care providers who want to know the technical workings of their measuring instruments.

Chapter 1 reviews the methods of transport of oxygen to the tissue by ventilation, perfusion to the blood, binding to hemoglobin in the red blood cells, and transport through the blood circulation. Chapter 2 describes the problems and diseases that can occur in oxygen transport, which motivate us to measure oxygenation. In chapter 3, we review the many ways oxygenation has been measured in the past, the CO-oximeter used as the gold standard, and provide an introduction to the pulse oximeter.

Chapter 4 begins with Beer's law for the absorption of light by hemoglobin and oxyhemoglobin, and develops the equations required for converting measured light transmission through the tissue to display the hemoglobin oxygen saturation. The light-emitting diodes, which alternately emit red light at 660 nm and infrared light at 940 nm and require precise wavelength control, are described in chapter 5. Chapter 6 covers the variety of light sensors, with emphasis on the single photodiode typically used.

Chapter 7 details the design of reusable and disposable probes and their flexible cables. The probes can transmit light through either the finger or ear, or use reflected light from the scalp or other skin surface. Chapter 8 covers the hardware, with block diagrams showing how red and infrared signals are amplified to yield the ratio of pulse-added red absorbance to the pulse-added infrared absorbance. These signals are used to control light-emitting diode levels and the ratio is used to calculate oxygen saturation. The flow charts and algorithms to perform oxygen saturation calculations are given in chapter 9, with worked out examples. Synchronization with the electrocardiogram improves accuracy during patient movement.

Chapter 10 describes ways to test performance of pulse oximeters: the technician's finger, electronic simulators, *in vitro* test systems, and optoelectronic simulators. In chapter 11, we find the resulting accuracies and descriptions of the inaccuracies caused by alternative forms of hemoglobin, optical and electrical interference, colored nail polish, etc. Chapter 12 describes the interface between the pulse oximeter, the operator, and the external world. Chapter 13 covers the many applications for pulse oximetry in intensive care, operating rooms, emergency, patient transport, general wards, birth and delivery, neonatal care, sleep laboratories, home care, and in veterinary medicine.

A glossary provides definitions of terms from both the medical and the engineering world. We also provide instructional objectives as a means of provoking further thought toward learning the information. We gleaned much of the design information from operator's manuals and from patents; periodical literature provided more general information. Rather than giving an exhaustive list of references, we have included review articles and books that can serve as an entry into further study. All contributors are from the Department of Electrical and Computer Engineering at the University of Wisconsin, Madison, WI, USA, and worked as a team to write this book. We would welcome suggestions for improvement of subsequent printings and editions.

John G. Webster
Department of Electrical and Computer Engineering
University of Wisconsin-Madison
Madison WI, USA
August 1997

CHAPTER 1

NORMAL OXYGEN TRANSPORT

Susanne A Clark

Oxygen is vital to the functioning of each cell in the human body. In the absence of oxygen for a prolonged amount of time, cells will die. Thus, oxygen delivery to cells is an important indicator of a patient's health.

Several methods have been developed to analyze oxygen delivery. Pulse oximetry is a common, noninvasive method used in clinical environments. This book discusses pulse oximetry, from applications to signal processing. Before continuing, it is essential to understand normal oxygen transport, which is the subject of this chapter.

Oxygen delivery to cells requires the use of the respiratory system as well as the circulatory system. Ventilation is the initial step, moving air into and out of the lungs. Within the lungs, gas exchange occurs. Oxygen is diffused into the blood, while carbon dioxide, a byproduct of cellular respiration, diffuses into the lungs. The oxygenated blood circulates around the body until it reaches oxygen depleted areas, where oxygen is diffused to cells, and carbon dioxide is transferred to the blood returning to the lungs. The ventilatory process is controlled by neurons in the brain stem. The circulatory system also can modulate cardiac output to effect the oxygen delivery.

1.1 VENTILATORY CONTROL

Ventilation is the involuntary, rhythmic process of moving air in and out of the lungs. This process is controlled by respiratory neurons in the brain stem. The respiratory neurons excite motor neurons, which in turn cause the movement of respiratory muscles. The output of the respiratory neurons is modulated by *chemoreceptors* and *mechanoreceptors*.

1.1.1 Neural control

The respiratory neurons in the brain stem are responsible for the pattern generation in normal breathing. The rate and depth of *ventilation* are modulated by these neurons. The respiratory neurons excite motor neurons in the spinal cord. The excitation of the motor neurons causes the contraction of the diaphragm, pectoral muscles, and intercostal muscles. All of these muscles combine efforts pulling the ribcage up and out, expanding the lungs, causing inspiration. The activity of respiratory neurons is thought to occur spontaneously, with occasional inhibition allowing the respiratory muscles to relax. This causes the rib cage to contract which yields expiration.

1.1.2 Respiratory feedback

The brain stem receives feedback from many mechanical and chemical receptors. The input from these neurons is analyzed by the respiratory neurons to determine the appropriate rate and depth of ventilation. Mechanoreceptors give feedback related to mechanical aspects of breathing. For example, stretch receptors are mechanoreceptors that provide feedback on the expansion of the lung and chest during both inspiration and expiration. An inflation index is the level of feedback provided that causes inhibition of inspiration, preventing overinflation of the lungs. A deflation index serves a similar purpose in expiration, hindering the collapse of the lungs.

Chemoreceptors provide information on the level of carbon dioxide, oxygen, and hydrogen ions in the blood. Chemoreceptors are located in the carotid arteries, as the oxygenated blood is being sent to the brain, and in the aorta, shortly after the oxygenated blood is being pumped from the heart to the body. Oxygen levels under normal conditions are high in the systemic arteries, and carbon dioxide and hydrogen levels are low.

The brain stem must process all of the information it receives and no single factor controls ventilation. Under normal breathing conditions, the brain stem is most sensitive to the levels of carbon dioxide and hydrogen. The oxygen concentrations are only important when the level is extremely low. Consider an extremely high level of carbon dioxide present in the blood, such as would occur during maximal exercise. However, stretch receptors indicate that the lung and chest are at maximal expansion, meaning the inflation index has been reached. Thus, the rate of breathing increases to compensate without a proportional increase in chest and lung expansion.

An unusual feature of ventilation is that breathing can be brought under voluntary control to some extent. However, it is not possible to commit suicide by refusing to breathe. Once the individual loses consciousness, the input from chemoreceptors will cause ventilation to be restored.

1.2 VENTILATORY MECHANICS

Ventilatory mechanics are based on the principle of air flow from areas of high pressure to areas of lower pressure. The contraction of the intercostal muscles, pectoral muscles, and the diaphragm causes the thoracic cavity to expand, decreasing the pressure in the thoracic cavity. The atmospheric pressure is higher than the pressure inside the lungs, causing air to flow into the lungs, which is termed inspiration. The relaxation of the intercostal muscles and the diaphragm causes the volume of the lungs to decrease, increasing the pressure in the thoracic cavity. As the pressure in the lungs increases reaching levels above the atmospheric pressure, air flows out of the lungs, which is referred to as expiration.

1.2.1 Inspiration

As discussed previously, the brain stem excites motor neurons in the spinal cord, which, in turn, causes the contraction of the diaphragm, the pectoral muscles, and intercostal muscles, located between the ribs. The contraction of the diaphragm causes the flattening and lengthening of the thoracic cavity. The intercostal muscles and pectoral muscles pull the ribcage up and out. Both of these sets of muscles work to expand the lungs. This means that pressure will be reduced within the lungs, since the air present will have a greater volume to expand in. This will create a pressure differential between the air outside the body and the air inside the body. Thus, air flows into the body (see figure 1.1(*a*)).

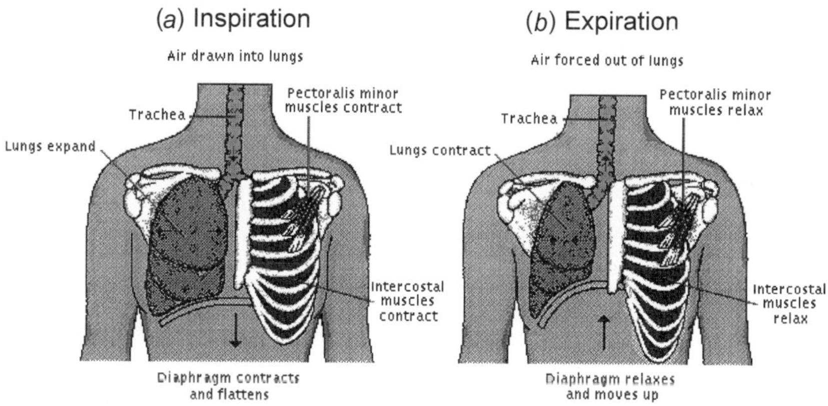

Figure 1.1 During inspiration, (*a*), the diaphragm, intercostal muscles and pectoralis minor muscles contract, causing the lungs to expand and air to enter the lungs. As the diaphragm, intercostal muscles and pectoralis minor relax, the lungs contract, causing air to leave the lungs (*b*), which is referred to as expiration (from Microsoft Encarta).

Air travels through the nasal cavity. Cilia are microscopic hairs within the nasal cavity that act to eliminate pollutants from entering the respiratory tract. Air and food both go through the *pharynx*. When food is swallowed, the epiglottis (part of the *larynx*), pharynx, and mouth cavity work together to shut off the opening to the trachea to avoid the entry of food particles into the lungs.

The larynx is commonly referred to as the voice box. Besides assisting with separation of food particles from air, the larynx contains the cricoid cartilage which reinforces the airway and assists in keeping it open. The larynx also contains the vocal cords. As air vibrates over the vocal cords, a sound is produced. The variation in elasticity and tension of the vocal cords determines the pitch of the sound.

The trachea is composed of ribbed cartilage which extends 10 cm to the bronchi. The trachea also contain cilia which act to filter out further pollutants. Two *bronchi* provide a path to each *lung* (see figure 1.2).

Each bronchus divides into even narrower *bronchioles*. Each bronchiole has five or more *alveolar ducts* at the end, which, in turn, end in *alveolar sacs*. Each alveolar sac contain several *alveoli* (see figure 1.3). Alveoli are the site of gas exchange.

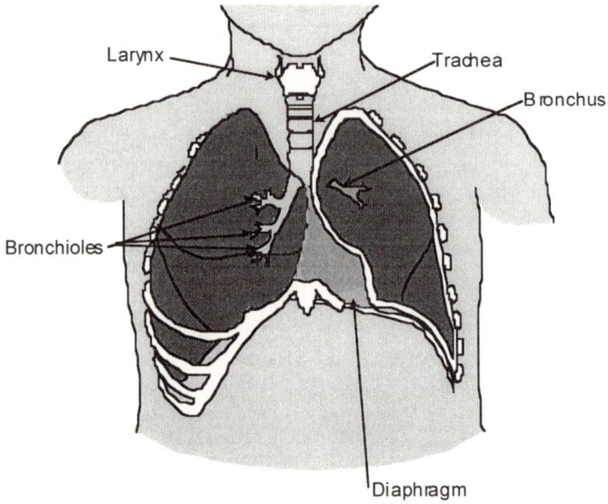

Figure 1.2 Air travels through the nasal cavity, into the pharynx, trachea, bronchi, and finally the lungs. The bronchi, bronchioles, alveolar ducts and alveoli compose the pulmonary tree with its branch like system (adapted from Corel Corporation).

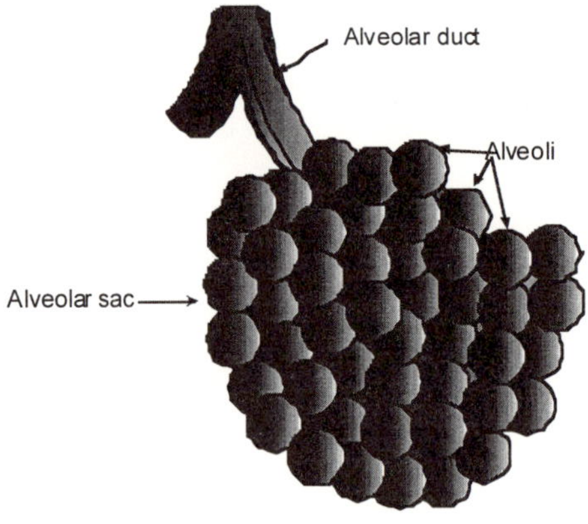

Figure 1.3 Ten or more alveoli are in one alveolar sac (adapted from Corel Corporation).

1.2.2 Expiration

Neurons in the brain stem cyclically inhibit the motor neurons in the spinal cord that cause muscle contraction in the diaphragm, the pectoral muscles, and intercostal muscles. The muscles then relax, causing the rib cage to contract, decreasing the amount of air space. This causes air to flow out of the lungs when the pressure inside the lungs is greater than the pressure outside the lungs (see figure 1.1(*b*)). Usually only 10% of the total lung volume is exchanged in normal breathing. With deeper, more rapid breathing, the turbulence of the air flow increases, causing greater resistance to airflow.

1.3 DIFFUSION TO BLOOD

The process of ventilation provides a continuous supply of fresh air in the lungs. After oxygenated blood has been circulated through the body, it is brought back to the lungs through arterial capillaries to exchange gases, receiving oxygen and ridding itself of carbon dioxide. Blood is reoxygenated and is then recirculated through the body.

Gas exchange occurs through the process of diffusion. *Diffusion* is the net movement of particles from an area of higher *partial pressure* to a region of lower partial pressure through a process of random motion. The actual gas exchange to the blood takes place through the process of diffusion in the *alveoli.*

1.3.1 The alveoli

The alveoli are surrounded by large *pulmonary* capillary beds. Since diffusion can only occur over a distance of 1 mm, the gas exchange takes between the two cells between the capillary and the alveolus, a distance of only 0.5 μm. The 600 million alveoli each adult has provide 70 m^2 of surface area for gas exchange (Curtis and Barnes 1989).

1.3.2 Gas exchange

Air in the alveoli has a higher partial pressure of oxygen and a lower partial pressure of carbon dioxide than the aortic blood. The pressure gradient causes diffusion to occur. The net movement of carbon dioxide will be towards the alveoli and the net movement of oxygen towards the blood (see figure 1.4). The blood then returns to the heart via a pulmonary venule to be pumped out to the rest of the body. Other gases may diffuse as a result of the partial pressure gradient between the air in the alveoli and the pulmonary arterial blood.

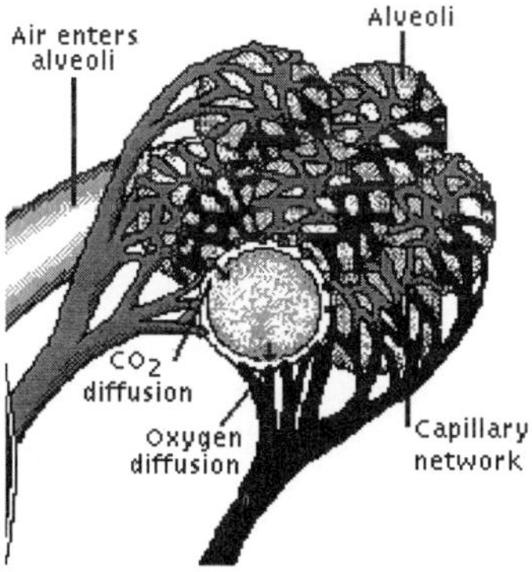

Figure 1.4 The capillaries surround the alveoli, providing the close proximity necessary for diffusion. Carbon dioxide diffuses from the capillary into the alveoli and oxygen diffuses into the blood (from Microsoft Encarta).

The partial pressure gradient of arterial (a) versus alveolar (A) pressure is affected by the concentration of carbon dioxide and water in the alveoli. The alveolar partial pressure of oxygen is

$$P_AO_2 = (P_{atm} - PH_2O)F_iO_2 - P_aCO_2/0.8 \quad (1.1)$$

where atmospheric pressure P_{atm} is typically 760 mm Hg (101 kPa), the water vapor pressure PH_2O is 47 mmHg (6.3 kPa) at 37 °C, the fraction of inspired O_2, F_iO_2 is 0.21 with room air, P_aCO_2 is the arterial carbon dioxide partial pressure, and 0.8 is the normal *respiratory quotient*. The respiratory quotient is the ratio of volume of CO_2 produced per volume of O_2 consumed.

The rate of gas movement is determined by the pressure gradient, temperature and path length over which the gas exchange occurs. This is defined as

$$\dot{V} = \frac{T\Delta P}{L} \times D \quad (1.2)$$

where $\dot{V}$ is the volume rate of gas exchange, ΔP is the pressure gradient, T is the absolute temperature, L is the path length, and D is a diffusion coefficient for a specific material.

In a living adult, it is not realistic to measure the path length or surface area. Thus, a diffusing capacity of the lung is defined to provide a quantitative measure of the effectiveness of *respiration*. This is defined as

$$D_L = \frac{\dot{V}}{\Delta P} \quad (1.3)$$

where D_L is the diffusing capacity of the lung (Ruch and Patton 1965).

1.4 BIND TO HEMOGLOBIN

Gases are not particularly soluble in blood, which is composed mostly of water. Thus, for effective oxygen transport, a secondary method of transport is required. The compound hemoglobin provides a binding mechanism that allows oxygen to be transported through the blood. *Hemoglobin* plays an essential role in transporting the necessary amount of oxygen to the body. For the same amount of plasma, 65 times more oxygen can be transported with hemoglobin than would be possible without hemoglobin.

1.4.1 Characteristics of hemoglobin

Hemoglobin is a respiratory pigment contained within red blood cells. One red blood cell contains approximately 265 million molecules of hemoglobin (Curtis and Barnes, 1989). Hemoglobin is composed of *heme* units, which are molecules containing iron, and *globin* units, *polypeptide* chains. One hemoglobin molecule contains four heme and four globin units. Each hemo and globin unit can carry one molecule of oxygen. Thus, one hemoglobin molecule can carry four molecules of oxygen (see figure 1.5).

As respiratory pigment, hemoglobin changes color when oxygenated. An oxygenated hemoglobin molecule is bright red, while a deoxygenated hemoglobin molecule, a hemoglobin molecule without oxygen, is dark red. This color change is used in the application of pulse oximetry to measure hemoglobin oxygen saturation.

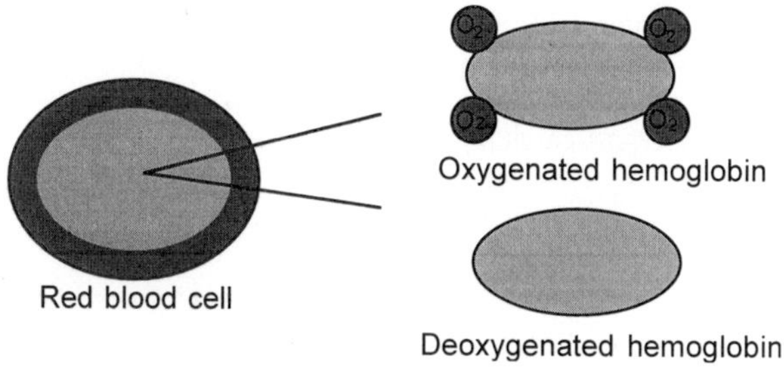

Figure 1.5 Hemoglobin molecules are contained within red blood cells. Each red blood cell contains approximately 265 million molecules of hemoglobin.

After a completely deoxygenated hemoglobin molecule combines with one oxygen molecule, it has a greater affinity for the second oxygen molecule. This is true for each additional oxygen molecule. The reverse process is also true. After the first oxygen molecule is released from the hemoglobin molecule, it is more likely to release the second oxygen molecule. Therefore, the oxyhemoglobin dissociation curve, which relates to the partial pressure of oxygen in the blood, is not a straight line, but a sigmoid.

1.4.2 Oxyhemoglobin dissociation curves

The oxyhemoglobin dissociation curve is the relationship between the partial pressure of oxygen in the blood and the percentage of oxygen bound to hemoglobin compared to the maximum (see figure 1.6). Factors such as decreasing carbon dioxide concentration, increasing pH, and decreasing temperature will shift the curve toward the left. A left-shifted curve implies that the hemoglobin molecules will be more saturated at a lower partial pressure of oxygen.

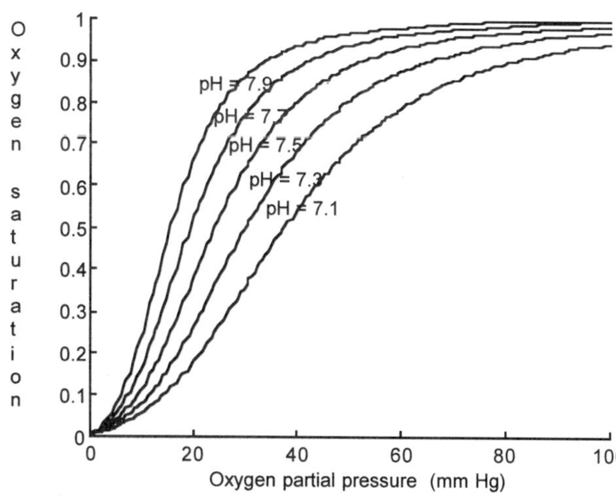

Figure 1.6 Increasing pH causes the oxyhemoglobin dissociation curve to shift to the left at a constant temperature of 37 °C.

A fetus has a oxyhemoglobin dissociation curve that is to the left of the mother's. This means that the fetus has a greater affinity for oxygen and will take oxygen from the mother's blood to meet its own need.

The volume of oxygen carried by hemoglobin per 100 mL of blood can be defined as follows:

$$C_{HbO_2} = 1.37 \times \mathrm{Hb} \times S_aO_2 \tag{1.4}$$

where C_{HbO_2} is the volume of oxygen carried by hemoglobin per unit of 100 mL of blood and is typically around 19 mL O_2/100 mL blood, 1.37 is the number of mL of oxygen bound to 1 g of fully saturated hemoglobin, Hb is the weight of hemoglobin, typically around 14 g Hb/100 mL blood, and S_aO_2 is the percentage of saturation of hemoglobin in arterial blood (Payne and Severinghaus 1986).

1.5 DISSOLVED IN PLASMA

Most of the oxygen transported by the body is bound to hemoglobin, but some oxygen is also dissolved in plasma. The total oxygen content of the blood is the sum of the bound oxygen and the dissolved oxygen.

Hemoglobin increases the amount of oxygen transported to the body by 65 times the amount carried by a specified volume of blood, but some oxygen is still carried dissolved in plasma. The volume of dissolved oxygen per 100 mL of blood is defined as

$$C_DO_2 = 0.003P_aO_2 \tag{1.5}$$

where C_DO_2 is the volume of dissolved oxygen in blood, 0.003 is the solubility of oxygen in blood as the percent by volume per mmHg, and P_aO_2 is the partial pressure of oxygen in the arteries. C_DO_2 is typically around 0.3 mL O_2/100 mL blood. It is significantly smaller than the bound oxygen content of blood, typically 19 mL O_2/100 mL blood. The oxygen content of the blood is the sum of the oxygen bound to hemoglobin and the oxygen dissolved in the plasma:

$$C_AO_2 = C_{HbO_2} + C_DO_2. \tag{1.6}$$

1.6 CIRCULATION

Once oxygen has been diffused to the blood, it is returned the heart. The *circulatory system* serves to transport oxygenated blood to the cells in the body. The heart is the primary pumping mechanism for transporting blood through the body.

1.6.1 The heart

Blood is pumped through body by the heart. The contraction of the heart is controled by a series of electrical impulses, originating from the sinoatrial node (SA node) and travels to the atrioventricular node (AV node), causing the polarization and depolarization of the muscle fibers of the heart. These electrical impulses can be recorded as the electrocardiogram (see figure 1.7).

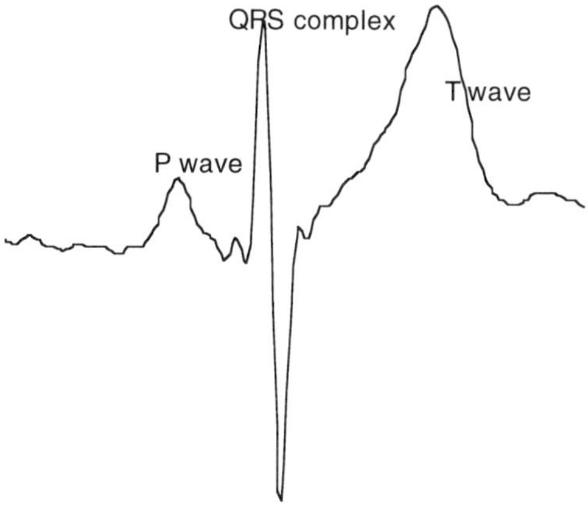

Figure 1.7 The P wave is caused by the depolarization of the atrial fibers just prior to contraction. The QRS complex is caused by the depolarization of the ventricles, causing the contraction of the ventricles. The T wave is caused by the polarization of the ventricles as the muscles relax. The polarization of the atrial fibers occurs simultaneously with the QRS complex and is obscured by the contraction of the larger muscle fibers in the ventricles. The peak of the QRS complex is the R wave. The R–R interval between consecutive heartbeats is used to calculate the heart rate.

1.6.2 Pulmonary circulation

The heart serves as the pumping mechanism for the blood. Blood that is oxygen depleted is pumped from the right *ventricle* of the heart to the lungs. The pulmonary arteries branch into smaller arterioles and eventually into arterial capillaries, which have a thickness of only one cell. This is where gas exchange occurs between the alveoli and the capillaries and blood is reoxygenated. Blood is then returned via pulmonary ventral capillaries to larger venules and eventually pulmonary veins. The pulmonary veins return blood to the left *atrium* of the heart.

1.6.3 Systemic circulation

Reoxygenated blood is returned to the heart in the left atrium. It is then pumped from the left ventricle via the *systemic* arteries to the body. Blood pressure within the arteries varies throughout a single heartbeat, reaching a maximum at systole, caused by the contraction of the ventricles, and a low at diastole, after the ventricles have relaxed. The systemic arteries also branch into smaller arterioles and even smaller capillaries. Oxygen is then exchanged with the tissues of the body. The blood, depleted of oxygen, is then returned via venal capillaries, venules, and veins to the right atrium of the heart where it is again reoxygenated (see figure 1.8).

1.6.4 Cardiac output

Mechanical and chemical stimuli are processed in the brain and provide feedback to the SA node and the AV node which control heart rate and stroke volume. The *cardiac output* (CO) of the heart is the product of the stroke volume (SV) and the heart rate (HR). The *cardiac index* (CI) is the cardiac output normalized by *body surface area* (BSA). A typical CI is in the range of 3 to 3.4 L/(min m^2). Normalization of the cardiac output allows comparisons related to circulation of people of varying sizes.

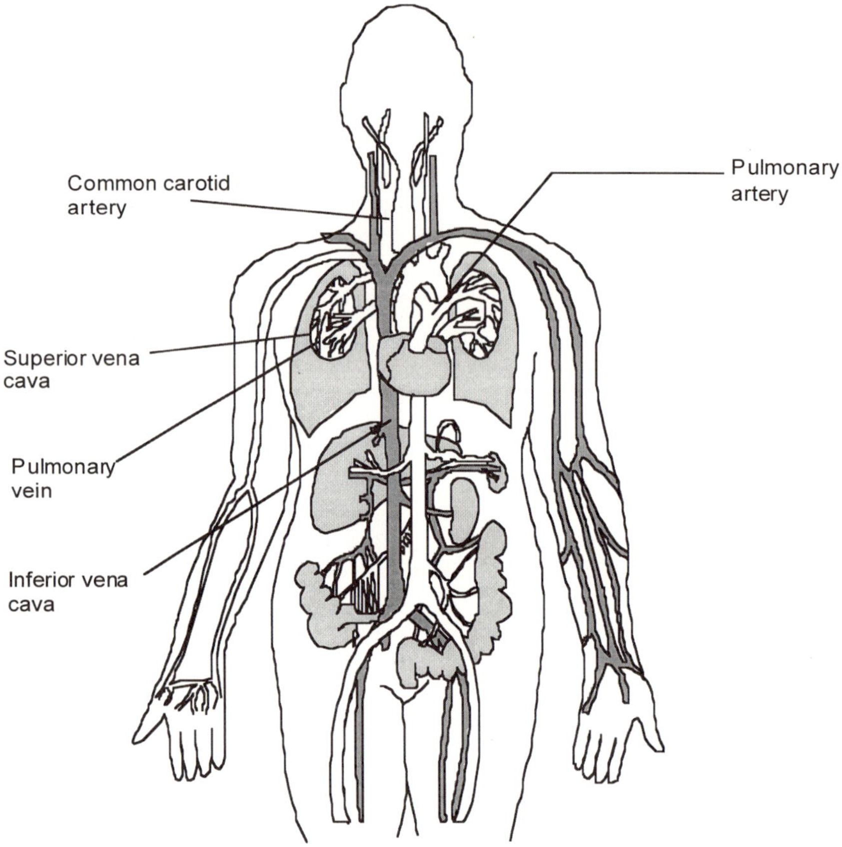

Figure 1.8 The right atrium receives blood from two veins, the superior vena cava and the inferior vena cava. The right ventricle pumps blood through the pulmonary artery, which sends the blood to the lungs to be oxygenated. The oxygenated blood returns to the heart via the pulmonary veins, where it pumped by the left ventricle to be distributed to the rest of the body (adapted from Corel Corporation).

1.7 DIFFUSION TO TISSUE

Diffusion occurs over a distance of about 1 mm. Thus, once blood is oxygenated, although it may pass through oxygen depleted tissue, oxygen does not diffuse until it reaches the capillaries with one cell thickness in the wall. Oxygen diffuses into the *interstitial fluid* and into the cells.

1.7.1 Diffusion into interstitial fluid and cell

Once blood reaches the systemic capillaries, the surrounding tissue usually has a lower partial pressure of oxygen than that of the blood. Oxygen diffuses into the surrounding tissue. When the tissue has a higher metabolic rate, the difference in partial pressure is greater, and more oxygen is released using the steep part of the oxyhemoglobin dissociation curve. While oxygen diffuses into the interstitial fluid, carbon dioxide diffuses into the blood. Once the oxygen is near the cell, it diffuses through the cell membrane.

1.7.2 Oxygen delivered

The oxygen delivery index (D_IO_2) is defined as

$$D_IO_2 = C_aO_2 \times \mathrm{CI} \times 10 \tag{1.7}$$

which is typically 550 to 650 mL/(min m^2). This is a measure of the amount of oxygen available to tissue. The oxygen consumption is a measure of the oxygen diffused into the tissue. It is defined as

$$\mathrm{CI} \times (C_aO_2 - C_vO_2) \tag{1.8}$$

where C_vO_2 is the oxygen content of the venous blood. A normal value for oxygen delivery is 115 to 165 mL/(min m^2). This means that not all of the available oxygen diffuses into the tissue (Payne and Severinghaus 1986).

1.7.3 Myoglobin

Myoglobin is a respiratory pigment found in muscles, which is responsible for the reddish brown color of the muscle cells. It has a greater affinity for oxygen than hemoglobin, its oxyhemoglobin dissociation curve is left-shifted, and will not release oxygen under the same conditions as hemoglobin in the blood. Only when the partial pressure of oxygen in the surrounding tissue is below 20 mmHg, such as in exercise, does myoglobin release its stored oxygen. Thus, myoglobin reduces the need for oxygen delivery to muscle tissue beds under extreme conditions, but can only supply a limited amount of oxygen for a short period of time (Hole 1981).

1.8 USE IN CELL

The purpose of respiration is to bring oxygen to cells for cellular respiration. The cells then use oxygen to in turn generate energy. Although a cell may survive for a short time without oxygen, producing energy through anaerobic methods, each individual cell must have oxygen.

Cellular respiration involves the breakdown of molecules, glucose, and releasing energy from them. This process involves oxidation and reduction chemical reactions. *Oxidation* is the loss of an electron, releasing energy, and *reduction* is the gain of an electron. Oxygen atoms serve to attract electrons. Oxygen is needed by the cell to oxidize glucose to release energy. The simplified equation for the complex chemical reactions, involving the *Kreb's Cycle*, taking place is

$$\text{glucose} + \text{oxygen} \rightarrow \text{carbon dioxide} + \text{water} + \text{energy} \qquad (1.9)$$

where carbon dioxide and water are byproducts of the chemical reaction. Energy is released in the form of *ATP*, a source of cellular energy used in various metabolic processes, and heat, which is lost.

REFERENCES

Curtis H and Barnes N S 1989 *Biology*. 5th edn (New York: Worth)
Hole J W Jr 1981 *Human Anatomy and Physiology* 2nd edn (Iowa: Brown)
Payne J P and Severinghaus J W (eds) 1986 *Pulse Oximetry* (New York: Springer)
Ruch T C and Patton H D (eds) 1965 *Physiology and Biophysics* 19th edn (Philadelphia, PA: Saunders)

INSTRUCTIONAL OBJECTIVES

1.1 Describe how the body accommodates for the increased demand for oxygen during exercise.
1.2 Describe how the respiratory system provides for gas exchange.
1.3 Describe the cardiovascular system and its role in transporting oxygen.
1.4 Explain the difference between oxygen content, oxygen saturation, and partial pressure of oxygen.
1.5 Describe the oxyhemoglobin dissociation curve, and factors which can shift the curve.
1.6 Describe the process of diffusion and its role in respiration.
1.7 Describe the neurological control of ventilation.
1.8 Explain why hemoglobin is required for oxygen transport.
1.9 Given alveolar gas concentration, calculate P_aO_2.
1.10 Given P_aO_2, S_aO_2, Hb, calculate C_aO_2.
1.11 Write the chemical equation for the use of oxygen in the cell.
1.12 Describe the muscles used for ventilation.
1.13 Describe the air flow resistance between the alveoli and the mouth.

CHAPTER 2

MOTIVATION OF PULSE OXIMETRY

Daniel J Sebald

Pulse oximeters have been commercially available for a little more than the last decade and have seen a tremendous growth in popularity becoming a quasi-standard, if not standard, monitoring device in hospital critical care units and surgical theaters. The instrument transcutaneously estimates oxygen saturation of arterial blood and provides vital information about the cardiorespiratory function of the patient. Pulse oximetry provides an empirical measure of arterial saturation. However, with state-of-the-art instrumentation and proper initial calibration, the correlation between the pulse oximeter measurement, S_pO_2, and arterial blood's actual oxygen saturation, S_aO_2, is adequate—generally less than 3% discrepancy provided S_aO_2 is above 70% (Severinghaus and Kelleher 1992)—for medical applications where detecting hypoxemia is essential. Quick acceptance of pulse oximetry as a monitoring device for surgery, recovery, critical care and research has shown that for determining hypoxemia any reasonably small loss in accuracy that may be attributed to measuring arterial oxygen saturation transcutaneously is outweighed by the advantages of noninvasiveness and continuous, immediate availability of data. In applications where accuracy is paramount, such as in detecting hyperoxia, the use of pulse oximetry is not so clear and remains to be decided in the medical community. However, mounting evidence suggests that the pulse oximeter is not very useful in these situations. Nonetheless, the importance of detecting hypoxemia, where pulse oximetry is best suited, is so great that the instrument plays a critical role in medicine despite its limitations.

2.1 PULSE OXIMETER PRINCIPLES

A pulse oximeter shines light of two wavelengths through a tissue bed such as the finger or earlobe and measures the transmitted light signal. The device operates on the following principles:

1. The light absorbance of oxygenated hemoglobin and deoxygenated hemoglobin at the two wavelengths is different. To be more precise, the set of associated extinction coefficients for the absorption of light for these wavelengths is linearly independent with great enough variation for adequate sensitivity but not so large that the blood appears opaque to either of the

light sources. This model assumes that only oxygenated and deoxygenated hemoglobin are present in the blood.

2. The pulsatile nature of arterial blood results in a waveform in the transmitted signal that allows the absorbance effects of arterial blood to be identified from those of nonpulsatile venous blood and other body tissue. By using a quotient of the two effects at different wavelengths it is possible to obtain a measure requiring no absolute calibration with respect to overall tissue absorbance. This is a clear advantage of pulse oximeters over previous types of oximeters.

3. With adequate light, scattering in blood and tissue will illuminate sufficient arterial blood, allowing reliable detection of the pulsatile signal. The scattering effect necessitates empirical calibration of the pulse oximeter. On the other hand, this effect allows a transmittance path around bone in the finger.

The principles above, associated issues and design and application of pulse oximeters comprise the better part of this text. The remainder of this chapter concentrates on the role and importance of pulse oximetry and limitations of the device.

2.2 S_pO_2 AS MONITOR OF HEMOGLOBIN OXYGENATION

Lack of oxygen can quickly lead to irreversible damage to cell tissue having a high metabolic rate, the heart and central nervous system being two examples. Although the human body is surprisingly robust in many ways, the physiological process of sustaining proper cell function via oxygen transport is a delicate and complex control system; one which if altered too significantly could become unstable and insufficient for meeting oxygen tissue demands. To emphasize the importance of proper tissue oxygenation, Table 2.1 lists survival times for different organ beds after the onset of anoxia, or cardiac arrest. Hence, it is important to safeguard against pathological conditions that might lead to improper tissue oxygenation.

Table 2.1 Organ robustness to anoxia (cardiac arrest), a consequence of metabolic rate and cellular oxygen stores. *Survival* time is the time before cellular damage occurs after total loss of oxygen delivery. *Revival* time is the time before function of the organ can no longer be restored. Revival times are generally four times longer than survival times in most organs except the brain, which has a revival time five times longer than its survival time (adapted from Nunn 1987).

Organ	Survival time after onset of anoxia
Cerebral cortex	less than 1 min
Heart	5 min
Liver and kidney	10 min
Skeletal muscle	2 h

2.2.1 Comprehensive approach

Arterial saturation, the variable which pulse oximetry is intended to measure, is just one of several variables a physician will consider when assessing the condition of a patient's cardiopulmonary system. In this regard, the clinician will address the fundamental issue of whether or not body tissue is being properly oxygenated (Vender 1992). This requires a comprehensive approach whereby arterial saturation plays a certain role. It is an extremely important one, but physicians typically do not use S_aO_2 as a sole monitor for pathological oxygenation conditions.

2.2.2 Arterial oxygen saturation

Arterial oxygen saturation pertains to blood in the arteries and arterioles throughout the body. This blood is of the same saturation throughout the arterial system. It is at the capillary level that saturation levels change. In a healthy adult, the normal operating range for S_aO_2 is greater than 90%, which corresponds to an arterial partial pressure, P_aO_2, of 60 to 100 mmHg (Ahrens and Rutherford 1993).

Owing to the complexity of the oxygenation process, it is difficult to address the wealth of uses for arterial saturation in critical care settings, operating rooms, and research laboratories. Physicians are interested in knowing S_aO_2 for a variety of reasons. Sometimes it is for quantitative assessment. Sometimes it serves as an important variable for safeguarding against, although it is not a direct indication of the dangerous condition of low cellular oxygenation. Table 2.2 gives several respiratory problems that might cause low S_aO_2, but this is by no means a complete list.

Table 2.2. Respiratory problems that might result in low S_aO_2 (adapted from Des Jardins 1990, Cherniack and Cherniack 1983, and Selecky 1982).

Respiratory problem	Example disease or possible source of problem
Poor lung compliance	Pneumonia, ARDS, fibrosis, emphysema
Increased airway resistance	Asthma, chronic bronchitis, cystic fibrosis
Low pulmonary diffusion capacity	Emphysema, pulmonary alveolar proteinosis
Airway obstruction	Choking, secretions from intubation, obstructive sleep apnea
Ventilatory muscle weakness	Lead poisoning, trauma to phrenic nerve
Increased true venous admixture	Congenital heart disease
Low inspired partial pressure of oxygen	Anesthesia equipment failure, high altitude
Hypoventilation	Acid-base imbalance

2.2.3 Hypoxia and hypoxemia

Hypoxia means lower than normal tissue oxygenation. *Hypoxemia* means lower than normal blood oxygenation. These are two quite different concepts. Hypoxia refers to the critically dangerous condition where cell function is in jeopardy. Table 2.3 shows different categories of hypoxia. The first category, hypoxic hypoxia, is a consequence of low arterial saturation. Hence, hypoxemia is a dangerous condition. However, it is not necessary that hypoxia exist under conditions of hypoxemia. Likewise, as table 2.3 suggests, hypoxia may occur when there is no evidence of hypoxemia. Therefore, a clinician carefully interprets results from monitoring blood oxygen content because S_aO_2 and, consequently, S_pO_2 provide *only* a measure of hypoxemia, *not* a measure of hypoxia.

Table 2.3 Different types of hypoxia (adapted from Bredle 1989, and Des Jardins 1990).

Type of hypoxia	Description
Hypoxic hypoxia	Arterial blood is poorly oxygenated due to low F_IO_2 or respiratory disease
Anemic hypoxia	Blood cannot transport adequate oxygen due to hemoglobin abnormalities
Circulatory hypoxia	Cardiac output is low or blood perfusion is inadequate
Histotoxic hypoxia	The tissue is incapable of using otherwise sufficient supplies of oxygen

2.2.4 Role of SpO$_2$ in avoiding hypoxia

Although monitoring blood oxygen saturation provides only clues to the oxygenation of cells, there is one variable which provides better evidence of hypoxia. That is lactate content of the blood. Energy utilization may take place in an anaerobic environment, and the byproduct of such a process is lactate. However, the anaerobic process is inefficient for generating energy, and cells cannot operate for long in this situation. The presence of lactate is not a problem initially because it may be broken down if oxygen stores are replenished soon enough (Ahrens and Rutherford 1993). Lactic acidosis occurs if this is not the case. This, in turn, affects the pH of blood which influences the cardiac and pulmonary control systems. However, if cardiac output and respiratory rate cannot increase the delivered oxygen, a dangerous situation results.

Although lactic acidosis may be a better indicator of hypoxia than arterial blood saturation, the problem is that lactic acidosis is an after-the-fact occurrence. As pointed out by Vender (1992), cell damage is likely occurring upon noting an increase in lactate. Herein lies the true value of blood saturation measurement. If monitored appropriately, it can help signal dangerous pathological conditions before cell damage occurs. However, as alluded to earlier S_pO_2 (i.e., S_aO_2) alone is not as helpful as when supplemented with measures of cardiac output, functional hemoglobin, blood pressure, heart rate, respiratory rate, urine output, patient comfort and a variety of other variables.

2.2.4.1 Anesthesiology. Tissue oxygenation and, consequently, blood saturation are of extreme importance to the anesthesiologist because the patient's cardiopulmonary system is placed in a state where it can no longer meet oxygen demands on its own. In a sense, the anesthetist becomes the controller for the patient's respiratory system, and S_pO_2 provides one of the better feedback variables. As a monitoring device to assist the anesthetist, pulse oximetry has literally revolutionized the field of anesthesiology because of its noninvasive nature, fast response and affordability (Fairley 1989). Note that the transition to pulse oximetry was not without controversy (Payne and Severinghaus 1985). Cyanosis, heart rate and blood pressure were generally what was available to the anesthesiologist for detecting hypoxia before the advent of pulse oximetry (Fairley 1989). Similar to lactate, all these variables are after-the-fact occurrences of hypoxia. Again, S_pO_2 does not give direct indication of hypoxia, which has its drawbacks, but it can be an early warning of its occurrence.

The most frequent use of pulse oximeters is by anesthesiologists during surgery and for about an hour afterwards in the recovery room. Anesthesiologists administer narcotics to the patient to suppress the central nervous system. This stops the patient's desire to breath. In addition, they administer muscle relaxants, which stops the ability to breathe and permits airways to collapse. Thus, it is necessary to restore breathing through intubation and artificial respiration. Anesthesiologists can monitor several variables, but most have limitations of late or unreliable response to an oxygenation problem.

Blood pressure declines long after oxygen declines, and the ECG indicates problems even later than blood pressure. An esophageal stethoscope indicates within one beat when the heart has stopped, but this is also long after oxygen has declined. The anesthesiologist can check for cyanosis. Again, this occurs long after oxygen has declined. Blood gas samples give an accurate measurement of oxygen, carbon dioxide, and pH but take about 5 min to process.

Pulse oximeters solved the problem of delay by continuously and noninvasively monitoring arterial oxygen saturation. Recall that adequate arterial saturation does not imply proper oxygenation. Furthermore, there is a delay between noting a drop in S_pO_2 and its cause. However, of the monitored variables, S_pO_2 is currently the best indication that an oxygenation problem exists or is about to occur, and it does so noninvasively.

The pulse oximeter probe is usually applied to the finger, since the body will decrease blood flow to the finger before more vital organs. It is more difficult to reliably secure probes to the ear, nose, and forehead. An arterial oxygen saturation drop from 98 to 96% alerts the anesthesiologist that something is going on. If the oxygen saturation drops to 90%, the default alarm sounds, which indicates that a serious problem may be at hand.

Continuously monitoring S_pO_2 catches several equipment malfunctions and improper placement of tracheal tubes, but naturally it does not identify the problem (Payne and Severinghaus 1985). The fact that S_pO_2 does not identify the source of the problem should not be viewed as a drawback to the pulse oximeter. Instead, this has implications in how to view pulse oximetry as a monitored variable. Fairley (1989) has figuratively stated the role of pulse oximetry in anesthesiology (Original metaphor attributed to Tremper and Barker (1989)):

> It was not until effective pulse oximetry became commercially available, for the first time, that large numbers of anesthesiologists could continuously monitor their patients' arterial oxygen levels. It is very important to recognize the nature of this monitoring. Since virtually every anesthetized patient breathes an oxygen enriched mixture, desaturation only occurs when there is a substantial increase in the difference between the (perceived) inspired oxygen tension and that in the arterial blood. Metaphorically, as the blindfolded anesthetist walks unknowingly towards the cliff of hypoxia—whether due to problems of inspired gas, equipment failure, underventilation, or abnormal pulmonary shunting—the protective hand of the pulse oximeter sentry stops him from falling over the edge. The oximeter will not tell him why he has been proceeding in that direction, or the direction back! On the other hand, should he start falling, the sentry functions on the vertical part of the dissociation curve and becomes an extremely

> sensitive (if not always accurate) indicator of progress during the drop. Interestingly, it is highly probable that many fewer blood gas samples are being drawn during anesthesia now that pulse oximeters are so universally available. Our detailed insight into our patients' pulmonary oxygen exchange is less than with P_aO_2 measurement but, because of the continuously available sentry, we believe our patients are safer. A prospective study to prove that important point with certainty may never be performed but, already, opinion seems overwhelmingly in favor of that belief.

Pulse oximetry has become a *de facto* standard for the American Society of Anesthesiologists (Eichhorn 1993). This means that, as alluded to in Fairley's description, although definitive statistical proof of the benefit of pulse oximetry may never be shown because of the rarity of complications due to anesthesiology in the operating theater, a large majority of those who use the device feel that it helps to reduce complications.

2.2.4.2 Postoperative and critical care. Pulse oximetry has proven very important to postoperative recovery because the patient's pulmonary control may still be compromised from the effects of anesthesia. For example, a randomized study by Lampe *et al.* (1990) found that of 141 patients having carotid endarterectomy 63% had episodes of S_pO_2 less than 90% and 21% had episodes of S_pO_2 less than 86% during the postoperative period. Similar studies also show large numbers of desaturation episodes, although variation in the data does exist (Severinghaus and Kelleher 1992).

The role of pulse oximetry in intensive and critical care units is similar to that for anesthesiology, although the patient's respiratory system may not be suppressed by narcotics and muscle relaxants. The instrument still acts as the sentry warning of desaturation from a variety of conditions, some of which were listed in table 2.2. In this setting, alarms and temporal records are very useful when constant surveillance of the patient is not possible.

2.2.4.3 Exams and research studies. The pulse oximeter is an excellent device for medical research studies such as sleep apnea and hypoxic ventilatory response (Severinghaus and Kelleher 1992). Medical exams such as stress tests also benefit from the noninvasive, continuous nature of pulse oximetry. In such cases, S_pO_2 may be used to catch hypoxemic events and also correlated with other variables to glean information about the patient's general health.

2.2.5 Photoplethysmography

Most pulse oximeters on the market feature a photoplethysmograph. The signal for the photoplethysmograph is derived from the same waveforms used to calculate S_pO_2. The photoplethysmograph may be used in a clinical setting in the same manner as a plethysmograph. However, the accuracy of the photoplethysmograph suffers from motion artifacts, and the patient must have adequate blood perfusion near placement of the pulse oximeter probe. Just as with the conventional plethysmogram, signal processing can derive heart rate from the photoplethysmogram waveform. Hence, most pulse oximeters also display heart rate. Similar to computing S_pO_2, temporal low-pass filtering abates the effect of motion artifacts on heart rate estimation.

2.2.6 Hyperoxia

Hyperoxia is the condition where blood in the system contains more than the normal amount of oxygen. Determining excessive levels of oxygen is important in many situations because of the toxic nature of oxygen radicals. Studies suggest that pulse oximetry is not useful for this type of application. For example, the role of S_aO_2 for determining retinopathy of prematurity in neonates is not quite clear, and furthermore the 2 to 3% inaccuracy of S_pO_2 for estimating S_aO_2 adds to this uncertainty (Severinghaus and Kelleher 1992).

2.3 LIMITATIONS

2.3.1 Instrument and operation limitations

Many of the limitations to pulse oximetry will come to light throughout the remainder of this text. However, table 2.4 summarizes some limitations given by Severinghaus and Kelleher (1992). These are described in detail in the original source.

Table 2.4 Limitations to pulse oximetry and its application in a clinical setting. Adapted from Severinghaus and Kelleher (1992).

Pulse oximetry limitations
Instrument incidence of failure
Low signal-to-noise ratio
Light shunting and poorly applied probe
Vasoconstrictors
Low-perfusion limits
Motion artifacts
Abnormal pulses
Ventilator-induced and venous pulse interference
Response times
Ambient light
Electrosurgery
Interference of MRI
Site selection for probe placement
Skin pigments, dyes, and nail polish
Dysfunctional hemoglobins
Burns and other dangers
False alarms and false nonalarms

2.3.2 Limitations in S_aO_2

Often in anesthesiology medical literature, articles regarding a limitation of pulse oximetry appear in which, if read more closely, what is actually meant is a limitation in monitoring arterial oxygen saturation, e.g., Hutton and Clutton-Brock (1993), and Mak (1993). Authors of such articles point out this fact. It is interesting how measuring S_pO_2 has become so associated with measuring S_aO_2.

Nonetheless, it follows that any limitations associated with S_aO_2 as a monitored variable are also associated with S_pO_2.

There are some caveats to using S_aO_2 to assess the condition of pulmonary function. It is difficult to regard any monitoring technique as foolproof, as there are usually misleading combinations of conditions that will result in the monitored variable appearing fine when, in fact, a potentially dangerous condition could exist for the patient. For example, Hutton and Clutton-Brock (1993) and Mak (1993) point out that pulse oximetry (i.e., S_aO_2) is a poor measure of hypoventilation when inspired oxygen concentration is high. It is in situations like this that a comprehensive approach to oxygenation assessment using other monitored variables is imperative.

REFERENCES

Ahrens T and Rutherford K 1993 *Essentials of Oxygenation* (Boston, MA: Jones and Bartlett)
Bredle D L 1989 Circulatory compensation as a response to hypoxia *Clinical Aspects of O_2 Transport and Tissue Oxygenation* ed K Reinhart and K Eyrich (New York: Springer)
Cherniack R M and Cherniack L 1983 *Respiration in Health and Disease* (Philadelphia, PA: Saunders)
Des Jardins T R 1984 *Cardiopulmonary Anatomy & Physiology: Essentials for Respiratory Care* (Albany, NY: Delmar)
Des Jardins T R 1990 *Clinical Manifestations of Respiratory Disease* (Chicago, IL: Year Book Medical)
Eichhorn J H 1993 Pulse oximetry as a standard of practice in anesthesia *Anesthesiology* **78** 423–6
Fairley H B 1989 Changing perspectives in monitoring oxygenation *Anesthesiology* **70** 2–4
Hutton P and Clutton-Brock T 1993 The benefits and pitfalls of pulse oximetry *Brit. Med. J.* **307** 457–8
Lampe G H, Wauk L Z, Whitendale P, Way W L, Kozmary S V, Donegan J H and Eger E I 1990 Postoperative hypoxemia after nonabdominal surgery: A frequent event not caused by nitrous oxide *Anesth. Analg.* **71** 597–601
Mak V 1993 False reassurance of pulse oximetry *Brit. Med. J.* **307** 732–3
Nunn J F 1987 *Applied Respiratory Physiology* (Boston, MA: Butterworths)
Payne J P and Severinghaus J W (eds) 1985 *Pulse Oximetry* (New York: Springer)
Selecky P A (ed) 1982 *Pulmonary Disease* (New York: Wiley)
Severinghaus J W and Kelleher J F 1992 Recent developments in pulse oximetry *Anesthesiology* **76** 1018–38
Tremper K K and Barker S J 1989 Pulse oximetry *Anesthesiology* **70** 98–108
Vender J S 1992 *Mixed Venous Oximetry* (video recording) (Secaucus, NJ: Network for Continuing Education)

INSTRUCTIONAL OBJECTIVES

2.1 State the fundamental question a clinician should ask when assessing the cardiopulmonary condition of a critically ill patient.
2.2 Give the normal values for S_aO_2 and P_aO_2 in the healthy adult.
2.3 State the difference between hypoxia, hypoxemia and hyperoxia.
2.4 Give several common problems that result in hypoxemia.
2.5 Describe the role of lactate as an indicator of improper oxygen transport.
2.6 Describe the role of S_pO_2 as an indicator of improper oxygen transport.
2.7 List several physiologic variables that may be used in conjunction with S_aO_2 for assessing a patient's cardiopulmonary condition.
2.8 Describe why pulse oximetry data are of importance to anesthesiology.
2.9 State how useful pulse oximetry is in detecting hyperoxia.
2.10 List several limitations to pulse oximetry.

CHAPTER 3

BLOOD OXYGEN MEASUREMENT

James Farmer

Oximetry is a general term that refers to the optical measurement of oxyhemoglobin saturation in the blood (Peterson 1986). Pulse oximetry is only one of those technologies. There are other methods of measuring oxygen content of the blood as well. Gradwohl (1948) describes two colorimetric methods of estimating the HbO_2 of the blood by direct comparison to a color chart. The Dare method used a thin layer of undiluted blood which was matched against a standard series of colored disks. The Tallqvist method used a drop of undiluted blood placed on absorbent paper. The absorbent paper was compared with a graded scale of colored blocks printed on paper. The Tallqvist method was reported to be inaccurate and not recommended. No information on the reliability of the Dare method was given.

This chapter describes several different chemical and optical methods of determining the oxygen saturation of the blood which are more deterministic than the ones above. The chapter examines the development of oximetry from a historical perspective. The final section of the chapter gives an overview of the design of a pulse oximeter.

Some of the methods described in this chapter find the partial pressure of oxygen (PO_2) and some find the oxygen saturation (SO_2). Chapter 1 describes the relationship between these two. It is interesting to note from a historical perspective that SO_2 was not always an accepted means of reporting blood oxygenation. Gradwohl (1948) stated, 'Hemoglobin estimations are reported in terms of percentage, but this incorrect. They should always be reported in terms of grams per 100 mL.'

3.1 CHEMICAL METHODS

The oxygen content of blood can be determined from a sample by using chemical reactions to remove the oxygen from the blood. These measurements can be done with varying degrees of success. The chemical reactions can be slow also. The Van Slyke method can take up to 20 min.

3.1.1 Van Slyke method

The Van Slyke apparatus (figure 3.1) is used in a method of measuring the oxygen content of a blood sample. A sample of blood is introduced to the apparatus anaerobically with a sample of potassium ferricyanide. Potassium ferricyanide is a releasing agent that releases the oxygen, carbon dioxide, and other gases from the blood sample. After removing the carbon dioxide from the mixture, the remaining gases are compressed into a fixed volume and the resulting pressure (P_1) is measured from the manometer. The oxygen is then absorbed with a reagent such as sodium hydrosulfite. The remaining gases are then recompressed into the same fixed volume and the final pressure (P_2) is measured (Hill 1966).

The difference of the two pressure measurements is a partial pressure due to the oxygen that was contained in the blood sample. The oxygen content of the blood sample is calculated by

$$\text{mL O}_2/100 \text{ mL blood} = K(P_1 - P_2) \tag{3.1}$$

where K is a constant relating to the reagents, apparatus, and the volume of the blood sample (Adams and Hahn 1982). Alternatively, the oxygen can be extracted from the blood with the Van Slyke apparatus and analyzed with a gas chromatograph (Hill 1966).

The technique is not simple to perform. Technical expertise and experience with chemical reactions are required to obtain accurate, reproducible results. However, the Van Slyke apparatus can provide measurements accurate to ±0.03% (Adams and Hahn 1982). The Van Slyke technique has been in the past a standard by which blood oxygen measurements were made (Miller 1966, Dennis and Valeri 1980).

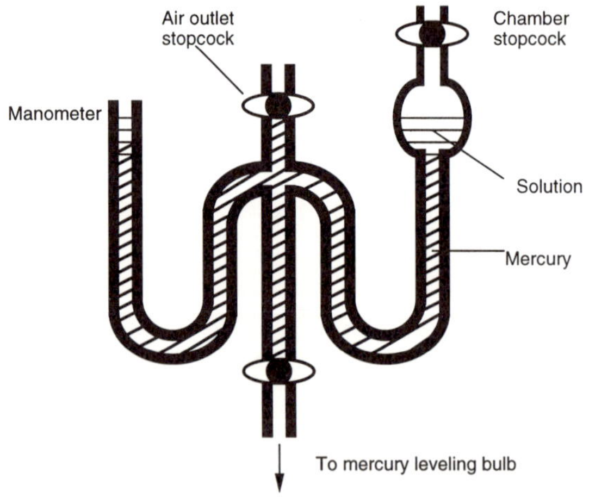

Figure 3.1 The Van Slyke apparatus (adapted from Adams and Hahn 1982).

3.1.2 Mixing syringe method

The mixing syringe method also measures the amount of oxygen released from a blood sample by a chemical reagent. The apparatus consists of two Luer-lock syringes joined to a manometer tap. One of the syringes is a precision automatic syringe which is able to accept and deliver a fixed volume of reagent. The automatic syringe is filled with the oxygen releasing agent and then emptied. This coats the inside of the syringe with the reagent and keeps the blood from any contact with the air. The oxygen releasing agent has a known oxygen partial pressure (P_r). The automatic syringe then draws a volume of blood (V_b) from the mixing syringe. The volume of blood and a known volume of the reagent (V_r) are mixed back and forth between the syringes. The partial pressure of oxygen of the blood–reagent solution (P_s) is then measured by a blood-gas analyzer. The oxygen content is calculated from the equation

$$\text{mL O}_2/100\text{ mL blood} = \alpha\frac{V_r + V_b}{V_b}\left[P_s - \left(\frac{V_r}{V_r + V_b}P_r\right)\right] \qquad (3.2)$$

where α is the solubility coefficient of oxygen in the blood–reagent solution at the temperature at which the measurement was made. Its value is obtained from either a separate experiment or from reference tables (Adams and Hahn 1982).

3.1.3 The Clark electrode

The Clark electrode uses the basic chemistry principles of oxidation and reduction to measure the PO_2 (partial pressure of oxygen) in a solution. When oxygen is dissolved in an aqueous solution and exposed to a 0.7 V polarizing voltage, the following reaction occurs

$$O_2 + 2H_2O + 4e^- \rightarrow 4OH^-. \qquad (3.3)$$

A silver anode immersed in a potassium chloride electrolyte bath will attract anions (Cl^-) to form silver chloride. This oxidation reaction produces a constant flow of electrons. A nearby platinum electrode undergoes a reduction reaction turning oxygen to hydroxyl ions (OH^-) as in equation (3.3). Figure 3.2 shows that the number of electrons used in the platinum cathode reaction is directly proportional to the PO_2 present in the bath. Therefore, by measuring the current between the two electrodes, the PO_2 in the solution is determined.

The entire Clark electrode system (figure 3.3) has a polypropylene sheath which slows the diffusion of oxygen from the blood to the electrode. This prevents the electrode from depleting the PO_2 in a particular place and eliminates the need to stir the blood *in vitro*.

The Clark electrode is the common sensing device used by blood gas analyzers to determine the PO_2 of the blood (Shapiro *et al* 1989). Using a variety of different electrodes, blood gas analyzers also determine the pH and PCO_2 of blood samples as small as 65 µL. The blood gas analyzers are very useful for *in vitro* measurements because they self-calibrate and self-diagnose malfunctions. Thus, interfacing blood gas analyzers with computers allows for automated measurements, patient data storage, and billing.

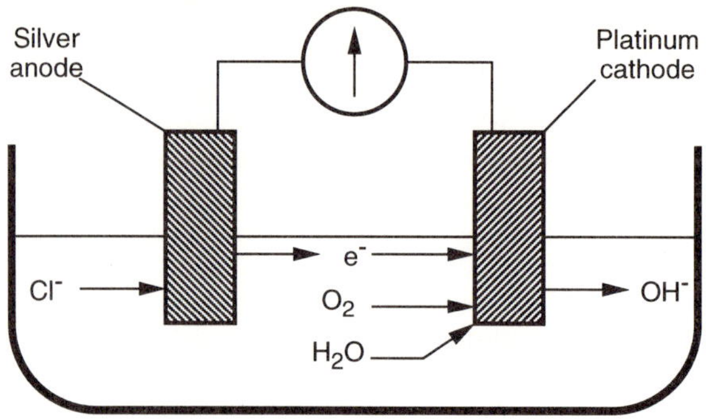

Figure 3.2 Since an aqueous solution has plenty of H_2O and the silver anode is able to supply an abundance of electrons, equation (3.3) is limited by the amount of oxygen present. Thus, the amount of current between the anode and the cathode is determined by the PO_2 present. This reaction shows the chemical reaction that occurs in a Clark electrode.

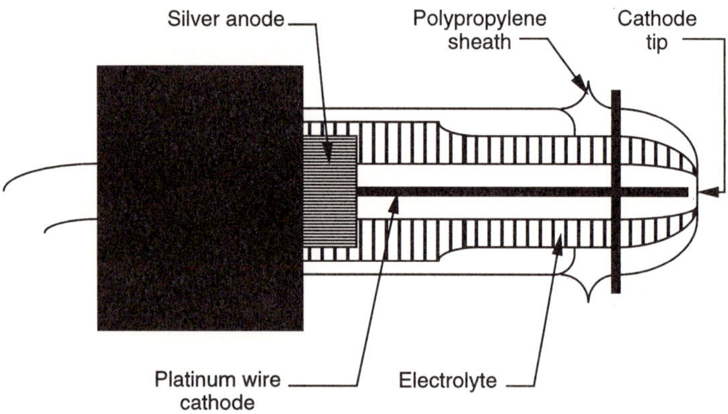

Figure 3.3 A Clark electrode (adapted from Shapiro *et al* 1989).

The Clark electrode can also be used to make *in vivo* measurements when designed to be used as catheter electrodes (Adams and Hahn 1982). Many catheter electrodes are designed specifically to be used with infants and are very small in diameter. Several different versions exist. Some versions have both the anode and cathode within the single electrode, as pictured in figure 3.3. But others have an external anode reference electrode on the skin.

One of the downfalls of the Clark electrode catheter system is calibration. Calibration takes place by drawing a blood sample near the end of the catheter and analyzing the sample with an *in vitro* blood gas analyzer. Another potential problem with the Clark electrode is keeping the tip clean. Though the polypropylene sheath helps to some extent, failure to keep the catheter in the flow of blood can cause errors as blood coagulates on the surface of the electrode.

3.1.4 The galvanic electrode

The galvanic electrode is similar in operation to the Clark electrode. As oxygen passes across the electrode, a chemical reaction occurs that produces a small electric current. But in this case the cathode is made of gold, the anode of lead, and the electrolyte solution is potassium hydroxide. In the Clark electrode, the silver anode and the platinum cathode participated in the chemical reaction. This made sure that the electrolyte solution was always replenished. But, the galvanic electrode has no means to replenish the electrolyte solution in the electrode and so it has a limited lifetime which depends on the PO_2 and exposure (Shapiro et al 1982).

3.2 TRANSCUTANEOUS PO_2 SENSOR

The Clark electrode can be used noninvasively to determine PO_2 of the blood. Under normal conditions, the PO_2 of the blood near the skin's surface ($P_{tc}O_2$, for the transcutaneous partial pressure of oxygen) is atmospheric. But, hyperemia of the skin can cause the $P_{tc}O_2$ to approach P_aO_2, the arterial partial pressure. Hyperemia of the skin can be induced by drugs, creams, abrasions, or heating the skin. In other words, by placing a Clark electrode on the skin with a heating element, the skin begins to diffuse oxygen so that the $P_{tc}O_2$ is nearly equal to the P_aO_2. The measurements given by the transcutaneous PO_2 electrode are stable with little drift and are widely accepted (Gothgen and Jacobsen 1987).

Heating is the easiest method of inducing hyperemia to control. With a heating element and a thermistor, the skin is heated to between 43 °C and 44 °C. This is the optimal temperature range for the $P_{tc}O_2$ to approach the P_aO_2 with minimal skin damage. The heat causes increased blood flow to the skin at the heating element site. This increased perfusion causes more O_2 to be delivered to this area and the excess O_2 diffuses through the skin more easily (Peura 1998). Figure 3.4 shows a cross sectional view of a transcutaneous PO_2 electrode, showing the heater and the thermistor.

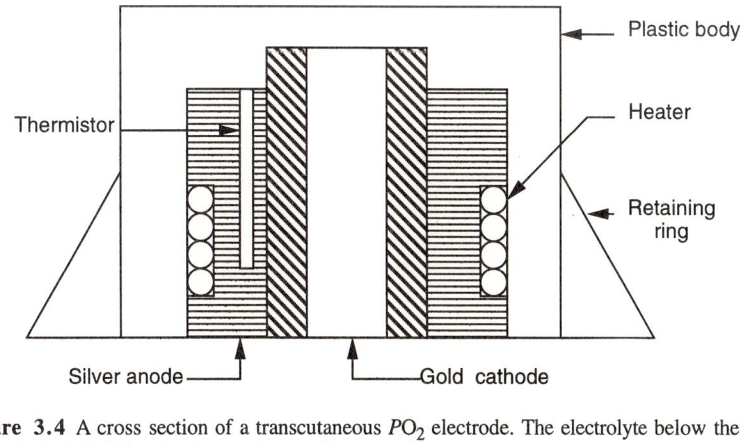

Figure 3.4 A cross section of a transcutaneous PO_2 electrode. The electrolyte below the anode and cathode is held in place by a polypropylene membrane.

One advantage of the transcutaneous PO_2 electrode is that it measures the *real* $P_{tc}O_2$ and is not an empirical calculation as with the pulse oximeter (Gothgen and Jacobsen 1987). $P_{tc}O_2$ can be thought of as a new PO_2 variable and not an estimation of P_aO_2. This contrasts with the pulse oximeter, which is an estimate of S_aO_2, and whose accuracy is dependent on its ability to predict S_aO_2 (Barker and Tremper 1984). But again, one of the disadvantages of the transcutaneous PO_2 sensor is the calibration. Like the Clark electrodes used with catheters and blood gas analyzers, the transcutaneous PO_2 measurement is based on an electrochemical reaction that needs to be calibrated frequently with some gas mixtures.

The transcutaneous PO_2 electrode has other disadvantages. There is a warm up time of 10 min for the heating element to induce enough blood flow to the measurement site. And even with the thermistor regulating the heating element, there is a risk of burns, especially in infants. It is recommended that the electrode be moved every 4 hours (Burtis and Ashwood 1994).

The $P_{tc}O_2$ does not vary more than 5% from the P_aO_2 in infants but is more dependent on blood flow in adults. The heating is not as effective in adults and so the $P_{tc}O_2$ is usually lower than the P_aO_2 (Barker and Tremper 1984). Also, transcutaneous PO_2 electrodes are unreliable when the blood pressure falls below 100 mmHg or when some anesthetics such as nitrous oxide are administered (Burtis and Ashwood 1994). Even with these problems, the transcutaneous PO_2 electrode has been found useful in clinical situations in the operating room, intensive care units, and emergency rooms (Waxman *et al* 1983).

3.3 *IN VITRO* OXIMETERS

3.3.1 Spectrophotometers

Spectrophotometry is the basis for all oximetry. The atoms of all molecules vibrate in specific patterns for each unique substance. As light passes through a substance, the frequencies of light similar to the vibrational frequencies of the substance are absorbed. A spectrophotometer measures the intensity of light transmitted through a particular substance at particular wavelengths. The fraction of light absorbed at a specific wavelength is determined by the absorptivity, or extinction coefficient, of the substance. The extinction coefficient of a substance can be graphed at various wavelengths as a spectrum. This spectrum is unique for every substance.

A photodetector is a device that converts light intensity into an electric current. A given intensity of light transmitted through a substance produces an electric current proportional to the intensity. By measuring the intensity of incident light on a substance (I_0) and measuring the intensity of light transmitted through the substance (I), the transmittance (T) of the substance can be calculated:

$$T = \frac{I}{I_0}. \tag{3.4}$$

Because each molecule absorbs an equal portion of light, the absorbance of light through a substance is linearly related to the concentration of substance present. From the measured transmittance (T), the absorbance (A) can be calculated from

$$A = 2 - \log(\%T). \quad (3.5)$$

Beer's law can now be used to find the amount of substance in a solution. Beer's law can be stated as

$$A = \varepsilon(\lambda)\, c\, d \quad (3.6)$$

where $\varepsilon(\lambda)$ is the extinction coefficient of the substance at a given wavelength λ of light, d is the length of the light path, and c is the concentration of the substance. For all substances, the linear relationship between absorbance and concentration only holds up to a certain concentration. Below this limit we can determine a calibration constant. The calibration constant can then be used as a standard to determine the unknown concentration of a substance with the same extinction coefficient as the standard.

For a solution with two unknown compounds, the absorbances at two wavelengths can be used to calculate the concentrations of both compounds. At the isosbestic point where the two extinction coefficients are equal, Beer's Law for the two samples can be written as

$$d = \frac{A_{ec}}{[c_1 + c_2]\varepsilon(\lambda_{ec})} \quad (3.7)$$

where A_{ec} is the absorbance at the isosbestic point and $\varepsilon(\lambda_{ec})$ is the extinction coefficient of the two substances at the isosbestic point. At the second wavelength Beer's Law gives

$$A_0 = d[c_1\varepsilon_1(\lambda_0) + c_2\varepsilon_2(\lambda_0)] \quad (3.8)$$

where A_0 is the absorbance and $\varepsilon_1(\lambda_0)$ and $\varepsilon_2(\lambda_0)$ are the extinction coefficients for the two compounds at the second wavelength. Because the sum of the concentrations of the two compounds is 1, we can solve equations (3.7) and (3.8) for the two concentrations.

If the solution contains more than just the two compounds as is the case with oximetry, solving equations (3.7) and (3.8) will give the relative concentration of c_1 to c_2 if the assumption can be made that none of the other compounds will absorb light at the two wavelengths used for the measurement. This assumption is sufficient for oximetry where the relative concentrations of Hb and HbO_2 are used to estimate S_aO_2.

Note that measuring the absorbance at the isosbestic point is not necessary to solve for c_1 and c_2. The absorbance at any two wavelengths can be used to solve for the concentrations with equally good results. The motives for the choice of the isosbestic point as one of the wavelengths used in the earliest oximeters are not clear. But the simplified mathematics may have been a reason (Nilsson 1960).

With the concentrations of Hb and HbO_2, an estimation of S_aO_2 is made from

$$S_pO_2 = \frac{HbO_2}{HbO_2 + Hb} \times 100\% \tag{3.9}$$

This assumes that any other substance present in the solution being measured has no effect on the absorbance of light at the chosen wavelengths. For example, it does not take into account the effect of the other types of hemoglobin present in the blood. These hemoglobin species do absorb light at certain wavelengths, but their relative concentration with respect to Hb and HbO_2 is small enough that for many applications equation (3.9) is an accurate estimate.

3.3.2 The CO-oximeter

CO-oximeters are spectrophotometers specifically designed to analyze the concentrations of several different types of hemoglobin including reduced hemoglobin (Hb), oxyhemoglobin (HbO_2), carboxyhemoglobin (COHb), and methemoglobin (MetHb). Each of these various forms of hemoglobin has its own extinction coefficient curve (figure 4.2). By using as few as four wavelengths of light, the amount of each of these forms of hemoglobin can be determined from a sample.

Instrumentation Laboratories Inc. coined, but did not copyright, the term CO-oximeter and released the first commercial CO-oximeter in 1966 (Moyle 1994). The CO-oximeter was originally introduced to measure COHb using three different wavelengths, 548 nm, 568 nm and 578 nm (Adams and Hahn 1982). The IL-282 CO-oximeter pictured in figure 3.5 uses four wavelengths of light, 535.0 nm, 585.2 nm, 594.5 nm, and 626.6 nm, to measure all four of the relevant forms of hemoglobin. These wavelengths are obtained by four interference filters mounted on a rotating wheel each selecting wavelengths of light from a Tl–Ne hollow cathode lamp (Zwart *et al* 1981). The CO-oximeter is able to operate in this narrow range of light because it only works with diluted plasma samples and like pulse oximeters does not have to deal with skin, muscle, or other tissue (Moyle 1994).

A four wavelength CO-oximeter would obtain absorbance readings on a blank solution at all four different wavelengths (λ_{1-4}). Then a reading is obtained at each wavelength for a diluted, hemolyzed sample. CO-oximeters use hemolyzed samples, blood samples with the red blood cell membranes removed, to reduce the amount of light scattering, which reduces the accuracy of the measurement.

The absorbance readings of the blank solution are subtracted from the readings from the samples at each wavelength to give the absorbance of the blood at each wavelength. From these absorbances of the blood, the concentration of each type of hemoglobin can be calculated from the equations

$$C_{Hb} = K[\varepsilon_{Hb}(\lambda_1)A_1 + \varepsilon_{Hb}(\lambda_2)A_2 + \varepsilon_{Hb}(\lambda_3)A_3 + \varepsilon_{Hb}(\lambda_4)A_4] \tag{3.10}$$

$$C_{HbO_2} = K[\varepsilon_{HbO_2}(\lambda_1)A_1 + \varepsilon_{HbO_2}(\lambda_2)A_2 + \varepsilon_{HbO_2}(\lambda_3)A_3 + \varepsilon_{HbO_2}(\lambda_4)A_4] \tag{3.11}$$

$$C_{MetHb} = K[\varepsilon_{MetHb}(\lambda_1)A_1 + \varepsilon_{MetHb}(\lambda_2)A_2 + \varepsilon_{MetHb}(\lambda_3)A_3 + \varepsilon_{MetHb}(\lambda_4)A_4] \tag{3.12}$$

$$C_{COHb} = K[\varepsilon_{COHb}(\lambda_1)A_1 + \varepsilon_{COHb}(\lambda_2)A_2 + \varepsilon_{COHb}(\lambda_3)A_3 + \varepsilon_{COHb}(\lambda_4)A_4] \tag{3.13}$$

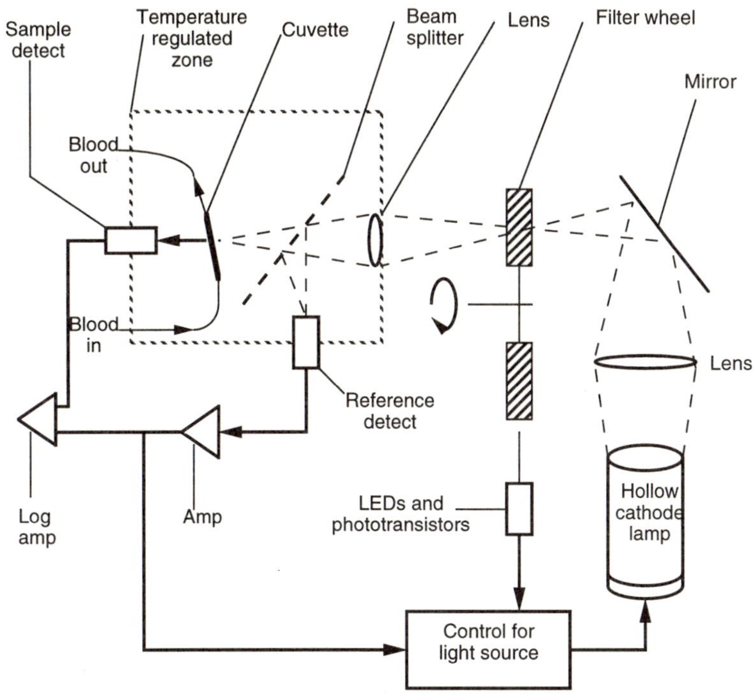

Figure 3.5 A schematic diagram of the IL-282 CO oximeter (adapted from Zwart *et al* 1981).

where C_x is the concentration of hemoglobin type x, $\varepsilon_x(\lambda_1)$ is the extinction coefficient of hemoglobin type x at the first wavelength, A_1 is the difference between the absorbance value of the blood and the blank solution at the first wavelength, and K is a constant set by the calibration procedure (Shapiro *et al* 1989).

CO-oximeters are subject to many sources of error. Any substances in the sample that scatter light affect the measurements because the amount of light transmitted is no longer solely a function of the light absorbed by hemoglobin species. Samples infected with small portions of lipids or cell fragments are common causes of light scattering. There are also errors associated with fetal hemoglobin samples. Results from CO-oximeters have been known to give falsely high COHb readings in fetal hemoglobin (Zwart *et al* 1981). Some CO-oximeters try to compensate for these errors by using more wavelengths of light. For example the AVL 912 uses 17 wavelengths to try to compensate for other light absorbing fragments that might be present in the solution (Moyle 1994).

Because CO-oximeters make measurements *in vitro* with discrete samples, they provide accurate oxygen saturation readings for only the times at which the samples are drawn. They do have their uses, notably as a standard for calibration of *in vivo* oximeters (Moyle 1994). The CO-oximeter is one of the most accurate methods available for measuring the four clinically relevant hemoglobin species. It is a standard against which other methods of measurement are compared (Shapiro *et al* 1989).

3.4 *IN VIVO* TWO-WAVELENGTH OXIMETERS

3.4.1 The first in vivo oximeters

In vivo oximetry originated in Germany in the 1930s when the use of the selenium photovoltaic cell became accepted (Peterson 1986). In 1934, Kramer showed that the absorbance of red light depended on oxygen saturation, but his implementation only used one wavelength of light (Payne and Severinghaus 1986). At about the same time, Matthes designed an oximeter which measured the transmission of light through the ear by a lamp with a photocell attached to the earlobe. At the time, regions of optical spectra were broadly defined and depended greatly on the lamp, photocell, and filters used. Matthes used wavelengths of red light, which varied the transmission measurements as the oxygen saturation varied, and compared them to measurements using green light, which did not vary with saturation. He later discovered that infrared light was a better choice than green (Nilsson 1960).

Glen Millikan is credited with coining the term oximeter while, during World War II, attempting to design a hemoglobin saturation meter for the ears of pilots to control the amount of oxygen they received (Severinghaus 1987). Millikan's ear piece was improved by Wood and Geraci in 1949. The biggest improvement was in the infrared filter. They also had the idea of using an inflatable balloon to cut off the circulation to the ear and make it bloodless. This made for a zero setting which tried to account for the other tissue present in the ear. Wood thought he had succeeded in the first absolute reading oximeter but he later showed that inconsistencies in the photocells used for light detection caused the device to be inaccurate (Payne and Severinghaus 1986).

3.4.2 The Cyclops

The Cyclops was a commercially available reflectance oximeter. It was named the Cyclops because of the large sensing device that was placed on the forehead of the patient. It used red and green light to determine SO_2. Limitations in the technology of photocells limited the Cyclops from using infrared light. The theory of the device was based on the fact that any reflection of the green light was due to nonblood reflection. The reflection of the red light was due to both the blood and noncomponents. So, by subtracting the green light reflection from the red light reflection, the reflection due to the blood was found.

The Cyclops was successful in producing trending information about SO_2. And if it was calibrated against two or more arterial blood samples, the Cyclops could produce accurate SO_2 values (Zijlstra 1958).

3.5 FIBER OPTIC OXIMETERS

3.5.1 In vitro reflectance oximeter

Polanyi and Hehir (1962) first described the design of a fiber optic oximeter to use as a catheter measurement device. They also used two wavelengths of light to measure the concentrations of Hb and HbO_2. They chose the specific values for their wavelengths of light to be 660 nm and 805 nm; 805 nm is the isosbestic point of HbO_2 and Hb. They used a filter wheel, similar to the one used in the CO-oximeter in figure 3.5, to obtain their wavelengths.

The concentrations of Hb and HbO_2 can be calculated in the same manner as in section 3.3.1. The only difference with this device is that this fiber optic oximeter was a reflectance device and measured an absorbance directly from the backscattered light in the blood. In Section 3.3.1, the measurement was a transmittance of light which was converted to an absorbance.

Polanyi and Hehir presented successful results with their fiber optic oximeter with *in vitro* experiments. But although they intended their device to be used *in vivo*, it did not come to be (Barker 1991).

3.5.2 In vivo reflectance catheter oximeter

In vivo catheter oximeters were not in widespread use until about 1980. These *in vivo* fiber optic catheter oximeters use much the same technology as the pulse oximeter. Many catheter oximeters are two-wavelength devices like the pulse oximeter. Some of them use three wavelengths to try to compensate for changes in pH or other variables. Figure 3.6 shows the basic configuration of a fiber optic reflectance oximeter.

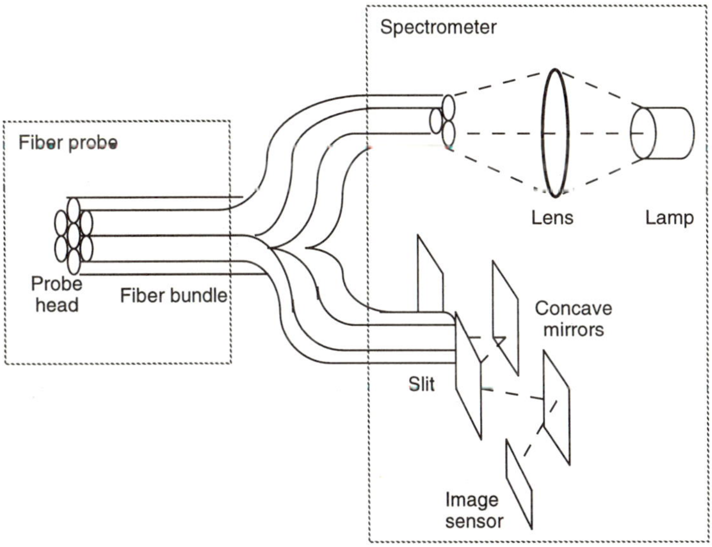

Figure 3.6 A fiber optic reflectance catheter oximeter (adapted from Ono *et al* 1991).

These devices do require user calibration though. The calibration can be done with a CO-oximeter or by the preferred method of *in vivo* calibration. Drift can occur after several hours and the catheter may need to be recalibrated.

Some of the early fiber optic catheters had the problem of wall artifacts, where reflections of light from a vessel wall would cause erroneous values of S_vO_2. New digital filtering techniques have helped to reduce that problem. Early catheters had a reputation for being stiff and hard to insert, but the use of plastic fiber optics has helped this issue (Barker 1991).

The early fiber optic oximeters were designed for cardiac catheterization to measure S_vO_2. Features have been added to some fiber optic probes for other uses. For example, some probes have a contact sensor or a pressure sensor to sense contact with tissue. This allows more stable data from *in vivo* tissue because the probe head can avoid excessive pressure which would affect microcirculation. Another application for the fiber optic oximeter is as a dental tool to diagnose SO_2 of gingiva (Ono *et al* 1991).

3.5.3 In vivo chemical oximeter

Peterson and Fitzgerald (1984) describe a chemical fiber optic oximeter suitable for measuring P_aO_2. A fluorescent dye in the tip of the probe reflects light sent by the oximeter back to a sensor. The dye has a unique property that it loses its luminescence in the presence of oxygen. Figure 3.7 shows the chemical fiber optic probe. The difference between this probe and the reflectance fiber optic probe is a small one. The reflectance fiber optic probe measures the change in color of the blood by reflecting light from it. This change in color indicates the degree of saturation. This device measures the change in color of a substance that changes color in the presence of oxygen.

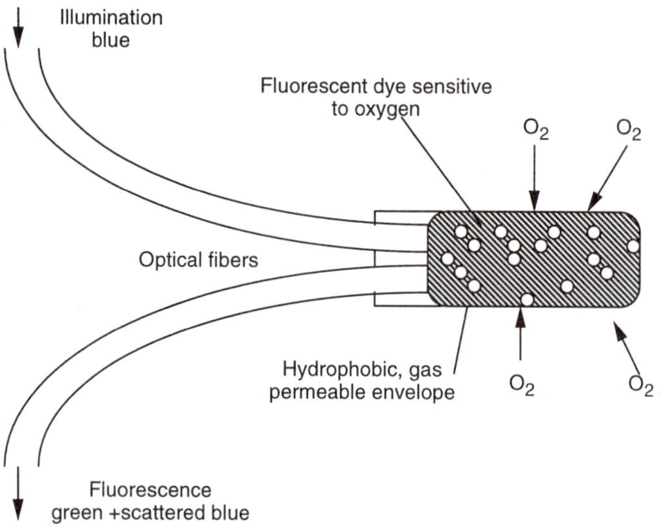

Figure 3.7 This figure shows the probe end of a chemical fiber optic oximeter. The coloration of the probe end changes in response to the amount of oxygen present.

3.6 *IN VIVO* EIGHT-WAVELENGTH OXIMETER

In 1970, Hewlett-Packard marketed an eight-wavelength oximeter, model 47201A. The device was designed to overcome some of the problems of the two-wavelength oximeter. It was designed to be self calibrating, accounted for factors like skin pigmentation, and claimed to be unaffected by motion (Merrick and Hayes 1976). The device also claimed to be precalibrated requiring no test samples. The only calibration necessary was an infrequent procedure that gave reference values of light intensity to the device, and did not involve the patient.

Figure 3.8 shows a block diagram of the 47201A. The lamp is a tungsten–iodine lamp which has a high output of light in the wavelengths of interest (650 nm to 1050 nm). The desired wavelengths are obtained with light filters. The filters are mounted on a rotating wheel so they cut the light beam sequentially. The wheel spins at 1300 rpm so about 20 samples at each wavelength are obtained every second.

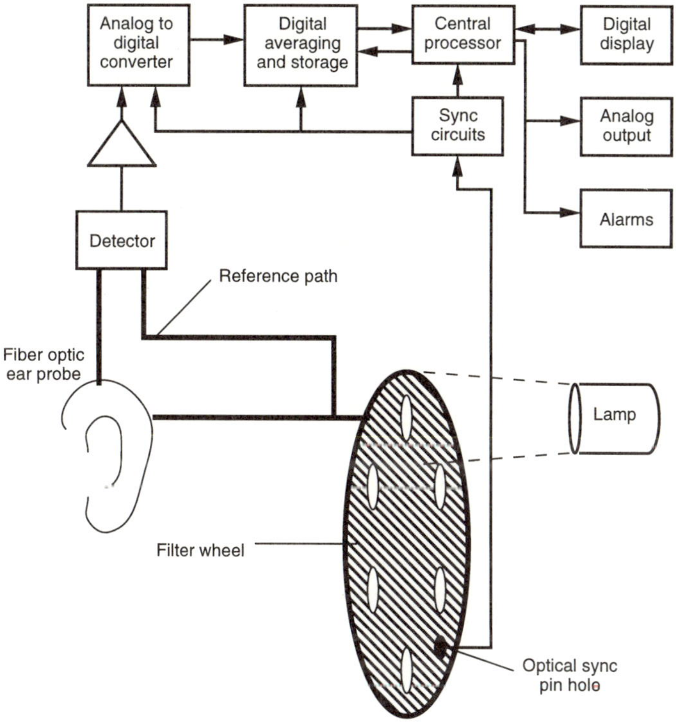

Figure 3.8 Block diagram of the Hewlett-Packard Model 47201A eight-wavelength ear oximeter.

The filtered light travels down two fiber optic paths. The first path leads directly to the photodetector and acts as a reference. This prevents any variations in the measurements due to changes in the light source or slight variations in the filters. The second fiber optic path goes to the ear probe. The filtered light is transmitted through the ear to a fiber optic cable which carries it to the photodetector. The current produced by the photodetector may be on the order of 0.5 nA so the output of the photodetector is amplified by a factor of 10^8. The absorbance is derived from the difference between the reference intensity and the intensity of the light transmitted through the ear.

For a time, the Hewlett-Packard device was the gold standard for oximeters. It worked fairly well and was the first introduction of noninvasive oximetry into a clinical environment. But it was found to be inaccurate for saturations less than 70%. And though it was a large improvement over previous devices, the Hewlett-Packard device was not totally immune to motion artifacts or skin pigmentation as it claimed. Also, it still required that the probe heat the skin. Devices that heat the skin put the patient at risk of burns, especially infants who have sensitive skin.

Although giving an improvement in performance, the device was huge, weighing almost 17 kg. The ear probe was also quite large and the fiber optics were fragile. Though it was the *gold standard* of oximeters in its time (Moyle 1994) it was used clinically only in sleep studies, pulmonary medicine, and physiology. The HP eight-wavelength oximeter was never used in anesthesiology or critical care as the pulse oximeter is today. Its use declined even more with improvements in the Clark transcutaneous PO_2 electrode (Severinghaus 1987).

3.7 PULSE OXIMETERS

The idea of exploiting the pulsatile nature of arterial blood in oximetry first belonged to Takuo Aoyagi while working in Japan for Nihon Kohden Corporation (Severinghaus 1987). Nihon Kohden's device used analog circuitry, had bulky fiber optic cables, and still had some of the instability problems of the Hewlett-Packard device. Other companies such as Minolta came up with similar products with similar problems (Santamaria and Williams 1994).

An anesthesiologist named William New saw the pulse oximeter marketed by Minolta and saw how to improve it. New, also an electrical engineer, teamed with Jack Lloyd to found Nellcor, Inc. Nellcor produced a microprocessor-based pulse oximeter, the N100, which was smaller, less expensive, needed no user calibration, and was accurate enough for clinical use. Nellcor is still the market leader in pulse oximetry (Santamaria and Williams, 1994). About the same time, Ohmeda came up with a similar device, the Biox II, which had similar success (Wukitsch *et al* 1988). Today, pulse oximeters exist in every intensive care unit, surgical suite, and in many emergency rooms in the United States (Santamaria and Williams 1994)

This section gives a brief description of the major parts of a pulse oximeter. Further detail of each of these parts can be found in later chapters.

3.7.1 Overview

By taking advantage of the pulsatile flow of blood, the pulse oximeter is able to overcome many of the problems of earlier technologies. The pulse oximeter tracks the change in light absorbance as the blood pulses. By tracking this peak-to-peak ac component, the absorbance due to venous blood or tissue does not have any effect on the measurement.

Light scattering is still a source of inaccuracy in pulse oximeters. Beer's law does not account for the scattering of light. So a direct calculation of S_aO_2 is not possible. The pulse oximeter measures absorbances at the two wavelengths and uses data from CO-oximeters to empirically look up a value for S_pO_2, an estimation of S_aO_2.

3.7.2 LEDs

One of the large improvements of the pulse oximeter over earlier oximeters is the use of LEDs as the light source. The LEDs can transmit large intensities of light proportional to the amount of drive current. The LED control block in figure 3.9 controls the amount of drive current and the timing of the LEDs. The timing of the pulsations is critical because the photodiode cannot distinguish between different wavelengths of light. The pulse oximeter relies on the microprocessor system to synchronize the pulsations of the LEDs with the samples taken by the ADC so that the absorbances detected by the photodiode can be attributed to the correct LED.

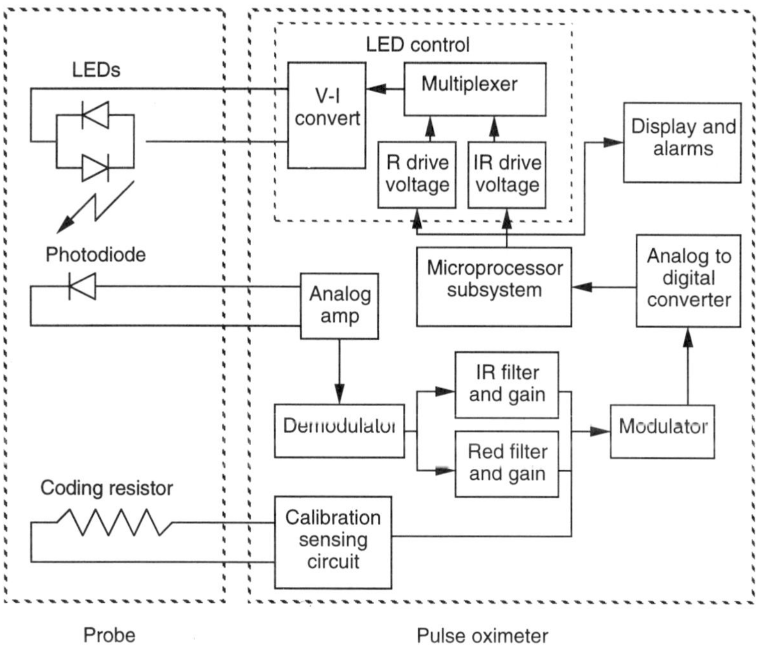

Figure 3.9 Block diagram of a pulse oximeter system. The arrows show the flow of data. The microprocessor also provides control and timing for the demodulator, modulator, and LED control circuits.

The two wavelengths chosen for pulse oximetry are 660 nm and 940 nm. These wavelengths were chosen based on the availability of LEDs at these wavelengths and because the extinction coefficients of Hb and HbO_2 vary as much as possible. HbO_2 has a higher extinction coefficient than Hb at 940 nm and a lower extinction coefficient at 660 nm. In other words, as S_aO_2 increases, the absorbance of light increases at 940 nm and decreases at 660 nm.

One disadvantage of using LEDs as a light source is that the exact wavelength of any single LED can vary by as much as ±15 nm. This would cause significant errors if unaccounted for. To account for this, some manufacturers characterize each LED and code it with a resistor value. By driving the coding resistor (figure 3.9) with a constant current source, the pulse oximeter can measure the voltage and take the characterization of the LEDs into account when empirically calculating S_pO_2 (Pologe 1987).

3.7.3 Photodiode

The photodetector is a silicon photodiode that produces current linearly proportional to the intensity of light striking it. Advances in silicon technology allow the photodiode to be small enough to fit in small, finger tip probes. These advances have helped make the pulse oximeter much more accurate and convenient than earlier devices. Early oximeter devices needed frequent calibration because the photoelectric devices used as sensors were often inconsistent (Miller 1966).

A photodiode cannot distinguish between red and infrared light, but to accommodate this, the microprocessor system alternately turns each LED on and off. The pulse oximeter repeatedly samples the photodiode output while the red LED is on, while the infrared LED is on, and while both are off. By sampling with both LEDs off, the pulse oximeter is able to subtract any ambient light that may be present (Pologe 1987).

3.7.4 Probes

Improved technology in photodiodes and LEDs has another benefit. They allow the probe to be small and attach to the pulse oximeter with conventional wires. The Hewlett-Packard eight wavelength oximeter was considered an accurate device, but because bulky fiber optic cables were needed to carry the light source to the patient and the transmitted light back to a light sensor, it was impractical (Rebuck *et al* 1983). Probes for the pulse oximeter are not only smaller, but can be disposable.

Figure 3.10 shows a transmission pulse oximeter and a reflectance pulse oximeter. As the names indicate, a transmission pulse oximeter measures the amount of light that passes through the tissue as in a finger probe. A reflectance pulse oximeter measures the amount of light reflected back to the probe. Both types use the same technology differing only in positioning of the probes and calibration.

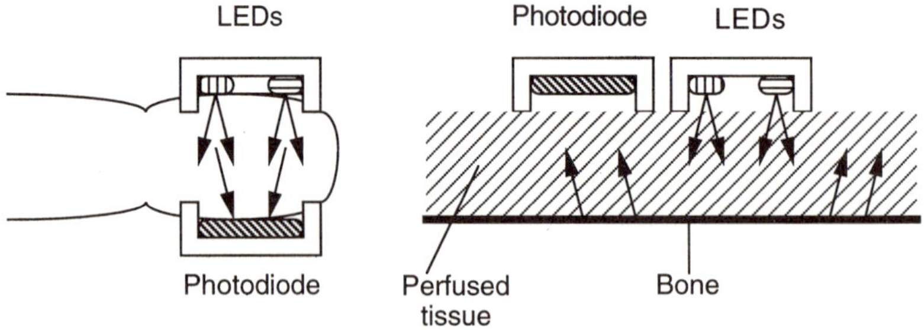

Figure 3.10 On the left is a transmission pulse oximeter measuring the transmission of light by two LEDs through the finger of a patient. On the right is a reflectance pulse oximeter measuring the amount of light reflected back to the probe.

3.7.5 Analog amplifier and signal processing

The photodiode generates a current proportional to the intensity of light. The analog amplifier converts this current to a voltage. Because the change in voltage due to the pulsations of the arteries is small in comparison to the dc portion of the signal, the dc component of the signal is subtracted from the rest of the signal by the demodulator. The demodulator also uses a sample-and-hold timing circuit to separate samples from the red LED from the samples of the infrared LED. The ac portions of these signals are low-pass filtered to remove electromagnetic interference. Then each signal goes through a programmable gain circuit after which a multiplexer with another sample-and-hold circuit modulates the red and infrared signals back into one to go through an analog-to-digital converter (ADC) for use by the microprocessor.

Using the data gathered from the ADC, the microprocessor calculates what is called a ratio of ratios. From this ratio of ratios and the value of the coding resistor, the microprocessor goes to an empirical look up table for its S_pO_2 value. The empirical look up table is generated by the manufacturer through laboratory tests done with a CO-oximeter. The signal processing algorithms also provide some noise reduction. Some pulse oximeters use an ECG in their signal processing algorithm to minimize errors due to motion artifacts.

3.7.6 A three-wavelength pulse oximeter for COHb determination

Current pulse oximeters estimate the arterial oxygen saturation of the blood by measuring absorbances at two wavelengths of light. Because of this, the pulse oximeter is only able to account for Hb and HbO_2. Increased levels of COHb, for example, will cause an overestimation in S_aO_2 because the pulse oximeter cannot distinguish between HbO_2 and COHb. In cases of carbon-monoxide poisoning, this could have terrible consequences if the clinician is unaware.

Table 3.1 A comparison of pulse oximetry and transcutaneous PO_2 electrodes from New (1985), Barker and Tremper (1984), and Severinghaus (1987).

Pulse oximeters	Transcutaneous $P_{tc}O_2$ electrodes
Require no heating	Have internal heaters which can cause burns and must be moved periodically to avoid skin damage, especially in infants
Have no delay	Require skin and electrode preparation and a warm up period of up to ten minutes
Never require user calibration	Require recalibration
Probes are clipped or taped on	Require operator skill to place monitor
Measure a pure respiratory variable (S_aO_2) and a pure circulatory variable (plethysmograph)	Are a sensitive, but not specific, monitor of blood oxygenation; a drop in $P_{tc}O_2$ may be caused by respiratory deficiency, circulatory deficiency, or both
Give an accurate reading or none at all	Report low PO_2 when electrode may not be placed well
Require pulsating arteries; fails during cardiac arrest, cardiopulmonary bypass, or distal placement to blood pressure cuff.	Detect low cardiac output
Require hemoglobin in the bloodstream and may fail with severe anemia or hemodilution	Are not dependent on hemoglobin
Can be in error with high levels of dyshemoglobin species present in the blood	
Display pulse rate	Do not display pulse rate

Buinevicius (1987) designed a three wavelength pulse oximeter to solve this problem. An additional LED at 810 nm was used in an attempt to determine the amount of COHb in the blood. Buinevicius also presented a method to calibrate the pulse oximeter for three wavelengths using three-dimensional solutions to Beer's law.

3.7.7 Comparison of pulse oximetry to transcutaneous PO_2 electrodes

Pulse oximeters and transcutaneous PO_2 electrodes are the two main technologies used to provide continuous information about the supply of oxygen to the body. Table 3.1 provides a comparison between the two technologies.

REFERENCES

Adams A P and Hahn C E W 1982 *Principles and practice of blood-gas analysis* 2nd edn (New York: Churchill Livingstone)

Barker S J 1991 Pulmonary artery oximetry *Proc. Optical Fibers in Medicine VI*, (Los Angeles, CA 1991) (Bellingham, WA: SPIE Optical Engineering Press)

Barker S J and Tremper K K 1984 Transcutaneous oxygen tension: a physiological variable for measuring oxygenation *J. Clin. Monitoring* **1** (2) 130–4

Buinevicius R P 1987 A three wavelength pulse oximeter for carboxyhemoglobin determination *MSc thesis*, Department of Electrical and Computer Engineering, University of Wisconsin-Madison

Burtus C A and Ashwood E R 1994 *Tietz Textbook of Clinical Chemistry* 2nd edn (Philadelphia PA: Saunders)

Dennis R C and Valeri C R 1980 Measuring percent oxygen concentration of hemoglobin, percent carboxyhemoglobin and methemoglobin, and concentrations of total hemoglobin and oxygen in blood of man, dog and baboon *Clin. Chem.* **26** 1304–8

Gradwohl R B H 1948 *Clinical Laboratory Methods and Diagnosis* 4th edn (St. Louis, MO: Mosby)

Gothgen I H and Jacobsen E 1987 Computing the oxygen status of the blood from heated skin PO_2 *Continuous transcutaneous monitoring* ed A Huch, R Huch and G Rooth (New York: Plenum)

Hill D W 1966 Methods of measuring oxygen content of blood *Oxygen Measurements in Blood and Tissues and their Significance* ed J P Payne and D W Hill (Boston: Little, Brown) **4** (1) 63–80

Merrick E B and Hayes T J 1976 Continuous non-invasive measurements of arterial blood oxygen levels *Hewlett-Packard J.* **28** (2) 2–9

Miller S E 1966 *A Textbook of Clinical Pathology* 7th edn (Baltimore, MD: Williams and Wilkins)

Moyle J T B 1994 *Pulse Oximeters* (London: BMJ)

New W Jr 1985 Pulse oximetry *J. Clin. Monitoring* **1** (2) 126–9

Nilsson N J 1960 Oximetry *Physiol. Rev.* **40** 1–22

Ono K, Masahiko K, Hiramoto J, Yotsuya K and Sato N 1991 Fiber optic reflectance oximeter spectrophotometry system for in vivo tissue diagnosis *Appl. Opt.* **30** 98–104

Payne J P and Severinghaus J W (eds) 1986 *Pulse Oximetry* (London: Springer)

Peterson J F 1986 The development of pulse oximetry *Science* **232** G135–40

Peterson J I and Fitzgerald R V 1984 Fiber-optic probe for in vivo measurement of oxygen partial pressure *Anal. Chem.* **56** 62–7

Peura R A 1998 Chemical biosensors *Medical Instrumentation: Application and Design* 3rd edn, ed J G Webster (New York: Wiley)

Polanyi M L and Hehir R M 1962 In vivo oximeter with fast dynamic response *Rev. Sci. Instrum.* **33** 1050–4 (Reprinted 1990 *Selected Papers on Optical Fibers in Medicine* ed B Thompson (Bellingham WA: SPIE Optical Engineering Press))

Pologe J A 1987 Pulse oximetry: technical aspects of machine design *Int. Anesthesiol. Clinics* **25** (3) 137–53

Rebuck A S, Chapman K R and D'Urzo A 1983 The accuracy and response characteristics of a simplified ear oximeter *Chest* **83** 860–4

Santamaria T and Williams J S 1994 Pulse oximetry *Med. Dev. Res. Rep.* **1** (2) 8–10

Severinghaus J and Astrup P 1987 History of blood gas analysis *Int. Anesthesiol. Clinics* **25** (4) 1–225

Severinghaus J W 1987 History, status, and future of pulse oximetry *Continuous transcutaneous monitoring* ed A Huch, R Huch and G Rooth (New York: Plenum)

Shapiro B A, Harrison R A, Cane R D and Templin R 1989 *Clinical Application of Blood Gases* 4th edn (Chicago: Year Book Medical)

Waxman K, Sadler R, Eisner M E, Applebaum R, Tremper K K and Mason G R 1983 Transcutaneous oxygen monitoring of emergency department patients *Am. J. Surg.* **146** 35–7

Wukitsch M W, Peterson M T, Tobler D R and Pologe J A 1988 Pulse oximetry: analysis of theory, technology and practice *J. Clin. Monitoring* **4** (4) 290–301

Zijlstra W G 1958 *A Manual on Reflection Oximetry* (Assen: Van Gorcum)

Zwart A, Buursma A, Oeseburg B and Zijlstra W G 1981 Determination of hemoglobin derivatives with the IL-282 CO-oximeter as compared with a manual spectrophotometric five wavelength method *Clin. Chem.* **27** (11) 1903–6

INSTRUCTIONAL OBJECTIVES

3.1 Explain why transcutaneous PO_2 electrodes require the skin to be heated.
3.2 Explain why a CO-oximeter uses hemolyzed blood samples to determine the hemoglobin components of the blood.
3.3 Explain the difference between absorptivity and absorbance.
3.4 Describe a noninvasive two-wavelength oximeter and its problems.
3.5 Describe a two-wavelength fiber optic oximeter.
3.6 Describe an eight-wavelength oximeter.
3.7 Describe how pulse oximeters overcome some of the problems of earlier oximeters.
3.8 Explain the need for a coding resistor in a pulse oximeter probe.
3.9 Explain why Beer's law cannot be used for direct computation of S_pO_2 and empirical lookup tables are used instead.
3.10 Given the concentrations of oxyhemoglobin and reduced hemoglobin for blood, calculate S_pO_2.
3.11 Explain why a pulse oximeter might not be as accurate for a patient who is a smoker.
3.12 Describe a three-wavelength pulse oximeter to determine COHb concentration and explain it might be more accurate for a patient who is a smoker than a two-wavelength pulse oximeter.

CHAPTER 4

LIGHT ABSORBANCE IN PULSE OXIMETRY

Oliver Wieben

This chapter describes the theoretical background for the measurement of light absorbance in pulse oximetry as a basis for determining arterial oxygen saturation. Beer's law and the derivation of a theoretical calibration curve for measured light absorbances in pulse oximeters is explained, although this curve is not valid in practice due to the scattering of light. Beer's law is used accurately to determine the oxygen saturation of hemoglobin solutions but does not apply for whole blood because of the scattering effects. Nevertheless, this model helps to develop an understanding of the absorbance of light as it passes through living tissue and why and how pulse oximetry works. The normalization of the measured signals and the calibration curves used in pulse oximeters are explained after an introduction of the theoretical model. The final part of the chapter describes mathematical approaches to incorporate light scattering in the model and describe its effects qualitatively and quantitatively.

4.1 BEER'S LAW

Beer's law (also referred to as Beer–Lambert's or Bouguer's law) describes the attenuation of light traveling through a uniform medium containing an absorbing substance. If monochromatic incident light of an intensity I_0 enters the medium, a part of this light is transmitted through the medium while another part is absorbed. The intensity I of light traveling through the medium decreases exponentially with distance

$$I = I_0 \ e^{-\varepsilon(\lambda)cd} \tag{4.1}$$

where $\varepsilon(\lambda)$ is the *extinction coefficient* or absorptivity of the absorbing substance at a specific wavelength, c the concentration of the absorbing substance which is constant in the medium, and d the optical path length through the medium (see figure 4.1). The concentration c is measured in mmol L^{-1} and the extinction coefficient is expressed in L $mmol^{-1}$ cm^{-1}.

Beer's law is based on the property that the sum of transmitted and absorbed light equals the incident light. It does not account for physical processes which include reflection of the light at the surface of the medium or scattering of light in the medium.

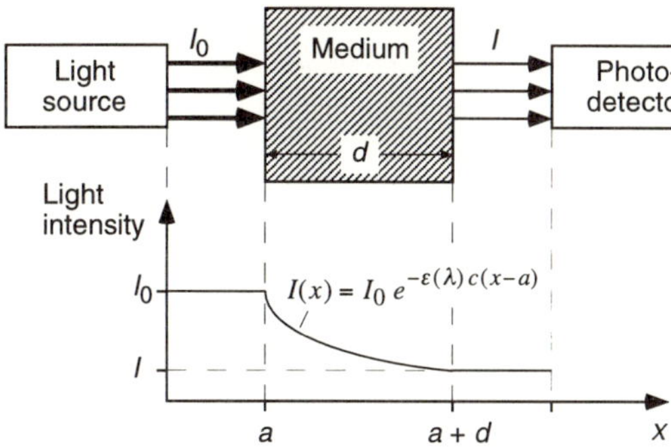

Figure 4.1 Beer's law: Incident light of intensity I_0 travels the distance a from a light source to the medium without being absorbed in the air. The light intensity decreases exponentially with distance in the absorbing medium. The intensity of the transmitted light I is determined by Beer's law. It stays constant after exiting the medium with optical path length d and can be measured by a photodetector.

4.1.1 Transmittance and absorbance of light

The *transmittance* (T) of light traveling through a medium with an absorbing substance is defined as the ratio of transmitted light I to the incident light I_0

$$T = \frac{I}{I_0} = e^{-\varepsilon(\lambda)cd}. \tag{4.2}$$

The *unscattered absorbance* (A) of this process is defined as the negative natural logarithm of the transmittance of light

$$A = -\ln T = \varepsilon(\lambda)cd. \tag{4.3}$$

The absorbance is sometimes referred to as the optical density of a medium.

4.1.2 Multiple absorbers

The properties of Beer's law are valid even if more than one substance absorbs light in the medium. Each absorber contributes its part to the total absorbance. The mathematical representation of this system of absorbers is a superposition of the individual absorbing processes. The resulting *total absorbance* A_t of light in a medium with n absorbing substances is the sum of their n independent absorbances

$$A_t = \varepsilon_1(\lambda)c_1 d_1 + \varepsilon_2(\lambda)c_2 d_2 + \ldots + \varepsilon_n(\lambda)c_n d_n = \sum_{i=1}^{n} \varepsilon_i(\lambda)c_i d_i \tag{4.4}$$

where $\varepsilon_i(\lambda)$ and c_i represent the extinction coefficient and concentration of the substance i and d_i represents the optical path length through the absorbing substance, which may differ from substance to substance in the medium.

Therefore, Beer's law allows us to determine the unknown concentrations of n different absorbing substances in a homogeneous medium if the absorbance of light is measured at n different wavelengths and the extinction coefficients of the substances are known.

4.2 HEMOGLOBIN EXTINCTION COEFFICIENTS

Hemoglobin is the main light absorber in human blood at wavelengths used in pulse oximetry. The absorbing characteristics of hemoglobin change with its chemical binding and the wavelength of the incident light. Although oxygenated and reduced hemoglobin absorb most of the light passing through blood, they do not represent the only two hemoglobin species present in human blood. Hemoglobin may combine with other substances such as carbon monoxide or hydrogen sulfide as well, which changes its color.

4.2.1 Functional hemoglobins

Binding oxygen in the pulmonary capillaries and releasing it in the systemic capillaries is the main purpose of hemoglobin. Hemoglobins that are able to bind reversibly with molecular oxygen are called *functional hemoglobins.*

When hemoglobin is fully saturated with oxygen (carrying four oxygen molecules), it is called *oxyhemoglobin* (HbO_2). If it is not fully saturated with oxygen it is called *reduced hemoglobin* (Hb). Therefore oxyhemoglobin and reduced hemoglobin are functional hemoglobins.

Most of the hemoglobins in a healthy individual are functional hemoglobins. The *functional oxygen saturation* (functional SO_2) is measured in percentage and determined by the amount of oxygenated hemoglobin (HbO_2) as compared to the sum of oxygenated and reduced hemoglobin (Hb). Another way to define this ratio is to use the concentrations of oxygenated hemoglobin (c_{HbO_2}) and reduced hemoglobin (c_{Hb})

$$\text{Functional } SO_2 = \frac{HbO_2}{Hb + HbO_2} \times 100\% = \frac{c_{HbO_2}}{c_{HbO_2} + c_{Hb}} \times 100\%. \qquad (4.5)$$

The functional oxygen saturation of explicitly arterial blood is called functional arterial oxygen saturation (functional S_aO_2) and is referred to as functional hemoglobin saturation as well.

4.2.2 Dysfunctional hemoglobins

Dysfunctional hemoglobins (or dyshemoglobins) do not support the transport of oxygen to the tissues. They are either unable to bind reversibly to oxygen or interfere with the ability of oxyhemoglobin to release its oxygen to the tissue.

The four most common dyshemoglobins are methemoglobin (MetHb), carboxyhemoglobin (COHb), sulfhemoglobin, and carboxysulfhemoglobin.

4.2.2.1 Methemoglobin. Methemoglobin is oxidized hemoglobin. It is a result of oxidation of a free heme iron (Fe^{2+}) instead of the reversible binding of oxygen to heme inserted into globin subunits.

$$HbFe^{2+} \Leftrightarrow HbFe^{3+} + e^{-}. \quad (4.6)$$

An enzyme system (including cytochrome b_5) is responsible for reducing the methemoglobin in the red cells by maintaining hemoglobin in the reduced state (Fe^{2+}).

Oxidized hemoglobin subunits are not capable of binding oxygen and altering the oxygen binding of the remaining ferrous hemes. Therefore, methemoglobin has a great influence on the functionality of hemoglobin. Under physiological circumstances the amount of methemoglobin remains below 0.6% of the total hemoglobin and this amount varies at a rate of 2 to 3% during the day. The absorbance spectrum of methemoglobin is strongly pH-dependent (Bunn 1986).

4.2.2.2 Carboxyhemoglobin. Carboxyhemoglobin is formed when hemoglobin combines with carbon monoxide (CO). The carbon atom of carbon monoxide is bonded to the iron atom of heme.

The affinity of hemoglobin binding with carbon monoxide is approximately 210 times larger than that of oxygen. Therefore, the presence of a high level of carbon monoxide will reduce the amount of oxygenated hemoglobin significantly. The level of carboxyhemoglobin in the blood varies with the habits and surroundings of the individual. Smoking, working in underground garages, traffic tunnels, mines, etc. increases the amount of CO in the blood. In nonsmokers the level of COHb is usually below 2% but this value varies with the local environment (Wukitsch *et al* 1988).

4.2.2.3 Sulfhemoglobin and carboxysulfhemoglobin. The reaction of oxyhemoglobin and hydrogen sulfide produces sulfhemoglobin. The relevant chemical reactions are complex and not thoroughly understood, although the absorbance spectrum of sulfhemoglobin is known.

The oxygen affinity of the heme iron in sulfhemoglobin is 100-fold lower than the oxygen affinity of unmodified hemoglobin (Bunn 1986). This chemical reaction is irreversible (Nellcor 1993). Carboxysulfhemoglobin results from a reaction of sulfhemoglobin with carbon monoxide. The concentrations of sulfhemoglobin and carboxysulfhemoglobin in human blood are usually not significant.

4.2.2.4 Fractional hemoglobin saturation. The *fractional oxygen saturation* is the fraction of oxygenated hemoglobin to the total hemoglobin. It is usually measured in percentage as well and is determined by the ratio of the concentrations of oxygenated hemoglobin to total hemoglobin

$$\text{Fractional } SO_2 = \frac{c_{\text{HbO}_2}}{c_{\text{total hemoglobin}}} \times 100\% \quad (4.7)$$

where total hemoglobin represents all different species of hemoglobin present in the blood.

4.2.3 Hemoglobin absorbance spectra

The chemical binding of the different hemoglobin species changes the physical properties of the hemoglobin as well. Figure 4.2 shows the extinction coefficients of oxyhemoglobin, reduced hemoglobin, methemoglobin and carboxyhemoglobin at wavelengths in the range of interest in pulse oximetry.

The absorbance of light in the red region of the spectrum is much higher for reduced hemoglobin than for oxyhemoglobin. The extinction coefficients of both hemoglobin species are equal at the point isosbestic point (805 nm). The reduced hemoglobin is more transparent to light from the infrared region than oxyhemoglobin.

The extinction coefficient of carboxyhemoglobin is about the same as that of oxyhemoglobin at the wavelength of 660 nm while it is almost transparent in the infrared region. Methemoglobin absorbs much light in the red region of the spectrum and its extinction coefficient remains higher than that of oxyhemoglobin in the infrared region.

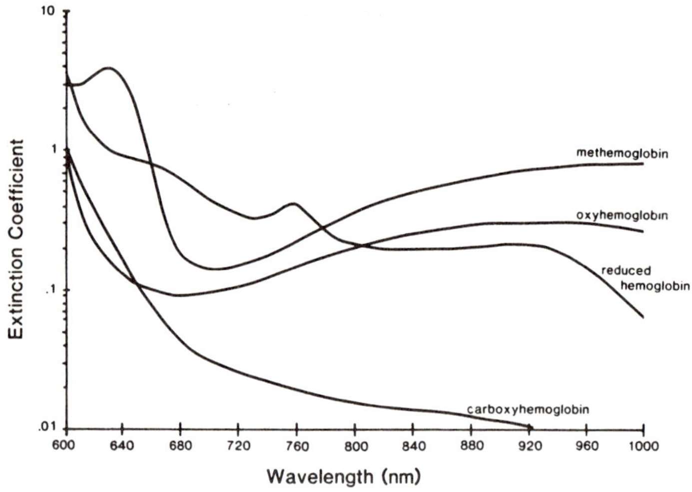

Figure 4.2 Extinction coefficients of the four most common hemoglobin species oxyhemoglobin, reduced hemoglobin, carboxyhemoglobin, and methemoglobin at the wavelengths of interest in pulse oximetry (courtesy of Susan Manson, Biox/Ohmeda, Boulder, CO).

4.3 BEER'S LAW IN PULSE OXIMETRY

Pulse oximeters determine the oxygen saturation of arterial blood by measuring the light absorbance of living tissue at two different wavelengths and using the arterial pulsation to differentiate between absorbance of arterial blood and other absorbers.

4.3.1 Criteria for the choice of wavelengths

Different reasons lead to the most common choice for wavelengths used in pulse oximetry. The red skin pigmentation absorbs a great amount of light at wavelengths shorter than 600 nm and therefore it is not desirable to measure light absorbance in this range. Large differences in the extinction coefficients of reduced hemoglobin and oxygenated hemoglobin change the absorbance of light significantly, even when the oxygen saturation changes slightly. A good choice for a wavelength in the red region is 660 nm because of a large difference in the extinction coefficients.

Another issue for the wavelength choice is flatness of the absorption spectra shown in figure 4.2 around the chosen wavelength. Otherwise shifts in the peak wavelength of the LEDs (see section 5.3) will result in a larger error. The absorbance spectra of reduced hemoglobin and oxygenated hemoglobin are relatively flat at 660 and 940 nm (Moyle 1994).

Mannheimer *et al* (1997) have shown that sensors fabricated with 735 and 890 nm emitters read more accurately at low saturations under a variety of conditions, while 660 and 990 nm emitters read more accurately at high saturations.

4.3.2 Absorbance in hemoglobin solutions

The different species of hemoglobin are the main light absorbers in arterial and venous blood. Most of the hemoglobin in human blood is either oxygenated or reduced hemoglobin which determine the functional oxygen saturation SO_2 (equation (4.5)). The concentrations of oxygenated hemoglobin (c_{HbO_2}) and reduced hemoglobin (c_{Hb}) can be expressed as a function of SO_2 as a fraction and the sum of the concentrations c_{HbO_2} and c_{Hb}

$$c_{HbO_2} = SO_2(c_{HbO_2} + c_{Hb}) \tag{4.8}$$

$$c_{Hb} = (1 - SO_2)(c_{HbO_2} + c_{Hb}). \tag{4.9}$$

According to Beer's law we derive the total absorbance A_t of a solution containing only reduced and oxygenated hemoglobin as absorbing substances from equation (4.4)

$$A_t = \varepsilon_{HbO_2}(\lambda) c_{HbO_2} d_{HbO_2} + \varepsilon_{Hb}(\lambda) c_{Hb} d_{Hb}. \tag{4.10}$$

Assuming that the optical path length d is the same for the oxygenated hemoglobin (d_{HbO_2}) and reduced hemoglobin (d_{Hb}) and using equations (4.8), (4.9), and (4.10), we derive

$$A_t = \left[\varepsilon_{HbO_2}(\lambda) SO_2 + \varepsilon_{Hb}(\lambda)(1 - SO_2)\right](c_{Hb} + c_{HbO_2})d. \tag{4.11}$$

Thus A_t can be expressed for known concentrations of hemoglobin in terms of functional oxygen saturation as a fraction, the extinction coefficients of hemoglobin, and the length of the optical path. Values for the extinction coefficients of adult reduced hemoglobin (ε_{Hb}) and adult oxygenated hemoglobin

(ε_{HbO_2}) at the two wavelengths most commonly used in pulse oximetry (660 nm and 940 nm) have been measured by Zijlstra *et al* (1991) (see table 4.1).

Table 4.1 Table of extinction coefficients of reduced and oxygenated hemoglobin in adults at the wavelengths of 660 nm and 940 nm (values from Zijlstra *et al* 1991).

Wavelength, nm	Extinction coefficient, L mmol^{-1} cm^{-1}	
	Hb	HbO_2
660	0.81	0.08
940	0.18	0.29

Figure 4.3 shows the characteristics of light absorbance for a sample with a fixed concentration of total functional hemoglobin ($c_{HbO_2} + c_{Hb}$) of 1 mmol L^{-1}, a fixed path length d of 1 cm and varying functional oxygen saturations. The two lines shown in figure 4.4 represent the properties for the two most commonly used wavelengths in pulse oximetry (660 nm and 940 nm). The absorbance of light at a wavelength of 940 nm increases with an increased oxygen saturation. At 660 nm the absorbance of light decreases rapidly with an increasing functional oxygen saturation (Pologe 1987).

It is possible to determine the concentrations of hemoglobins in hemoglobin solutions or hemolized blood by using a device such as a spectrophotometer (see section 3.3).

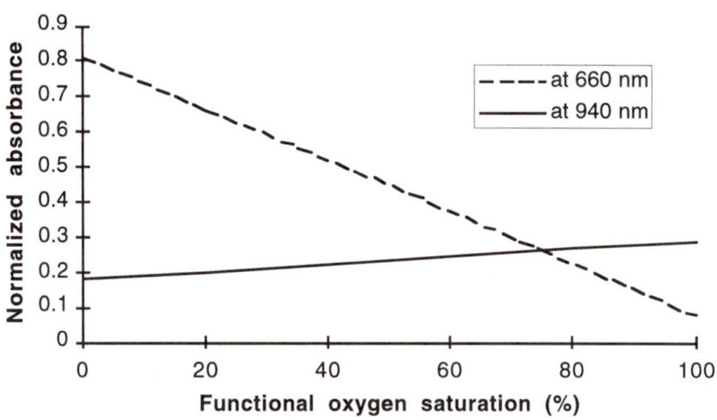

Figure 4.3 Changes in light absorbance in hemoglobin solutions as a function of functional oxygen saturation for the wavelengths used in pulse oximetry. Absorbance decreases rapidly with increasing oxygen saturation at 660 nm (dashed line) but increases slightly with increasing oxygen saturation at 940 nm (solid line).

4.3.3 Pulsation of the blood

Light traveling through biological tissue (e.g. the finger or earlobe) is absorbed by different absorbing substances. Primary absorbers of light in the region of interest are the skin pigmentation, bones, and the arterial and venous blood. Instead of measuring the arterial oxygen saturation of the blood *in vitro* with a sample of arterial blood and a spectrophotometer, or at a wide range of different wavelengths as with the Hewlett-Packard ear oximeter, pulse oximeters take advantage of *arterial pulsation*. Figure 4.5 shows the amount of absorbed and transmitted light in living tissue as a function of time.

The arteries contain more blood during systole than during diastole, and therefore, their diameter increases due to increased pressure. This effect occurs only in the arteries and arterioles but not in the veins. The absorbance of light in tissues with arteries increases during systole mainly because of the larger amount of absorbing substances (hemoglobin), due to the fact that the optical path length d in the arteries increases. This alternating part of the total absorbance allows us to differentiate between the absorbance due to venous blood, a constant amount of arterial blood, and other nonpulsatile components such as skin pigmentation (dc component of the total absorbance) and the absorbance due to the pulsatile component of the arterial blood (ac component). The alternating part of the light absorbed by the living tissue usually does not exceed 1% to 2% of the constant absorbance of the dc components. The time varying signal of transmitted light is referred to as the plethysmographic (or photoplethysmographic) signal.

The intensity of the light passing through the tissue during diastole is high (I_H). The absorbers that are present during diastole are the DC components. All DC components except the nonpulsating arterial blood are collectively represented by $\varepsilon_{DC}(\lambda)$, c_{DC}, and d_{DC}. The diameter of the arterial vessels is minimal (d_{min}) and therefore the absorbance due to arterial hemoglobin is minimal and the amount of transmitted light is high (I_H) and has a peak (see figures 4.4 and 4.5)

$$I_H = I_0 \, e^{-\varepsilon_{DC}(\lambda) c_{DC} d_{DC}} e^{-[\varepsilon_{Hb}(\lambda) c_{Hb} + \varepsilon_{HbO_2}(\lambda) c_{HbO_2}] d_{min}} \qquad (4.12)$$

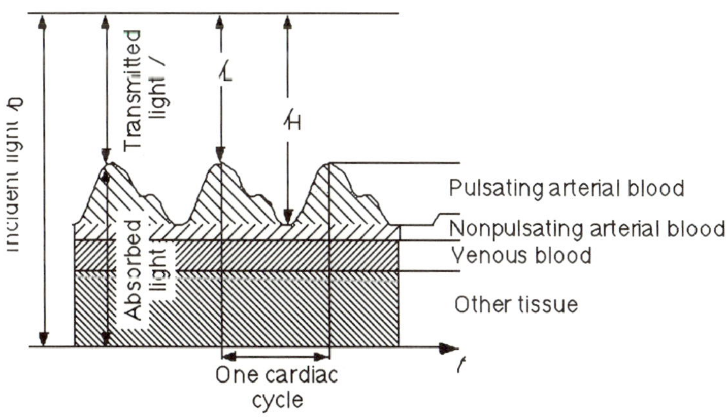

Figure 4.4 Absorbed and transmitted light in living tissue. The amount of absorbed light correlates with the pulsation of arterial blood. A constant amount of light is absorbed by the skin pigmentation, bone, other tissue, venous blood and the nonpulsating part of the arterial blood. More blood is present in the arteries during systole and therefore more light is absorbed. The intensity of the transmitted light varies from I_H (maximum) to I_L (minimum) within one cardiac cycle.

The optical path length in the arteries increases during the systole to d_{max}. The amount of absorbed light reaches a maximum peak and therefore the transmitted light reaches the low peak I_L:

$$I_L = I_0 \, e^{-\varepsilon_{DC}(\lambda) c_{DC} d_{DC}} e^{-[\varepsilon_{Hb}(\lambda) c_{Hb} + \varepsilon_{HbO_2}(\lambda) c_{HbO_2}] d_{max}}. \tag{4.13}$$

The light intensity I of the light arriving at the photodetector is a function of the diameter d of the arteries and arterioles. During one cardiac cycle this diameter changes from d_{min} to d_{max}. By substituting d with $d_{min} + \Delta d$ we derive the following expression from Beer's law, where I is expressed as a function of I_H and Δd, the part of the diameter that changes from 0 to $d_{max} - d_{min}$ with time

$$I = I_H e^{-[\varepsilon_{Hb}(\lambda) c_{Hb} + \varepsilon_{HbO_2}(\lambda) c_{HbO_2}] \Delta d}. \tag{4.14}$$

Figure 4.5 shows these properties in a simplified model.

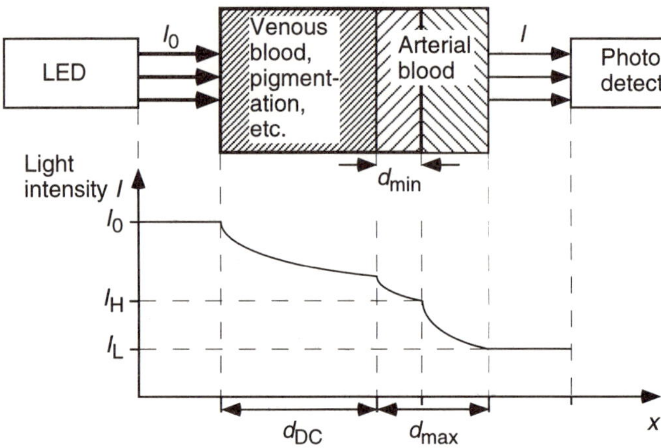

Figure 4.5 Beer's law in pulse oximetry. The DC components of the tissue (e.g. skin pigmentation, bone, venous blood and the nonpulsating part of the arterial blood) absorb a constant amount of the incident light I_0. The effective optical path length in the DC components without the constant level of arterial blood is represented by d_{DC}. During diastole the optical path length through the arteries has a minimum length of d_{min} and the light intensity at the photodetector is maximal (I_H). The optical path length reaches a maximum d_{max} during systole and the hemoglobin in the arteries absorbs a maximum amount, causing I to decrease to a minimum level of I_L.

4.3.4 Measurement of pulse oximeters

The reading of the pulse oximeter S_pO_2 is an estimation of the arterial oxygen saturation S_aO_2. Measuring at two wavelengths allows us to distinguish the concentrations of only two different absorbers (Hb and HbO_2). But in humans more species of hemoglobin, such as carboxyhemoglobin and methemoglobin, are present. These other hemoglobins absorb light as the functional hemoglobins do and therefore influence our measurements. As long as we do not measure at as many wavelengths as absorbers are present in the blood, we can not determine the concentrations of Hb and HbO_2 and therefore the arterial oxygen saturation correctly (Barker and Tremper 1987).

Due to the fact that Hb and HbO_2 are the main absorbers, the error may be small. Nevertheless, the results of determining either the actual functional or fractional oxygen saturation (see equations (4.5) and (4.7)) of the arterial blood are not exact. This problem is also discussed in sections 10.1.1 and 11.1.1. The oximeter reading becomes less accurate if the concentrations of dyshemoglobins are larger than in normal humans. Section 11.7 deals with the presence of high concentrations of dysfunctional hemoglobins.

4.4 SATURATION VERSUS NORMALIZED RATIO

The arterial oxygen saturation can be derived based on Beer's law as a function of the ratio of absorbances at two wavelengths. Due to nonlinearities in the LEDs, the photodetector, and light absorbance in the tissue, the absorbances have to be normalized in the ratio. This model results in a theoretical calibration curve, but it is not used in practice as will be described in the following sections.

4.4.1 Normalization

The measured light intensities at the different wavelengths have to be *normalized* before they can be compared with each other due to the fact that the light-emitting diodes (LEDs) may emit light with different intensities. The absorbing characteristics of the DC components and the sensitivity of the photodetector differ for the two different wavelengths and the tissue absorption and path length varys widely from patient to patient and with the probe site (de Kock and Tarassenko 1991). The normalized signal I_n is calculated by dividing the transmitted light intensities (the *raw signals*) by their individual maximum peaks ($I_{H,R}$ for the red wavelength and $I_{H,IR}$ for the infrared wavelength). From equation (4.14) we derive

$$I_n = \frac{I}{I_H} = e^{-[\varepsilon_{Hb}(\lambda)c_{Hb} + \varepsilon_{HbO_2}(\lambda)c_{HbO_2}]\Delta d}. \tag{4.15}$$

This results in normalized signals with the same intensities $I_{H,n}$ during diastole. The normalized signals of the transmitted red and infrared light are independent of the incident light levels and photodetector nonlinearities as shown in figure 4.6. The AC components of the normalized signals represent only changes of transmitted light caused by the pulsation of blood in the arteries and can be compared with each other. They depend on the absorbers present in the arterial blood (ideally Hb and HbO_2) and the actual optical path length d through the volume changing part of the arteries.

4.4.2 Ratio of normalized signals

The absorbance of the light is derived by calculating the natural logarithm of the measured and normalized transmitted light level. Dividing the raw signal by the transmitted light during diastole I_H as in equation (4.15) and calculating the total absorbance then is comparable to calculating the total absorbance only due to the AC components in the pathway. The transmitted light during diastole represents the new nonchanging incident light level and the *ratio R* of these normalized absorbances at the red (R) and infrared (IR) wavelengths depends only on the light absorbers present in the arterial blood (see equation (4.3))

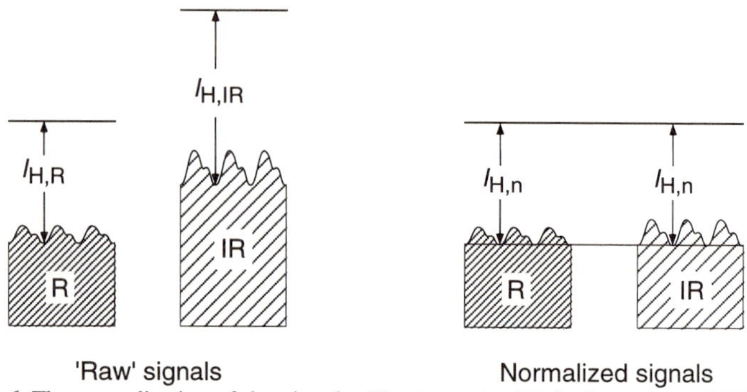

Figure 4.6 The normalization of the signals. The transmitted light from the red LED (R) and from the infrared LED (IR) is divided by its individual DC component. Thus, both normalized light intensities have the same magnitude during diastole. The normalized signals determine the basis for the calculation of the arterial oxygen saturation.

$$R = \frac{A_{t,R}}{A_{t,IR}} = \frac{\ln(I_{L,R} / I_{H,R})}{\ln(I_{L,IR} / I_{H,IR})}. \tag{4.16}$$

By using equation (4.15) the ratio can be derived as

$$R = \frac{[(\varepsilon_{Hb}(\lambda_R)c_{Hb} + (\varepsilon_{HbO_2}(\lambda_R)c_{HbO_2}]\Delta d_R}{[(\varepsilon_{Hb}(\lambda_{IR})c_{Hb} + (\varepsilon_{HbO_2}(\lambda_{IR})c_{HbO_2}]\Delta d_{IR}}. \tag{4.17}$$

Assuming that the optical path lengths d_R for red light and d_{IR} for the infrared light are equal, only the arteries change their diameter, and using equation (4.11)

$$R = \frac{\varepsilon_{Hb}(\lambda_R) + [\varepsilon_{HbO_2}(\lambda_R) - \varepsilon_{Hb}(\lambda_R)]SaO_2}{\varepsilon_{Hb}(\lambda_{IR}) + [\varepsilon_{HbO_2}(\lambda_{IR}) - \varepsilon_{Hb}(\lambda_{IR})]SaO_2}. \tag{4.18}$$

In this form the ratio R is not a function of the optical path length and can be derived from the arterial oxygen saturation instead of the concentration of the hemoglobins in the blood (see de Kock and Tarassenko 1993).

4.4.3 Theoretic calibration curve

Equation (4.18) can be rewritten in a form where S_aO_2 is a function of the measured and calculated ratio R

$$S_aO_2 = \frac{\varepsilon_{Hb}(\lambda_R) - \varepsilon_{Hb}(\lambda_{IR})R}{\varepsilon_{Hb}(\lambda_R) - \varepsilon_{HbO_2}(\lambda_R) + [\varepsilon_{HbO_2}(\lambda_{IR}) - \varepsilon_{Hb}(\lambda_{IR})]R} \times 100\%. \tag{4.19}$$

Therefore, the functional oxygen saturation in arterial blood can be derived theoretically by calculating the ratio R of measured and normalized total light absorbances in the red and infrared region and using equation (4.19). Figure 4.7 plots this relationship as the *theoretical calibration curve.*

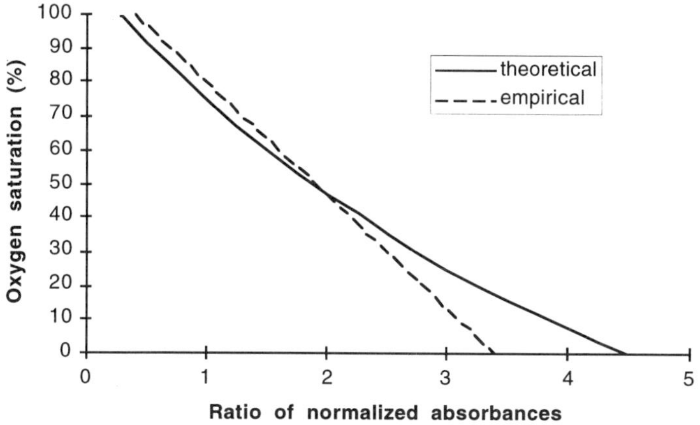

Figure 4.7 Calibration curves for pulse oximeters: the solid line is the theoretical curve by Beer's law and the dashed line is the empirical curve. The difference between these curves is due mainly to light scattering effects. This empirical calibration curve is derived by a second order polynomial.

4.5 VALIDITY OF BEER'S LAW IN PULSE OXIMETRY

Incident light passing through human tissue is not split only into absorbed light and transmitted light as proposed by Beer's law. Some parts of the light are reflected and others are scattered.

Light reflection at the skin surface and light absorbance due to tissue other than the pulsating arterial blood are overcome by using the plethysmographic waveform. However, the skin surface, tissue, muscle, bone and especially blood cause light scattering which increases the absorbance of light (see following section). Blood is a nonhomogeneous liquid, which is capable of nonlinear absorbance of light, e.g. as the concentration of hemoglobins varies (Wukitsch *et al* 1988).

The variation in light absorbance is not entirely due to the increased optical path length during systole. If the change in diameter were the only reason, the variation would be much less. The reason is a change in the axis of the red blood cells, which changes their absorbance as well. Red blood cells have the shape of a biconcave disk. Their major diameter is aligned parallel to the direction of blood flow during diastole and aligns perpendicular to the direction of flow during systole. Therefore, the optical path length is larger during systole and increases light absorbance. Even the light reflectance changes with the axis of the red blood cells, which is important for the use of reflectance probes. As a result of these properties, the absorbance and reflectance of blood in motion varys within the cardiac cycle and with the velocity of blood flow (Moyle 1994).

4.6 LIGHT SCATTERING

The results of oximetry measurements with whole blood differ from the results of the theory based on Beer's law. A physical phenomenon called *light scattering* highly increases the absorbance of light. Nevertheless, pulse oximeters read the arterial oxygen saturation of the blood accurately enough for clinical use under normal circumstances. This is due to the fact that most of the commercial pulse oximeters use a calibration curve based on empirical data, because modeling the problem of light scattering mathematically for different conditions is very complex. Several approaches have been made to create models which describe the real process within certain limits of accuracy.

4.6.1 Light absorbance in whole blood

Unfortunately Beer's law does not apply for whole blood. The absorbance of light is not simply proportional to the concentration of hemoglobin or to the length of the optical path. Beer's law assumes no light scattering, which is not true in whole blood, besides the fact that the LEDs do not emit monochromatic light.

Shymada and Yoshida (1984) verified that the influence of multiple scattering can not be overcome be subtracting the DC level as had been expected. Kramer *et al* (1951) stated that the absorbance of light due to oxyhemoglobin and reduced hemoglobin is increased in whole blood compared to hemolyzed blood by factors of the order of five.

The reasons for the increased absorbance are mainly *scattering* and *multiple scattering*. Light scattering causes the deviation of a light beam from its initial direction. It occurs when light is refracted by an object of a size similar to the magnitude of the wavelength of the light and a change in the index of refraction at the interface of this object. The wavelengths of red and infrared light do have the same order of magnitude as the geometric dimensions of red blood cells (approximately 7 μm in diameter). The discontinuity in the index of refraction at the interface between plasma and red blood cells and the great proportion of red blood cells in blood yield a highly light scattering medium. Light that is scattered once will likely be scattered again by cells and therefore multiple scattering occurs (Steinke and Shepherd 1986). Multiple scattering increases the optical path length and therefore increases the absorbance.

The intensity of the light scattered by the tissue depends on such factors as the red blood cell concentration in the blood; on the size, shape, orientation, and index of refraction of the scattering particles; on the tissue thickness; and on the aperture cone of the detector (Fine and Weinreb 1995). The thickness of the tissue, the distance between the LED and the photodiode, and the concentration of hemoglobin will vary from patient to patient and the shape and orientation of the red blood cells is irregular. Thus it is difficult to develop a physical model which can be used under different circumstances.

4.6.2 Models for light absorbance including scattering

It would be very useful to find a relationship between S_aO_2 and the ratio R of normalized absorbances for whole blood instead of only for hemoglobin solutions. An accurate scattering theory for whole blood could replace the empirical calibration curves used for the S_pO_2 readings. A few attempts are described below.

4.6.2.1 Twersky's multiple scattering theory. Twersky (1962, 1970a,b) has developed an analytical theory to describe the scattering of light by large, low-refracting, and absorbing particles. It is based on electromagnetic field theory and uses statistical averages to expand the theory for scattering and absorbing valid for a single particle, to find a formulation valid for multiple scattering (de Kock and Tarassenko 1993).

The total absorbance of whole blood can be expressed as the sum of absorbance as described by Beer's law and a second term representing the attenuation of light due to scattering. These two processes can be treated as independent processes. The intensity of scattering depends on variables such as those mentioned in section 4.6.1. The theory can be adapted for a special setting and will provide accurate results, but once the physiological conditions change, recalibration is required (Fine and Weinreb 1995). Hitachi, Ltd uses Twersky's approach in one of their US patents (Ito *et al* 1993).

4.6.2.2 Comparison of different models. Steinke and Sheperd (1986) compared Twersky's theory of radiation scattering and photon diffusion equations. They found Twersky's original equation to give the best fit for the measured data.

Marble *et al* (1994) found the three dimensional photon diffusion theory to be useful for modeling tissue optics although the pulse oximeter system violates many of the requirements of the model. However, they came to the conclusion that this theory can not replace clinical calibration studies.

De Kock and Tarassenko (1993) also found Twersky's theory to give the best fit to the experimental data. They compared results of this model with the photon diffusion theory and the Kubelka–Munk theory.

4.6.3 Influence of scattering on pulse oximeter readings

Although the assumptions of Beer's law are violated in pulse oximetry, the actual readings of the devices show a good correlation between the measurement and the actual arterial oxygen saturation.

Steinke and Sheperd (1986) found that the scattering effects of the light passing through whole blood depend on the wavelength of the light and the oxygen saturation. The relationship between oxygen saturation and total scattering effects (absorbance due to hemoglobin plus multiple scattering) is approximately linear and so scattering does not influence the linearity of the pulse oximeter in a negative way. In contrast, the total absorbance has a larger slope than that due only to the absorbance of hemoglobin following Beer's law. Therefore, light scattering increases the sensitivity of the whole blood oximeter.

Fine and Weinreb (1993, 1995) demonstrate that the ratio of total absorbances is a function of the effective blood layer thickness and the concentration of hemoglobin. Therefore physiological factors such as temperature or peripheral vasoconstriction reduce the accuracy of saturation readings. The error increases as the level of arterial oxygen saturation decreases. This is dangerous because the clinician has to question the readings of the oxygen saturation when it is most critical for the patient.

4.6.4 Calibration curves used for pulse oximeters

Commercial pulse oximeters are calibrated from *in vitro* data (see section 10.1). A large set of data obtained in clinical studies is collected containing information about the ratio R of the absorbances calculated by the pulse oximeter and the actual arterial oxygen saturation S_aO_2 measured by a very accurate method such as the CO-oximeter (see section 3.3). Lookup tables or equations are used to find the relationship of these two variables for a pulse oximeter reading.

To relate the measured values of the ratio R to the reading of the pulse oximeter, the equation of the theoretical calibration curve based on Beer's law can be modified as Mendelson and Kent (1989) described

$$S_pO_2 = \frac{k_1 - k_2 R}{k_3 - k_4 R}. \tag{4.20}$$

In this equation the extinction coefficients from equation (4.19) are replaced by constants k_i. These constants are determined by clinical studies to give the curve a best fit to the *in vitro* measured data. Another approach for a mathematical representation is the use of a polynomial such as found for example in the Ohmeda 3700 and Radiometer OX100 pulse oximeters (Fine and Weinreb 1995)

$$S_pO_2 = k_1 + k_2 R + k_3 R^2. \tag{4.21}$$

Figure 4.7 provides an example of a calibration curve used in pulse oximeters in comparison to the theoretical calibration curve.

REFERENCES

Barker S J and Tremper K K 1987 Pulse oximetry: applications and limitations *Int. Anesthesiol. Clinics* **25** 155–75

Bunn H F 1986 *Hemoglobin: Molecular, Genetic, and Clinical Aspects* (Philadelphia PA: Saunders)

Fine I and Weinreb A 1993 Multiple scattering effect in transmission oximetry *Med. Biol. Eng. Comput.* **31** 516–22

Fine I and Weinreb A 1995 Multiple scattering effect in transmission pulse oximetry *Med. Biol. Eng. Comput.* **33** 709–12

de Kock J P and Tarassenko L 1991 In vitro investigation of the factors affecting pulse oximetry *J. Biomed. Eng.* **13** 61–6

de Kock J P and Tarassenko L 1993 Pulse oximetry: theoretical and experimental models *Med. Biol. Eng. Comput.* **31** 291–300

Ito Y, Kawaguchi F Yoshida M and Kohida H 1993 Method and equipment for measuring absorptance of light scattering materials using plural wavelengths of light *US patent 5,239,185*

Kramer K, Elam J O, Saxton G A and Elam W N Jr 1951 Influence of oxygen saturation, erythrocyte concentration and optical depth upon the red and near-infrared light transmittance of whole blood *Am. J. Physiol.* **165** 229–46

Mannheimer P D, Casciana J R, Fein M E and Nierlich S L 1997 Wavelength selection for low-saturation pulse oximetry *IEEE Trans. Biomed. Eng.* **44** 148–58

Marble D R, Burns D H and Cheung P W 1994 Diffusion-based model of pulse oximetry: in vitro and in vivo comparisons *Appl. Opt.* **33** 1279–85

Mendelson Y and Kent J C 1989 Variations in optical absorption spectra of adult and fetal hemoglobins and its effect on pulse oximetry *IEEE Trans. Biomed. Eng.* **36** 844–8

Moyle J T B 1994 *Pulse Oximeters* (London: BMJ)

Nellcor 1993 Hemoglobin and the principles of pulse oximetry *Reference Note: Pulse Oximetry Note Number 1* (Pleasanton, CA: Nellcor)

Pologe J A 1987 Pulse oximetry: technical aspects of machine design *Int. Anesthesiol. Clinics* **25** (3) 137–53
Shymada Y and Yoshida I 1984 Effects of multiple scattering and peripheral circulation on arterial oxygen saturation measured with a pulse-type oximeter *Med. Biol. Eng. Comput.* **22** 475–8
Steinke J M and Shepherd A P 1986 Role of light scattering in whole blood oximetry *IEEE Trans. Biomed. Eng.* **33** 294–301
Twersky V 1962 Multiple scattering of waves and optical phenomena *J. Opt. Soc. Am.* **52** 145–71
Twersky V 1970a Interface effects in multiple scattering by large, low refracting, absorbing particles *J. Opt. Soc. Am.* **60** 908–14
Twersky V 1970b Absorption and multiple scattering by biological suspensions *J. Opt. Soc. Am.* **60** 1084–93
Wukitsch M W, Petterson M T, Tobler D R and Pologe J A 1988 Pulse oximetry: analysis of theory, technology, and practice *J. Clin. Monitoring* **4** 290–301
Zijlstra W G, Buursma A and Meeuwsen-van der Roest W P 1991 Absorption spectra of fetal and adult oxyhemoglobin, de-oxyhemoglobin, carboxyhemoglobin, and methemoglobin *Clin. Chem.* **37** 1633–8

INSTRUCTIONAL OBJECTIVES

4.1 Describe the properties and limitations of Beer's law.
4.2 Describe different species of hemoglobin and their effect on the oxygenation of blood.
4.3 Describe the functional and the fractional hemoglobin saturation and their difference.
4.4 Describe the properties and assumptions of the spectrophotometric method to determine oxygen saturation in hemoglobin solutions.
4.5 Describe the principles of pulse oximetry and what a pulse oximeter measures.
4.6 Describe why and how a pulse oximeter measures the absorbance in the arterial blood only.
4.7 Describe the normalization of the signals and the reasons for this normalization.
4.8 Explain how and why the ratio of the normalized signals is calculated.
4.9 Explain errors in the spectrophotometric method when used for whole blood samples.
4.10 Describe the different physical phenomena occurring when light travels through tissue and blood.
4.11 Describe what light scattering is and where it occurs in pulse oximetry.
4.12 Describe the influence of light scattering on the accuracy of a pulse oximeter.

CHAPTER 5

LIGHT-EMITTING DIODES AND THEIR CONTROL

Brad W J Bourgeois

In order to make pulse oximetry practical in the modern medical environment, a light source is required that is powerful enough to penetrate more than a centimeter of tissue yet diminutive enough to fit in a small probe. Chapter 4 shows that it also is desirable for the light source at each desired wavelength to have a very narrow *emission spectrum*, which minimizes error in the measurement of arterial oxygen saturation (S_aO_2). Fortuitously, light-emitting diodes (LEDs) fulfill all the requirements for the light source in a pulse oximeter.

However, LEDs are not without drawbacks. The primary problem faced by pulse oximeter designers is how to deal with variations and shifts in the *peak wavelength* of each LED. Because the main function of a pulse oximeter, measuring arterial oxygen saturation, is so heavily dependent upon accurate values for the two wavelengths of light, a design which does all it can to compensate for LED wavelength changes will outperform its competition.

This chapter discusses important characteristics of LEDs, a LED driver circuit in a pulse oximeter, and various problems with the use of LEDs in pulse oximetry.

5.1 AN INTRODUCTION TO LIGHT-EMITTING DIODES

Light–emitting diodes are the light source of choice for all pulse oximeters on the market today. Their small size, excellent drive characteristics, and large light output over a very narrow bandwidth make them the ideal choice for the source of light at both the red and infrared wavelengths used in pulse oximetry.

The fact that LEDs are available for use in pulse oximetry is due to a combination of science and luck. LEDs are only available over an approximately 700 nm range of wavelengths, from blue in the visible spectrum into the near infrared. By contrast, the electromagnetic spectrum extends over a range of 10^{14}. Another fortuitous fact is that the window of low absorption on the hemoglobin extinction curves occurs within the range of LED availability. The common LED wavelengths of 660 and 940 nm work very well in pulse oximetry, which allows for lower cost due to the off-the-shelf availability of these LEDs.

5.1.1 Description, materials, and operation

An LED is an optoelectronic semiconductor which produces light by *electroluminescence* (D.A.T.A. Handbook 1992). LEDs are characterized by high light emitting efficiency compared to other methods of light emission such as cathode, high-temperature, and photoluminescence. The electroluminescence occurs by the injection and recombination of minority carriers in the forward-biased *p–n* junction. Most LEDs are made from III–V, II–VI, and IV semiconductors, with the most common materials being gallium arsenide phosphide (GaAsP), gallium phosphide (GaP), and gallium arsenide (GaAs). GaAsP and GaP LEDs emit light in the visible spectrum (approximately 380 to 780 nm), while GaAs is used in infrared LEDs. Another material not as commonly used to make LEDs which can produce light in both the visible and IR regions of the spectrum is gallium aluminum arsenide, GaAlAs.

Figure 5.1 shows the light emission mechanism of an LED. When an electron gains enough energy to cross the forbidden energy gap E_g, it enters the conduction band. When an electron in this conduction band returns to the lower energy level of the valence band, the electron releases energy in the form of a photon of light. The wavelength of light emitted from an LED is determined by

$$E_g = hc/\lambda, \tag{5.1}$$

where E_g is the forbidden bandwidth in electron volts, h is Planck's constant (6.626×10^{-34} J s), c is the speed of light in a vacuum (3.00×10^8 m/s), and λ is the wavelength of the emitted photon. The value of E_g, which is a physical property of the LED material(s), determines the wavelength of emitted photons and is directly related to the forward voltage of an LED (see section 5.2.1).

5.1.2 Bandwidth considerations

Another factor considered in the use of LEDs in pulse oximetry is the emission spectrum of the LED. Because of the steep slope of the deoxyhemoglobin (Hb) extinction curve at 660 nm, it is extremely important that the red LEDs used in pulse oximeter probes emit a very narrow range of wavelengths centered at the desired 660 nm in order to minimize error in the S_pO_2 reading, which is the pulse oximeter's estimation of arterial oxygen saturation (New and Corenman 1987, 1988). The width of the wavelength range of the IR LED is not as important for accuracy due to the relative flatness of both the Hb and HbO_2 (oxyhemoglobin) extinction curves at 940 nm. LEDs again perform very well for this requirement. Typical LEDs have a *spectral bandwidth* in the range of 60 nm to less than 20 nm, with visible LEDs usually having smaller bandwidths of approximately 25 nm and IR LEDs typically having larger bandwidths near 50 nm.

5.2 LIGHT-EMITTING DIODE SPECIFICATIONS

Before discussing the specifications of LEDs available on the market, the performance desirable for LEDs in pulse oximetry will be given. The two predominant factors are the radiated power (or light output) and the size of the LEDs.

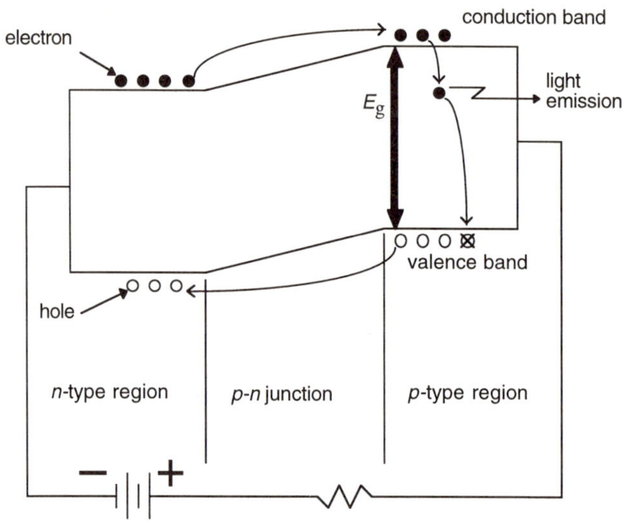

Figure 5.1 The light emission mechanism of an LED. Electrons gain energy moving to the conduction band. They emit light when dropping to the valence band.

The radiated power of an LED is measured in milliwatts. The typical radiated power of both the red and IR LEDs used in pulse oximetry is 1 mW at 20 mA dc. Brighter LEDs are available, but generally the radiated power does not exceed 10 mW.

Modern manufacturing techniques have shrunk LEDs to sizes smaller than a millimeter in length or diameter, while remaining bright enough to be used in devices such as pulse oximeters. LED size is not an obstacle in the design of pulse oximeter probes.

5.2.1 Forward voltage

The forward voltage is defined as the potential drop across the *p–n* junction of the diode from anode to cathode. While ordinary silicon diode forward voltages are near 0.7 V, LEDs forward voltages can range from 0.9 to 2.5 V typically. Equation (5.1) shows that an inverse relationship exists between a material's forbidden energy gap E_g and the wavelength of emitted photons. In addition, the forward voltage of an LED is directly related to E_g. Therefore, an LED with a relatively small forward voltage has a small E_g and a long emitted wavelength (e.g. in the infrared region). Conversely, an LED with a relatively large forward voltage has a large E_g and a short emitted wavelength (e.g. in the blue–green region).

5.2.2 Forward current

The forward current is defined as the current flowing through the LED in the direction from anode to cathode. With sufficient current, an LED will emit light. A very important property of LEDs is that radiated power, to a first approximation, varies linearly with forward current over the range of current found in pulse oximeters. Typical values for forward current have a large range, from 2 to 50 mA. Figure 5.2 shows the relationship between current and voltage for a typical 660 nm LED.

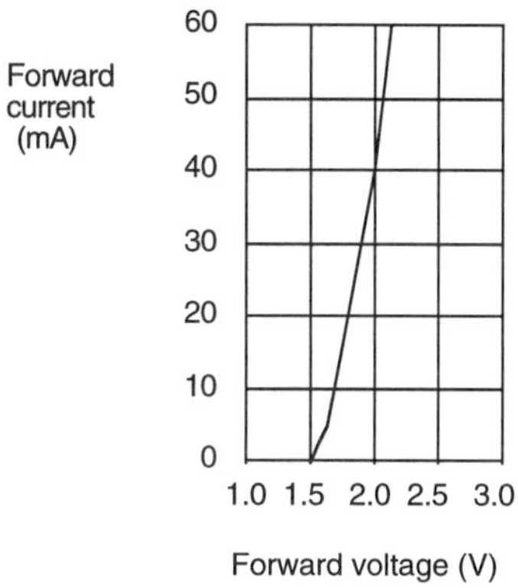

Figure 5.2 Forward current–voltage characteristic for a typical 660 nm GaP LED.

5.2.3 Power dissipation

Another consideration for LEDs used in pulse oximetry is power consumption. While the vast majority of pulse oximeters are used in a stationary environment where power is readily available from the nearest wall outlet, some are portable units used in a variety of emergency medical situations. These portable units may need to function for an extended period of time without a power supply recharge. It is therefore essential that LED power consumption be minimized while still providing adequate radiated power for pulse oximetry.

The maximum power dissipation rating for an LED can be defined as the largest amount of power that can be dissipated while still remaining within safe operating conditions. This power is a function of three parameters: ambient temperature, rated maximum junction temperature, and the increase in junction temperature above ambient per unit of power dissipation for the given LED's package and mounting configuration. The latter of these parameters is defined as the *thermal resistance* of the device, and is very important in reliable system design. The worst-case value for thermal resistance, that with no heat sink, can be calculated from

$$R_{TH} = (T_J - T_A)/P_D \ {}^\circ\mathrm{C/W}, \tag{5.2}$$

where R_{TH} is the thermal resistance, T_J is the junction temperature, T_A is the ambient temperature, and P_D is the rated power dissipation of the LED. Another method for calculating thermal resistance is to use the negative reciprocal of the slope of the forward current versus ambient temperature graph, figure 5.3. This value is in units of °C/mA, which can be converted to the thermal resistance in °C/W by multiplying the denominator by the LED forward voltage (D.A.T.A.

Handbook 1992). Because the skin is the primary sink for LED heat in pulse oximetry, the design engineer must consider power dissipation in order to prevent possible burns to the patient's skin.

Typical LEDs are 2 to 10% efficient, meaning that the majority of power dissipated by an LED becomes heat. The optical power absorbed by the tissue also becomes heat. As with forward current, a broad range of power ratings is available, typically from 20 to 300 mW. An interesting fact to note is that the typical IR LED, with its lower forward voltage (see section 5.2.1), required a greater forward current to dissipate the same optical power as a typical red LED. This is because red photons contain more energy than infrared photons.

5.2.4 Reverse breakdown voltage

As with all diodes, under reverse bias virtually no current will flow across the *p–n* junction until the reverse breakdown voltage has been reached. Above that voltage, large currents flow and damage the diode, unless a resistor limits the current. Most LEDs have a fairly small value for this specification, usually in the range of 3 to 5 V. This specification is important in pulse oximetry due to the arrangement of the LEDs in a probe. To minimize the number of wires in each probe (and hence cost), the LEDs are wired in a parallel arrangement with polarities reversed. This means that while one LED is ON, the other LED is under reverse bias. The typical LED has a reverse breakdown voltage that is larger than the forward voltage of most LEDs, minimizing the difficulty in dealing with this specification.

5.2.5 Reverse current

In an ideal diode, no current flows in the reverse direction when the *p–n* junction is reverse-biased. In reality, a minute amount of current actually does flow in the reverse direction. In LEDs, this current typically ranges from 0.01 to 10 μA. Since this current is extremely small compared to the forward current of the LED wired in parallel, this shunt current has a negligible effect.

5.2.6 Operating temperature

Pulse oximeters are usually used in a stable medical environment at room temperature. However, emergency situations may arise in which a pulse oximeter has to operate under extreme temperatures. Fortunately, LEDs are extremely rugged devices with a basic specified range of operating temperature from –40 to 85 °C. Many LEDs with an even larger operating temperature range are available.

Most LED parameters are specified at a given temperature. In addition, information is given for how some of these parameters vary over a given temperature range (see section 5.5). The most important of these parameters is maximum forward current versus temperature, which determines the thermal resistance of the LED (see section 5.2.3). Figure 5.3 shows the relationship between maximum forward current and temperature for a typical high-power 660 nm red LED.

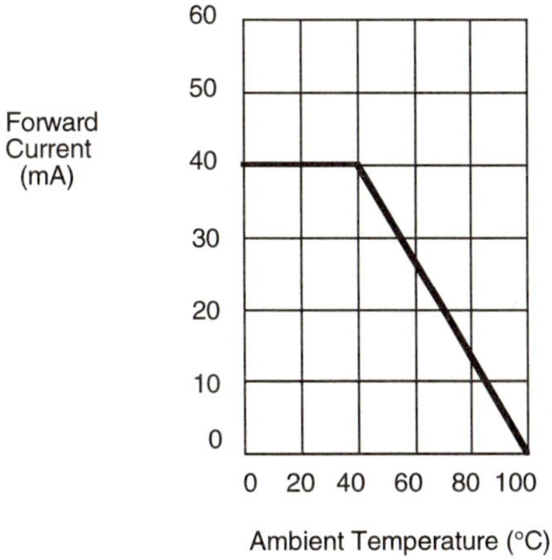

Figure 5.3 Maximum forward current versus temperature for a typical high-power red LED.

5.2.7 Switching times

Switching time is the time required for an LED to switch from its ON state to its OFF state or vice versa. Most LEDs have a switching time in the low hundreds of nanoseconds. In the application of pulse oximetry, this is much faster than required because of the extremely low frequency of the arterial pulsatile waveform (~1 Hz). For reasons which will be explained in chapter 8, in most pulse oximeters LED switching cycles occur at a rate of 480 Hz, much more slowly than the maximum switching capabilities of LEDs.

5.2.8 Beam angle

Beam angle is defined as the angular measure of radiated power measured on an axis from half-power point to half-power point. It is simply a measure of how focused the emitted light is. In LEDs on the market today, beam angles can range from a few degrees to a maximum of 180°. In pulse oximetry, the beam angle only needs to be narrow enough to ensure that the maximal light output enters the tissue. The scattering of light occurring in the tissue serves to ensure that the light spreads over the entire sensor area.

5.2.9 Pulse capability

Pulse capability is defined as the maximum allowable pulse current as a function of *duty cycle* and frequency. This parameter is important in pulse oximeters for two main reasons. The first reason is that, as discussed in chapter 8, LEDs are pulsed in pulse oximeters. The second reason is that the small LEDs used by some manufacturers in pulse oximeter probes may not be able to tolerate enough sustained current to sufficiently excite the photodiode. Since the allowable pulse current is always substantially higher than the maximum sustained current, smaller LEDs can be used than could be if the LEDs were constantly on. For example, the LEDs in Criticare pulse oximeters have a duty cycle near 5%, while in Nellcor devices it is 25%. Figure 5.4 shows the pulse capability of a typical 660 nm LED.

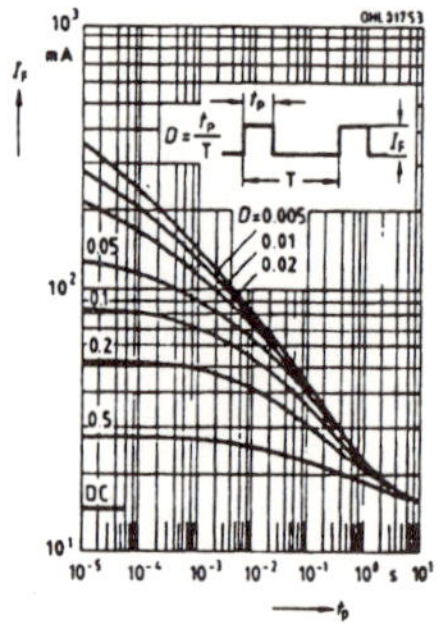

Figure 5.4 Pulse capability of a typical high-power 660 nm LED. The maximal pulse current is a function of the duty cycle *d* and frequency (from Siemens 1993).

5.2.10 Cost

Chapter 7 states that a disposable probe for use in pulse oximetry has some advantages over reusable probes, such as convenience and guaranteed sterility. With the widespread use of disposable probes, cost is the prohibitive factor in their manufacture. The cost of the two LEDs used in each probe is therefore important for the purpose of minimizing the overall expense of each probe. Today, both red and IR LEDs can be purchased in bulk for just a few cents each, making them a minor factor in the overall cost of a probe. (Allied Electronics, Inc. 1995, Digi-Key Corporation 1995). However, testing each LED to find its peak wavelength, as discussed in the following section, does increase the overall cost to the manufacturer.

5.3 MEASURING AND IDENTIFYING LED WAVELENGTHS

Chapter 4 notes that the choice of 660 and 940 nm for the light wavelengths was not arbitrary with respect to optical considerations. Because of the steep slope of the Hb extinction curve at 660 nm, it is important that the red LEDs used in pulse oximeter probes have a peak wavelength of exactly 660 nm in order to minimize error in the S_pO_2 reading (see chapter 11). Error in the peak wavelength of the IR LED is not as important for accuracy due to the relative flatness of both the Hb and HbO_2 extinction curves at 940 nm. An alternative to having LEDs with precise peak wavelengths of 660 and 940 nm is to have the pulse oximeter itself somehow compensate for any deviation from those nominal values. This section discusses these concerns.

As is the case with all mass manufacturing processes, imperfections occur in each lot of LEDs produced. For pulse oximetry, the most important of these is peak wavelength shift. Peak wavelength is defined as the wavelength at which the radiated power of the device is maximum. Although bulk LEDs theoretically all have the same peak wavelength, figure 5.5 shows that the actual peak wavelength of any LED may vary from the rated value by as much as 15 nm (Pologe 1987).

In order to solve this problem, pulse oximeters can compensate for a number of different LED peak wavelengths. This technique has the advantage of lowering cost by allowing probe manufacturers to buy and use LEDs in bulk instead of being able to use only LEDs with peak wavelengths of exactly 660 and 940 nm.

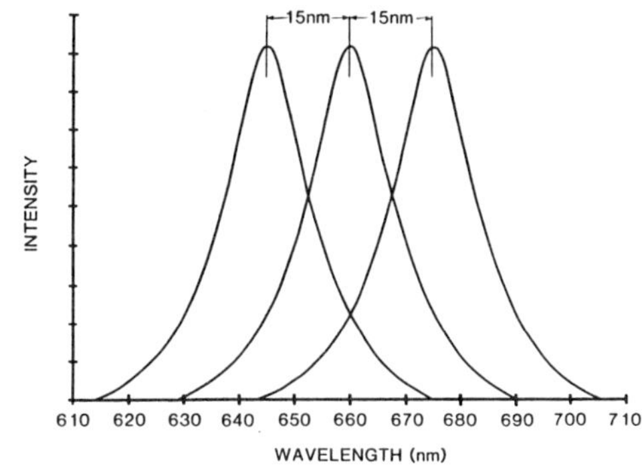

Figure 5.5 Center wavelength variation of LEDs of the same type from the same lot (from Pologe 1987).

While many methods exist to solve this problem, only the one most commonly used in pulse oximeters will be explained here.

The first step in the process for the probe manufacturer is to test each individual LED to find its exact peak wavelength. This is done by testing each individual LED with a spectrophotometer to experimentally determine the wavelength of light at which the LED has its highest power output. The LEDs are then separated into a certain number of groups, with each group having a small, distinct range of wavelengths, for example 660 to 661 nm.

Knowing the center wavelengths for a particular LED pair allows the proper set of calibration curves, specific to that wavelength combination, to be chosen from the entire family of curves that exist. This is most often done by developing a two-dimensional matrix with, for example, the red LED wavelength values in the heading row and the IR LED wavelength values in the heading column. Each matrix location then identifies the appropriate set of calibration curves for the given pair of LEDs.

The final problem to be solved is to have the pulse oximeter somehow interrogate each new probe to find out which calibration curve must be used to accurately determine arterial oxygen saturation. The most common technique is to include in the probe connector a coding resistor with a specific value. Each unique resistor value represents to the pulse oximeter those pairings of LED wavelengths that correspond to one calibration curve. The microprocessor simply sends a current through the resistor and measures the voltage drop across it, in effect finding the value of the coding resistor. By finding this voltage value in a lookup table, the microprocessor can indirectly determine the proper calibration curve to be used for that probe (New and Corenman 1987, 1988).

Chapter 8 provides details of how the pulse oximeter performs this interrogation of each probe.

Kästle *et al* (1997) describes how Hewlett-Packard avoided using a coding resistor by selecting red LEDs within a ± 1 nm variation of wavelength. A later sensor used new high-efficiency AlGaAs red LEDs to achieve a four-fold increase in intensity, with corresponding lowered heat dissipation. They also note that the red LED may emit an undesired secondary emission peak (<4% of maximum intensity) at about 800 to 850 nm, which may interfere with the IR LED. They place the LED in an integrating sphere to diffuse the light for wavelength measurement by an optical spectrometer having a wavelength resolution of 0.2 nm.

5.4 LED DRIVER CIRCUIT

This section presents an overview of the operation of a specific LED driver circuit used in many pulse oximeters. Greater detail about the microprocessor control, signal processing, and other hardware or software concerns can be found in chapters 8 and 9.

Figure 5.6 shows the LED driver circuit. This circuit, and the LED driver circuits in many of the pulse oximeters on the market today, provide up to 50 mA of pulse current to each LED. The microprocessor automatically alters the amount of current supplied to the LEDs according to the absorption of the tissue on which the pulse oximeter is being used. Factors such as skin pigmentation, skin thickness, and optical path length, among many others, determine the absorption of the tissue. The microprocessor first determines if the photodiode is receiving a proper amount of light, enough to adequately excite but not saturate the photodiode. The microprocessor then supplies voltage feedback to the LED driver circuit, which allows current to the LEDs to be adjusted as needed. No complex calculations are necessary to determine current adjustments, as radiated power varies nearly linearly with drive current over the range of current utilized in pulse oximetry.

The microprocessor controls how much current is provided to each LED by dynamically adjusting the reference voltage seen at the driver amplifier, U3A. U1 supplies the reference voltage, which is switched selectively for the red and IR LEDs using U4A and U4B. The microprocessor changes the reference voltage for the red or IR LEDs by changing the data supplied to U1, which is a multiplying DAC, before the voltage is switched to the amplifier. The microprocessor attempts to achieve and keep the optimal drive current without clipping the transducer signal.

The control signals REDLED/ and IRLED/ come from the microprocessor and control the switches U4A and U4B along with the transistor network that drives the LEDs. The LEDs are never on at the same time, although during part of the LED switching cycle they are both off to allow the photodiode to detect ambient light.

When REDLED/ is low, IRLED/ is high and U4A is closed, placing the red reference voltage at pin 3 of U3A. Since REDLED/ is low, transistor Q5 is off, which allows Q3 to turn on and conduct current from the LED through R1, the sense resistor. Also with REDLED/ low, Q2 turns on, allowing current to flow from the positive supply to the red LED anode, turning on the red LED. Since IRLED/ is high, Q1 is off, with no current conduction, and Q6 is on, pulling the base of Q4 to ground, which keeps Q4 off. The current path in this case is from VCC through Q2, the red LED, Q3, and R1, the sense resistor. The voltage drop across R1 is fed back to U3A, which compares it to the reference voltage and changes the drive current of Q3 accordingly.

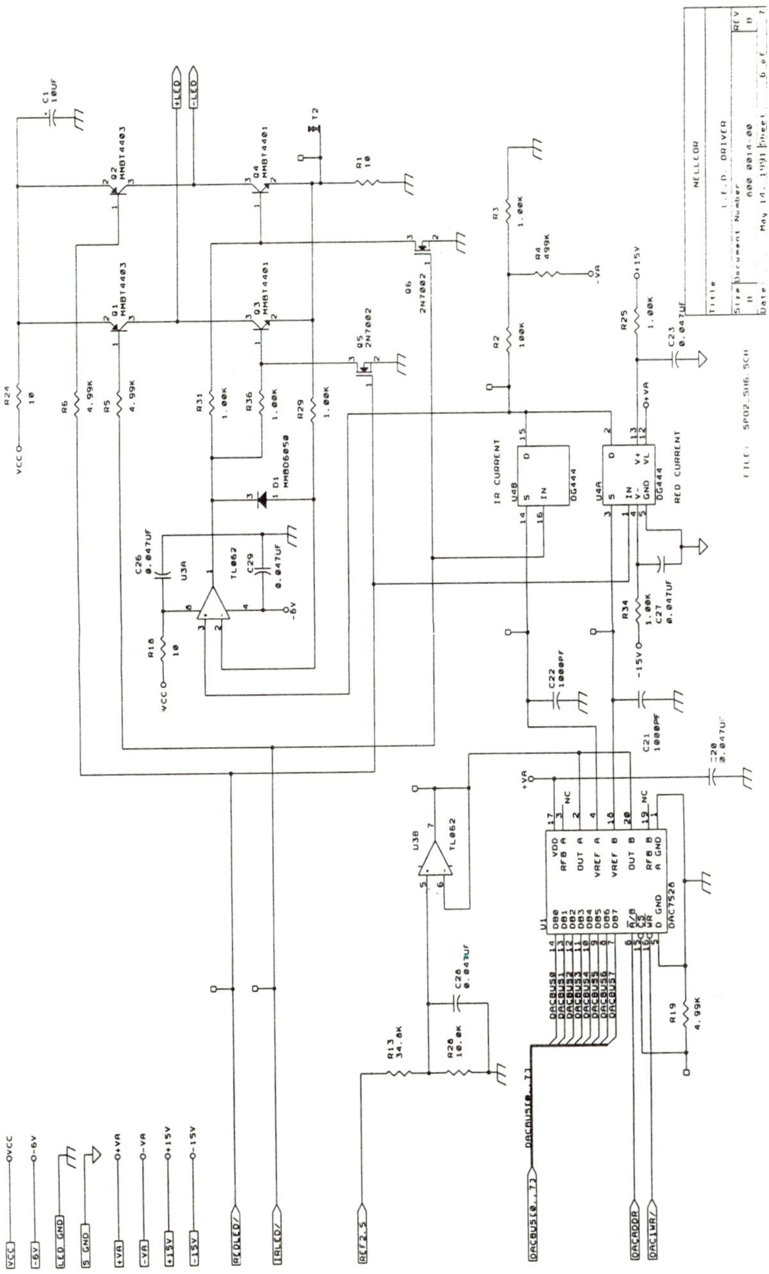

Figure 5.6 LED driver circuit. (From Protocol. 1994. Propaq® 100–Series Monitors *Schematics & Drawings Set*, Schematic 00950, 6 of 7, Section 2E, Protocol Systems, Inc., Beaverton, OR).

When REDLED/ is high, IRLED/ is low and the reference voltage is applied to U3A through U4B. Having REDLED/ high causes Q2 to be off and Q5 to be on, turning off Q3. Having IRLED/ low turns Q1 on, allowing current to flow to the IR LED anode, turning on the IR LED. Q6 is also turned off, which allows the base of Q4 to be pulled up in voltage by U3A until Q4 conducts. The current path in this case is from VCC through Q1, the IR LED, Q4, and R1. The voltage drop across R1 is again fed back to U3A, which compares it to the reference voltage and changes the drive current of Q4 accordingly.

In the case when both the IRLED/ and REDLED/ control signals are high, both switches, U4A and U4B, are open and all of the drive transistors are off. The resistors R2, R4, and R3 form a voltage divider network that makes the reference input of U3A slightly negative with respect to ground. Because of this, U3A drives its output negative. However, D1 will not allow U3A's output to drop below approximately –0.6 V so that the drive transistors Q3 and Q4 can be turned on quickly when needed (Protocol 1994, pp 2E5–6).

5.5 LED PEAK WAVELENGTH SHIFT WITH TEMPERATURE

As discussed in section 5.3, pulse oximeter probe manufacturers could use any of a number of methods to compensate for LED peak wavelengths which vary from the nominal values of 660 and 940 nm, with the method of choice being the use of a coding resistor to indicate to the microprocessor which set of calibration curves to use for a given probe.

However, the peak wavelength of an LED can shift during operation due to a change of the *p–n* junction temperature. It is more difficult to account for this wavelength shift when determining which set of calibration curves to use. The effect that LED peak wavelength shift due to temperature has upon S_pO_2 will be discussed in detail in chapter 11. Lastly, two methods of minimizing the negative effects of temperature changes upon S_pO_2 will be discussed.

5.5.1 p–n junction heating

Equation (5.1) shows that the wavelength of emitted light in an LED depends on the forbidden energy gap E_g. In turn, E_g is dependent upon temperature (Varshni 1967, Panish and Casey 1969). In GaAs, GaP, and most other common semiconductor materials, E_g decreases as temperature increases. Therefore, the peak wavelength of an LED should increase as the *p–n* junction temperature increases. Typically, the peak wavelength will increase by 0.35 to 0.6 nm/°C (Miller and Kaminow 1988).

The main factor affecting the *p–n* junction temperature of the two LEDs is drive current, which causes ohmic heating at the *p–n* junction. Although the LEDs are sequentially pulsed with a duty cycle of 2 to 50% depending on the make of the oximeter (Reynolds *et al* 1991), the resulting average current (duty cycle multiplied by drive current) is still sufficient to substantially heat the *p–n* junction, as power dissipated is directly proportional to the drive current I according to the equation $P = VI$.

5.5.2 Studies

de Kock *et al* (1991) tested the effect upon peak wavelength of driving a red and IR LED at 10% and 100% of the rated maximum drive current. The nominal wavelengths of the tested LEDs were 660 and 950 nm, with 30 min allotted for thermal equilibrium to be reached. At 380 Hz with a 25% duty cycle, they found that the increased drive current increased the center wavelength of the red LED by 8 nm, while the center wavelength of the IR LED did not shift at all. Figures 5.7 and 5.8 show their results.

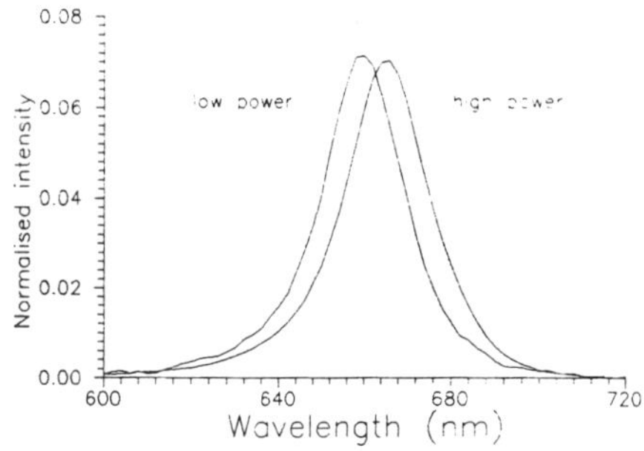

Figure 5.7 Normalized red LED spectra at low and high forward current (from de Kock *et al* 1991).

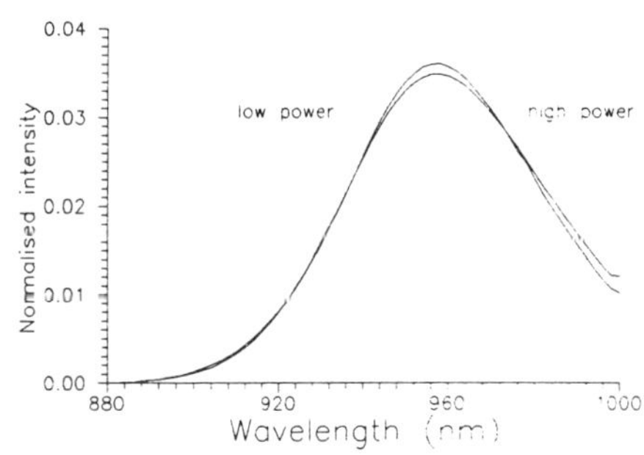

Figure 5.8 Normalized IR LED spectra at low and high forward current (from de Kock *et al* 1991).

The same group studied the effect of ambient temperature upon LED peak wavelength in 1991. As in the study mentioned above, the red and IR wavelengths were 660 and 950 nm. The spectrum of each LED was measured using a spectrophotometer at 2 nm intervals at ambient temperatures ranging from 0 to 50 °C in 10 °C steps. Ten minutes was given for thermal equilibrium to be established at each temperature step. The group found that over this range of ambient temperatures, the red LED had an increase of 5.5 nm in its peak wavelength, while the IR LED had an increase of 7.8 nm. In addition, no significant change was found in the spectral bandwidth of either LED over the temperature range. The measured bandwidths were found to be approximately 25 and 55 nm for the red and IR LEDs, respectively. Figures 5.9 and 5.10 show these results.

Kästle *et al* (1997) list a wavelength shift of about 0.12 nm/K but note that the red and IR LEDs tend to track temperatures, which compensates for errors in the ratio R.

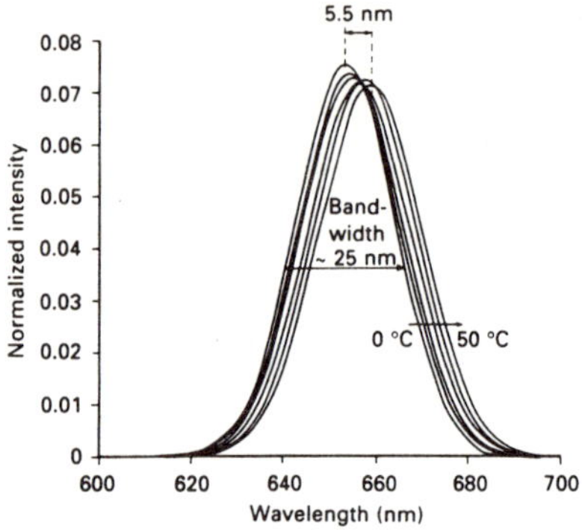

Figure 5.9 Shift in emission spectrum of a red LED as ambient temperature is increased from 0 to 50 °C in 10 °C intervals (from Reynolds *et al* 1991).

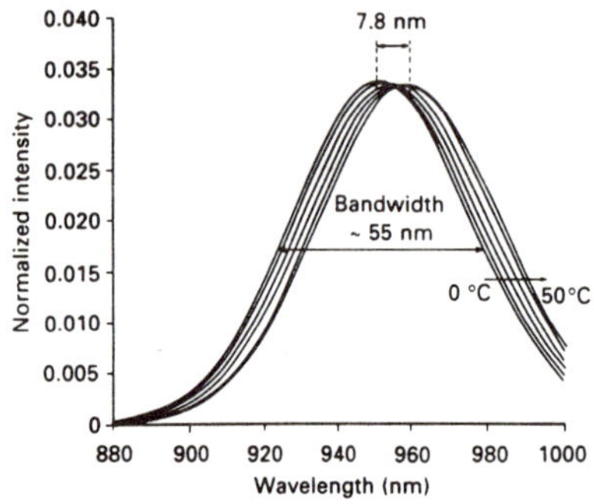

Figure 5.10 Shift in emission spectrum of an IR LED as ambient temperature is increased from 0 to 50 °C in 10 °C intervals (from Reynolds *et al* 1991).

5.5.3 Two methods to compensate for LED temperature changes

As expected, a shift in LED peak wavelength due to a change in temperature can cause erroneous S_pO_2 readings. A full discussion of this problem will be given in chapter 11.

One way to compensate for LED temperature changes is to have a temperature sensor built into the probe along with the LEDs and photodiode (Cheung *et al* 1993). Temperature information is fed back to the microprocessor, which then estimates how much the peak wavelength of each LED has changed from its rated value (which the microprocessor determined from the probe's coding resistor). The microprocessor then chooses the set of calibration curves to match the new set of LED wavelengths. One inherent problem with this method is that the temperature–peak wavelength relationship given as a specification by the manufacturer will not be exactly the same for each individual LED, making the microprocessor's calculation of new LED peak wavelengths potentially inaccurate. Another problem is the difference between the sensed temperature and the actual temperature of the *p–n* junctions of the LEDs. If the two LEDs are being driven with different currents, as is normally the case, they will probably be at different temperatures. The temperature sensor will read at best an average of the two LED temperatures, and at worst an average of the two LED temperatures along with the skin and ambient temperatures. In addition, the sensor and additional wires needed will add cost to the probes, making a cost–benefit analysis of this method necessary before its inclusion in a pulse oximeter design.

A second, similar method to compensate for LED temperature changes is to measure the LED drive current directly. The microprocessor would then use that drive current value to calculate the estimated temperature change, and from that, calculate the estimated peak wavelength shift. This method eliminates the second problem listed above for the temperature sensor solution, but still leaves the problem of variations in the relationship between temperature and peak wavelength among individual LEDs. Another advantage of this method is that no extra wires or other components need to be added to a probe, making this the less expensive of the two methods discussed here.

5.6 PREVENTION OF BURNS IN PULSE OXIMETRY

In order to prevent burns on a patient's skin due to LED heat, the Food and Drug Administration now requires that the contact region between the skin and the oximeter probe not exceed 41 °C. Given an average body temperature of 37 °C, a pulse oximeter system should be designed to yield a maximum temperature rise of 4 °C at the skin–probe contact region, which is the primary dissipator of the LED heat. The relevant LED specification is thermal resistance, discussed in section 5.2.3. In pulse oximetry, the thermal resistance of each LED is on the order of a standard LED mounted in a PC board, which is a specification given in LED product catalogs. As previously mentioned, many pulse oximeters on the market have a maximal LED pulse current of 50 mA. This is sufficiently small to prevent dangerous LED heating, while still providing adequate light to the photodiode.

Mills and Ralph (1992) tested the heating of six pulse oximeter probes over a span of 3 h. The probes were placed in an incubator kept at a constant temperature of 36.9 to 37 °C. The working temperatures of the probes were quite similar, with a range of 39.1 to 39.7 °C over the entire 3 h. One of the probes was monitored for 24 h, and during that time its temperature remained constant within a range of ±0.1 °C.

The conditions of this test, however, did not do anything to simulate the reaction of skin to heating of a few degrees for several hours. To ensure that no burning occurs, the probe's point of application should be inspected often. In addition, the position of the probe on the patient should be changed regularly, especially if the probe application area suffers from low perfusion, which limits the skin's ability to dissipate heat.

5.7 LED PACKAGING

Most LED packages are made of resin, offering superior mechanical strength and the ability to withstand vibration and shock. In some of today's pulse oximeter probes, the two LEDs can be found in one package, which has the distinct advantage of keeping costs down. In the Nellcor SCP–10 reusable and Oxisensor II D–25 disposable pulse oximeter probes, the two LEDs come encased in a transparent rectangular solid with approximate dimensions of 5 mm long by 4 mm wide by 2 mm thick. The LEDs themselves are flat squares with sides of approximately 0.25 mm.

Other probes have discrete LEDs inside, with the LEDs lying side by side and a mirror to reflect light at a 90° angle to the tissue. Some probes even have three or four LEDs in them to increase light output. The details of these and many other probes will be discussed in Chapter 7.

There is no wrong choice for LED packaging as long as the LEDs are small yet powerful enough to perform the task at hand. However, there is definitely a superior choice for packaging when trying to minimize costs, and that choice is almost always a package containing both LEDs.

REFERENCES

Allied Electronics, Inc 1995 *Catalog 956* (Fort Worth, TX: Allied Electronics)

Cheung P W, Gauglitz K F, Hunsaker S W, Prosser S J, Wagner D O and Smith R E 1993 Apparatus for the automatic calibration of signals employed in oximetry *US patent 5,259,381*

D.A.T.A. Handbook 1992 *LED Lamps and Displays* (Englewood, CO: D.A.T.A)

de Kock J P, Reynolds K J, Tarassenko L and Moyle J T B 1991 The effect of varying LED intensity on pulse oximeter accuracy *J. Med. Eng. Technol.* **15** (3) 111–6

Digi-Key Corporation 1995 *Catalog No. 956* (Thief River Falls, MN: Digi-Key)

Kästle S, Noller F, Falk S, Bukta A, Mayer E and Miller D 1997 A new family of sensors for pulse oximetry *Hewlett-Packard J.* **48** (1) 39–53

Miller S E and Kaminow I P 1988 *Optical Fibre Telecommunications* Vol II (New York: Academic) pp 487–8

Mills G H and Ralph S J 1992 Burns due to pulse oximetry *Anaesthesia* **47** 276–7

New W Jr and Corenman J E 1987 Calibrated optical oximeter probe *US patent 4,700,708*

New W Jr and Corenman J E 1988 Calibrated optical oximeter probe *US patent 4,770,179*

Panish M B and Casey H C Jr 1969 Temperature dependence of the energy gap in GaAs and GaP *J. Appl. Phys.* **40** 163–7

Pologe J A 1987 Pulse oximetry: technical aspects of machine design *Int. Anesthesiol. Clinics* **25** (3) 137–53

Protocol 1994 Propaq® 100–Series Monitors *Schematics & Drawings Set* (Beaverton OR: Protocol Systems)

Reynolds K J, de Kock J P, Tarassenko L and Moyle J T B 1991 Temperature dependence of LED and its theoretical effect on pulse oximetry *Br. J. Anaesthesia* **67** 638–43

Siemens 1993 *Optoelectronics Data Book* (Cupertino CA: Siemens)

Varshni Y P 1967 Temperature dependence of the energy gap in semiconductors *Physica* **34** 149–54

INSTRUCTIONAL OBJECTIVES

5.1 Sketch a current–voltage curve for an LED and indicate the approximate maximal current to prevent damage to the LED.
5.2 State the maximal LED current that will not cause burns to the patient.
5.3 Sketch and describe the current control system for LEDs in a pulse oximeter.
5.4 Describe how a pulse oximeter determines the LED wavelengths in a given probe.
5.5 Explain the process by which an LED emits light. Relate the explanation to the main point of interest on an LED I–V plot.
5.6 Discuss bandwidth characteristics of red and IR LEDs, including temperature and drive current effects.
5.7 Explain how a change in LED drive current indirectly affects the output spectra of red and IR LEDs.
5.8 Explain how a change in LED temperature affects the output spectra of red and IR LEDs.
5.9 Discuss two techniques which would automatically compensate for a shift in the peak wavelength of the LEDs.
5.10 Give reasons why LEDs are convenient to use in pulse oximetry.

CHAPTER 6

PHOTODETECTORS AND AMPLIFIERS

Jeffrey S Schowalter

The photodetector is the main input device of the pulse oximeter system. These devices, found in the probe, sense the intensity of light emitted by each LED after the light passes through the tissue. The photodetector produces a current which is linearly proportional to the intensity of incident light. This current is then converted to a voltage which is passed on to the pulse oximeter unit for processing. The choice of photodetector depends on factors such as performance, packaging, size, and cost. However, most pulse oximeters currently use silicon photodiodes. Transimpedance amplifiers, which convert the photodiode current to a voltage are also discussed.

6.1 PHOTODETECTION DEVICES

A variety of devices can be used to sense the intensity of a light source. These include photocells, photodiodes, phototransistors, and integrated circuit (IC) sensors. When choosing a photodetection sensor, several things need to be considered. First, since the pulse oximeter uses two specific wavelengths, *spectral response*, or the relative response of the device to different wavelengths must be considered. Another important consideration is the linearity of the output signal. With the pulse oximeter, an output linearly proportional to the intensity of incident light, also known as *illumination* (E), is highly desirable. A third important factor is *sensitivity*, or the ratio of the electrical output signal to the intensity of incident light. A related consideration is the response time, or how quickly the output of the device is able to respond to a change in the incident light. Size also becomes a consideration since many of these devices are mounted in disposable probes (as will be discussed in chapter 7). Finally, as is the case with any commercial device, the cost must be considered. Each type of device merits consideration although the photodiode is most frequently chosen for pulse oximeter applications.

6.1.1 Photocells

A *photocell* exhibits a change in resistance that is proportional to light intensity. In these semiconductor devices, also referred to as *photoconductors* and *photoresistors*, the electrical conductivity of the material is dependent on the number of carriers in the conduction band. Incident light increases the number of carriers generated and thus increases the conductivity. These devices have spectral responses dependent on the type(s) of materials used in their manufacture. In the visible/near infrared range (400 to 1400 nm), which includes the wavelengths used in pulse oximetry, the most common materials used are cadmium sulfide (CdS) and cadmium selenide (CdSe). Equation (6.1) shows the relationship between the resistance of a photocell and the illumination E:

$$R = AE^{-a} \tag{6.1}$$

where R is the resistance of the device and A and α are constants dependent on manufacturing process and material type (Pallas-Areny and Webster 1991). This equation shows that the relationship between resistance and light intensity is highly nonlinear. In addition, the resistance changes quite dramatically as a function of light intensity. For example, a typical CdS photocell can have its resistance change by a factor of 10^4 between an illuminated condition and a dark condition. Photocells are also temperature sensitive, exhibiting a changing resistance/incident light relationship with temperature. Increased temperature also causes increased thermal noise. The response time of the photocell is relatively slow. Time constants are on the order of 100 ms and these devices exhibit light memory, making their response dependent on the previous light level. In addition, photocells are relatively large in size with typical diameters of 5 to 25 mm (Vig 1986). Photocells are widely used and are relatively inexpensive (~$1), but are not typically used in pulse oximetry applications.

6.1.2 Photodiodes

A photodiode produces an output current or voltage which is proportional to the intensity of the incident light. The *p–n* junction photodiode consists of one layer of *n*-type semiconductor material along side a layer of *p*-type semiconductor material (see figure 6.1). When a photon is absorbed, it creates an electron-hole pair. Electrons from the *p*-side will move across the depletion region toward the *n*-side and holes from the *n*-side will be transported to the *p*-side. As a result, an electric current is generated.

Figure 6.2 shows a simple model for the photodiode. It is made up of the parallel combination of a current source, an ideal diode, and a junction capacitance.

For this photodiode the total current supplied (I) can be expressed as

$$I = I_P - I_D \tag{6.2}$$

where the photocurrent I_P can be expressed as

$$I_P = SE \tag{6.3}$$

and the diode current I_D is expressed as

$$I_D = I_0\left[\exp\left(\frac{qV}{kT}\right) - 1\right] \tag{6.4}$$

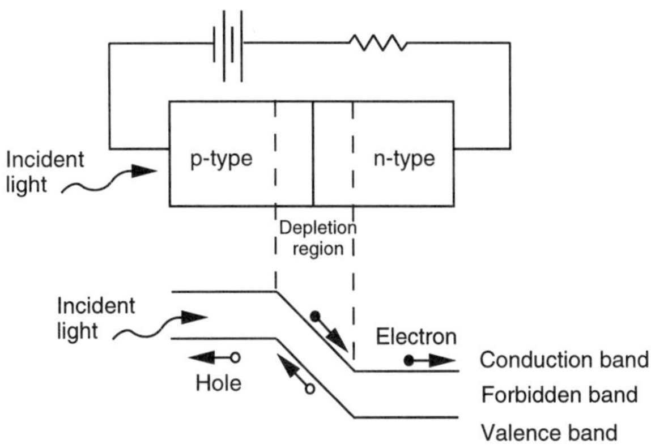

Figure 6.1 *p–n* junction of a photodiode. Electrons move towards the *n* layer and holes move towards the *p* layer (adapted from Hitachi 1992).

where S is the *sensitivity* or the unit of photocurrent produced per unit of input light, E is the illumination, I_0 is the inverse saturation current, V is the voltage applied to the diode, k is Boltzmann's constant, and T is absolute temperature.

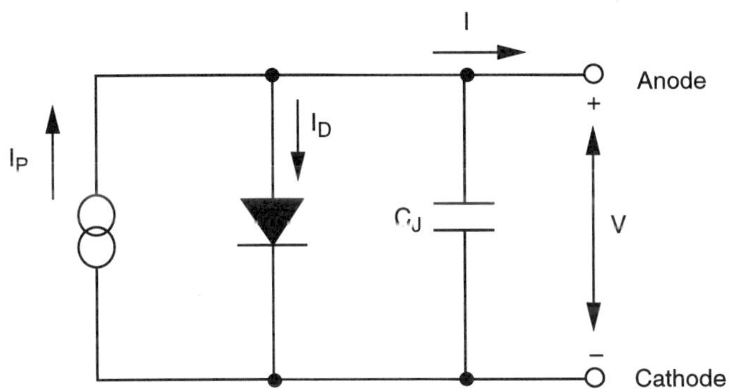

Figure 6.2 Simplified photodiode equivalent circuit model. The current induced by the incident light is denoted by I_P.

The photodiode operates in one of two modes. The photovoltaic operating mode generates a light induced voltage produced by an open-circuit photodiode. This output voltage however is not a linear function of incident light. In the open-circuit condition ($I = 0$), the output voltage is given by

$$V_{OC} = \frac{kT}{q}\ln\left(\frac{I_P}{I_D}+1\right). \tag{6.5}$$

The photoconductive operating mode generates a light-induced current produced by a photodiode connected so that the photodiode voltage is zero or constant with varying light intensity. In this mode, output current is linearly proportional to the level of incident light. In the short-circuit condition ($V = 0$), the output current is given by

$$I_{sc} = SE. \tag{6.6}$$

Figure 6.3 shows the current versus voltage characteristics of a photodiode for various levels of incident light.

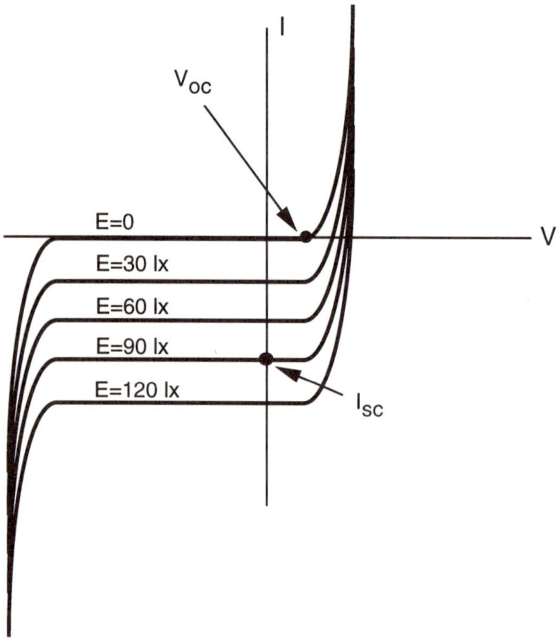

Figure 6.3 Current versus voltage characteristics for a photodiode. For an open circuit condition (I = 0), increasing light intensity results in a logarithmic increase in voltage. For a short circuit condition (V = 0), the photodiode's current varies linearly with increasing incident light intensity.

When the *p–n* photodiode is used in the photoconductive mode, a highly linear relationship exists between the incident light level and the output current. The sensitivity of a typical photodiode varies by only 0.05% over most of its range (which may span up to seven decades) but can increase to several percent at high levels of incident light/output current. Sensitivity, however, varies significantly with incident light wavelength (see figure 6.4). The spectral response is determined by the material used for fabrication and the physical depth of the *p–n* junction. The silicon photodiode, shown in figure 6.4, works well with the wavelengths of interest to pulse oximetry. Photodiodes, when used in the photoconductive mode, are also relatively insensitive to temperature variations with the typical sensitivity varying by approximately +0.2%/°C. These devices have response times much faster than that of the photocell with typical values running on the order of 20 μs. With radiant sensitive areas on the order of 1 to 7 mm^2, silicon photodiodes' prices are equivalent to photocells (~$1).

There are several different variants of the basic *p–n* photodiode. These include the *p–i–n* photodiode, the Shottky photodiode, the metal–semiconductor–metal photodetector and the avalanche photodiode. Of these, the *p–i–n* photodiode is frequently found in pulse oximetry applications.

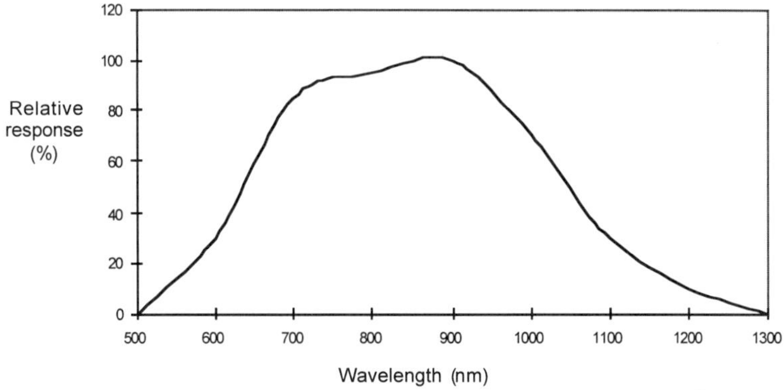

Figure 6.4 Spectral response of a Si photodiode as a function of wavelength.

6.1.2.1 p–i–n photodiodes. The *p–i–n* photodiode has an intrinsic (lightly doped) layer between the *n* and *p* layers. This modified structure typically results in lower junction capacitance than for *p-n* diodes of the same optical sensing area. As a result, *p–i–n* photodiode response times are faster than *p–n* junction photodiodes. Bandwidths typically run on the order of 10 MHz for these devices. Since cost and size are comparable to the *p–n* photodiode, these devices are also used in pulse oximetry applications.

6.1.2.2 Shottky photodiodes. In these devices, a thin metal layer deposited on a semiconductor can form a Shottky barrier and if thin enough can pass incident light. These devices are primarily used to detect ultraviolet (UV) light and have smaller junction capacitances than either *p–n* or *p–i–n* photodiodes. This results in lower response times with frequency responses exceeding 100 GHz (Sloan 1994), far exceeding the requirements of the typical pulse oximeter application. However, with a primary spectral response in the UV range, these devices are not suitable for pulse oximetry applications.

6.1.2.3 Metal–semiconductor–metal (MSM) photodetectors. These devices commonly use interdigitated metal fingers to form the two electrical contacts of the device. They have low capacitance because of a small active area and a relatively fast response time. These devices have lower sensitivity than *p–i–n* photodiodes due to large amounts of surface covered by metal and as such are not currently being used for pulse oximeter applications.

6.1.2.4 Avalanche photodetectors (APDs). Avalanche photodetectors are high speed photodetectors that make use of the avalanche multiplication effect of photons. APDs operate under large reverse bias voltage so the electric field is large. However, the multiplication effects result in increased noise, reduced bandwidth, avalanche buildup time (which slows response time), and noise multiplication. They are typically used as amplifiers for applications requiring detection of extremely low levels of light (Fraden 1997).

6.1.3 Phototransistors

A phototransistor can be thought of as a photodiode with a built-in current amplifier. Phototransistors typically have 100 to 500 times the sensitivity of a corresponding photodiode. In these devices, incident light on the base of the transistor induces a current. This current is then amplified by the transistor resulting in a significant increase in collector current. The sensitivity of these devices is not as linear as photodiodes with the sensitivity varying 10 to 20% over the useful range of the phototransistor (Siemens 1993). These devices do not have the light memory problems associated with photocells but sensitivity can vary as much as 50% among devices of the same type because of process and beta variations (Sprague Electric 1987). The response time for the typical phototransistor is 125 μs. Size, cost and signal-to-noise ratio (SNR) of a phototransistor are equivalent to that of a photodiode. Although early pulse oximeters used phototransistors (Schibli *et al* 1978), currently photodiodes are the sensor of choice in pulse oximetry applications.

6.1.4 Integrated circuit (IC) sensors

Integrated circuit sensors are becoming increasingly popular for sensing incident light levels. These devices incorporated the features of a photodiode along with a current-to-voltage converter so that the output is a voltage which is a direct function of the incident light. Since photodiodes and IC amplifiers are built out of semiconductor material, IC designers have been able to fabricate both the hybrid circuitry and the photodiode on the same silicon substrate. Combining these two devices onto one chip eliminates problems commonly encountered in discrete designs such as leakage current errors, noise pick-up, and gain peaking due to stray capacitances (Burr-Brown 1994a,b,c). These devices typically are four times as costly as equivalent photodiodes and as such do not appear to have gained much acceptance by pulse oximeter manufacturers. However, at least one manufacturer (Protocol Systems 1992) is using this integrated circuit photodiode/transimpedance amplifier configuration.

6.2 PHOTODIODE CHARACTERISTICS

Because of their relatively low cost and linear output current response to incident light, both the standard *p–n* diodes and *p–i–n* (New and Corenman 1987) diodes are currently in use today as the photodetectors of choice for use in pulse oximetry systems. The following sections describe several key parameters related to photodiode operation. These parameters serve as evaluation criteria for the designer when selecting a photodiode for use in a pulse oximeter.

6.2.1 Junction capacitance

Photodiode junction capacitance is an important parameter and is proportional to the junction area. It also decreases with increasing reverse bias voltage so it may be expressed at a specified reverse bias voltage across the photodiode. The response speed of the photodiode depends on the *RC* time constant of the junction capacitance and the load resistance. Therefore a higher response speed can be obtained by applying a larger reverse bias voltage to the photodiode. It should be noted however, that this technique is not typically used in pulse oximetry applications.

6.2.2 Dark current

Dark current is the reverse leakage current that flows in a photodiode in the absence of light. Dark current is usually specified at some specified reverse bias voltage or with zero voltage bias. Although technically, no dark current should flow with zero bias, most zero bias applications have a small voltage across the photodiode such as the offset voltage of the op amp. The dark current increases as the reverse voltage or ambient temperature increases.

6.2.3 Sensitivity

Since the output current of the photodiode is linear, the sensitivity is normally expressed as the output current level for a known incident light level at a specified temperature. The light source used to produce this specification varies among photodiode manufacturers though and can cause some confusion. In some cases, the sensitivity is determined with an LED optical source and thus the center frequency of the LED is specified. In this case incident light is expressed in mW/cm^2. If the light source is an International Commission on Illumination (CIE) standard light source (normally a tungsten lamp), it is expressed in lux (lx).

6.2.4 Spectral response

Figure 6.4 shows that photodiodes have a spectral response and as such care should be taken when selecting a photodiode. Normally photodiode manufacturers specify the spectral response by providing the wavelength of peak sensitivity. The designer should keep in mind the wavelengths of interest in pulse oximetry (660 nm and 940 nm) when deciding on an appropriate photodiode.

6.2.5 Packaging

Photodiodes used in pulse oximeter probes have some unusual mounting and mechanical assembly requirements. A variety of characteristics should be considered including cost, hermetic seal, package material and package type. In general, photodiodes are available in three types of packages: the can package, the ceramic stem package and the resin mold package.

6.2.5.1 Can package. In the can package (figure 6.5(*a*)), the photodiode chip is mounted on a metallic stem and is sealed with a cap that has a window to allow incident light to reach the semiconductor surface.

6.2.5.2 Ceramic stem package. With the ceramic stem package, the photodiode chip is mounted on a ceramic stem (figure 6.5(*b*)) and is coated with resin.

6.2.5.3 Resin mold package. For the resin mold package (figure 6.5(*c*)), the photodiode chip is mounted on a lead frame and molded with resin. Some of these devices use molding that is transparent only to certain wavelengths of light, thereby limiting the wavelength sensitivity of the photodiode. This is the most common package type used in pulse oximetry today.

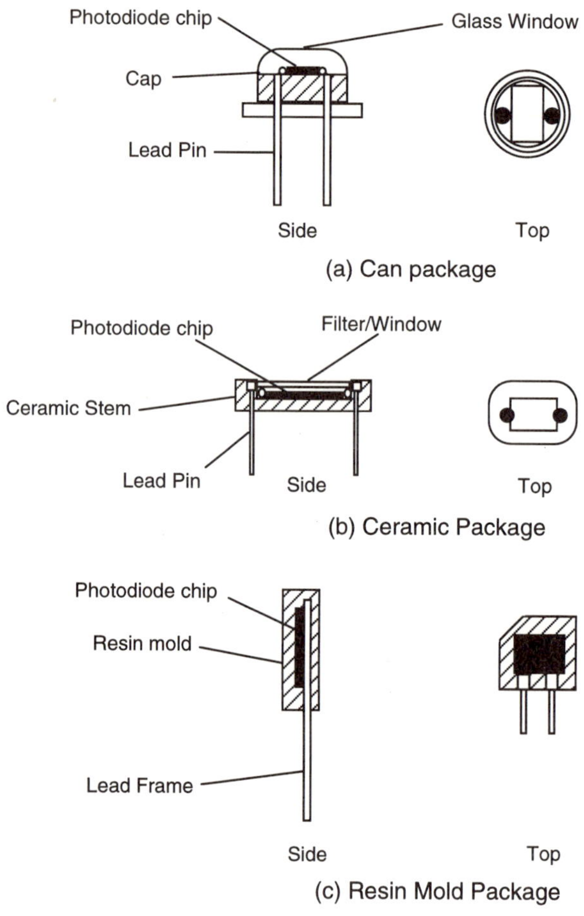

Figure 6.5 Typical photodiode packaging (adapted from Sharp 1988).

Table 6.1 shows the characteristics for several typical photodiodes.

Table 6.1 Electro-optic characteristics of two *p–i–n* photodiodes.

	Sharp PD4663PS	TRW OP913
Output current	120 nA @ 1000 lx	55 mA @ 5.0 mW/cm^2
Dark current	200 pA	25 nA
Peak sensitivity wavelength	840 nm	875 nm
Junction capacitance	2 pF	150 pF

6.3 OPTICAL CONCERNS

Since this is a system with an optical interface, it is important to minimize the effects from light other than the optical signals of interest. One way to minimize unwanted light incident upon the detector is to place some type of light filter over the detector. This allows light of wavelengths of interest to pass through the filter but does not allow light of other wavelengths to pass through the filter. For the pulse oximeter to work effectively, most of the light being transmitted from the LEDs must not reach the photodiode unless it has passed through tissue containing arterial blood.

6.3.1 Optical filtering

Optical filtering, placed between sources of light and the photodiode, is used to limit the spectral response of the photodiode. A number of optical filter types can be used. Cheung *et al* (1993) recommend a red Kodak No. 29 wratten gel filter to eliminate the flickering effect of fluorescent light. However, these external filters do not appear to be used much in actual pulse oximetry designs. In addition, the photodiode mounting package may contain filtering material. Many photodiodes are mounted in clear plastic which absorbs UV wavelengths (Burr-Brown 1994a,b,c). This is useful to filter out some of the unwanted effects of fluorescent lighting on the photodiode. Many photodiodes are available in a variety of packaging types each of which filters out selected wavelengths of light.

6.3.2 Optical interference

To minimize errors, the pulse oximeter designer must attempt to limit the light reaching the photodiode to that which has traveled through tissue containing arterial blood (Nellcor 1993). This can be accomplished through thoughtful LED/photodiode placement. Light impervious barriers should be placed between LEDs and the photodiode in all areas where the emitted light could reach the photodiode without passing through tissue (New and Corenman 1987). Two additional measures can be taken to ensure this (figure 6.6). One is to decrease the angle of incidence to the photodiode. The second is to coat the housing around the photodiode with a material that does not scatter or reflect light.

There are two types of optical interference that may cause problems for the photodiode. The first is excessive ambient light. The source of this type of error may be surgical lamps, fluorescent lights, infrared heat lamps and direct sunlight. Usually this type of interference will saturate the photodiode so that no pulse can be distinguished, however some of these sources may result in apparently normal but inaccurate readings. The second source of interference is optical cross-talk. This type of interference typically may occur when multiple probes are used in close proximity. In this case, light from one LED probe is sensed by the photodiode of another probe.

6.4 AMPLIFIERS

Since photodiodes generate an output current, an amplifier must be used to translate that current into a voltage for use by the pulse oximeter. Transimpedance amplifiers, or current-to-voltage converters, are amplifiers that convert an input current to an output voltage. These are the most common types of amplifiers used in pulse oximetry applications today.

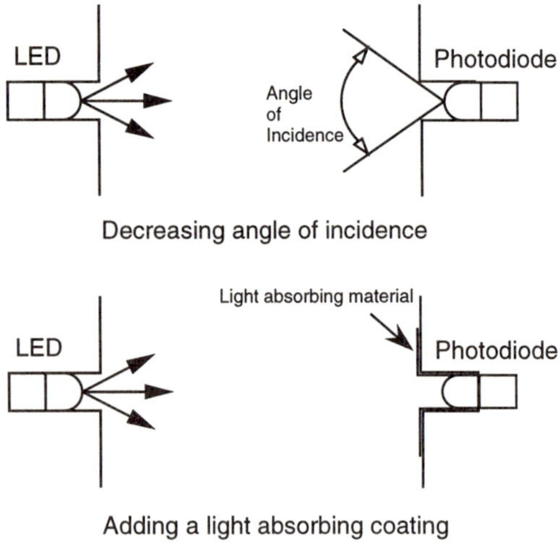

Figure 6.6 Minimizing photodiode optical interference (adapted from Marktech International 1993).

6.4.1 Standard transimpedance amplifier configuration

Figure 6.7 shows the standard transimpedance configuration.

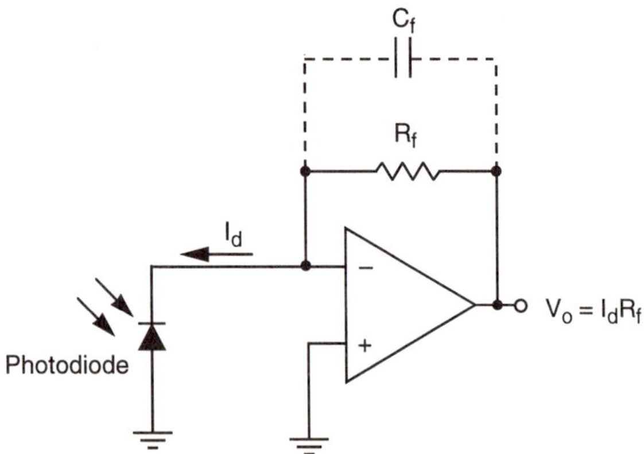

Figure 6.7 Typical transimpedance amplifier used with a photodiode.

In this configuration, the current generated by the photodiode is converted to a voltage. Because of the virtual ground, the op amp maintains zero voltage across the photodiode. Current flows through the feedback resistor and creates a voltage at the output that is proportional to the light intensity as given by

$$V_0 = I_d R_f. \tag{6.7}$$

The transimpedance gain is then equal to the value of the feedback resistor. Even though standard resistor values can give substantial gains, Cysewska-Sobusiak (1995) noted that the effective transmittance of light through the finger in pulse oximetry applications never exceeds 5%. Even with the use of superbright LEDs, relative light intensity can be expected to be fairly low.

Although the transimpedance amplifier appears to be a simple and straightforward design, it is subject to a number of multidimensional constraints. These constraints have been well documented (Burr-Brown 1994a,b,c, Graeme 1992, 1994, Wang and Ehrman 1994, Kirsten 1996), and several alternative configurations have been proposed. However, because this standard configuration is frequently used in pulse oximetry applications, several general guidelines are provided.

6.4.1.1 Photodiode capacitance. Photodiode junction capacitance should be as low as possible. The junction capacitance affects noise and bandwidth of the circuit.

6.4.1.2 Photodiode active area. The photodiode active area should be as small as possible for largest signal-to-noise ratio. The area of the photodiode is directly proportional to the junction capacitance. Kästle *et al* (1997) describe an active area of 1 to 2 mm^2 for the Hewlett-Packard sensor.

6.4.1.3 Feedback resistor. The feedback resistor should be made as large as possible to minimize noise. This is because the feedback resistor is the dominant source of noise in the circuit. This thermal (Johnson) noise increases as a function of the square root of the value of the feedback resistance.

$$\text{thermal noise} = \sqrt{4kTBR} \tag{6.8}$$

where k is Boltzmann's constant, T is absolute temperature, B is the noise bandwidth (Hz), R is the feedback resistance (Ω), while the signal voltage increases as a function R. Therefore the signal-to-noise ratio improves by the square root of the feedback resistance as the feedback resistance is increased.

In addition, a high resistor value of feedback resistance is preferred to an equivalent low resistance T network. Although the transimpedance gain is equivalent, the T network will have a lower signal-to-noise ratio due to current noise and offset voltage.

6.4.1.4 Op amp. An FET op amp is a requirement for this configuration. The lower the bias current of the op amp, the higher the sensitivity.

6.4.1.5 Feedback capacitor. The capacitor in the feedback loop minimizes gain peaking and improves stability. The choice of capacitor value is critical. Graeme (1992) analyzed this circuit configuration and provided several simplified formulas for determining the appropriate value of feedback capacitance, C_f. For relatively large area photodiodes, where the junction capacitance is much larger than the feedback capacitor

$$C_f = \sqrt{\frac{C_I}{2\pi R_f f_c}} \tag{6.9}$$

where f_c is the unity gain frequency of the op amp, C_I is the total input capacitance = photodiode junction capacitance + op amp input capacitance, R_f is the feedback resistance.

A more general formula, for use with small photodiode junction capacitances is

$$C_f = \frac{1}{4\pi R_f f_c}\left(1 + \sqrt{1 + 8\pi R_f C_I f_c}\right). \tag{6.10}$$

Note that larger values of capacitance can be used but this decreases signal bandwidth where the bandwidth can be calculated by

$$BW = 1.4 f_p$$

where

$$f_p = \sqrt{\frac{f_c}{2\pi R_f (C_I + C_f)}}. \tag{6.11}$$

6.4.1.6 Shielding. Since this circuit configuration has high sensitivity and high input impedance, the transimpedance amplifier is sensitive to noise coupling from electrostatic, magnetic, and radio frequency sources.

Electrostatic coupling, typically from ac voltage sources, can create noise in the photodiode/transimpedance amplifier circuit. To prevent this, some pulse oximeter manufacturers have completely enclosed their photodiodes in a metallic shielding such that only the detector surface is exposed. One item of concern is that the shield produces a capacitance between amplifier and ground that may, in some cases, affect the performance of the system.

Magnetically coupled noise is somewhat more difficult to control since it is unaffected by the electrostatic shielding. Sensitive loop areas need to be minimized. High value resistors are sensitive to magnetic coupling so connections between these resistors and op amp inputs should be as short as possible.

Radio frequency interference (RFI) sources should be expected both from the main processing unit of the pulse oximeter itself (as discussed in chapter 8) as well as from other patient monitoring devices. The best prevention is through the use of shielding and filtering. Shielded twisted pair cabling is typically used to send photodiode signals back to the pulse oximeter. The desired photodiode signals of interest are not in the radio frequency range so filtering, even before input to the amplifier, is also effective.

6.4.2 Differential transimpedance amplifier

Since the photodiode signal of interest is a current, it is available to drive two different inputs in differential fashion as shown in figure 6.8. Since these currents are in different directions, if the feedback resistances are equal, the differential output signal will be twice what it was for the single-ended transimpedance amplifier configuration (figure 6.8). An advantage of this configuration is that for a given gain level, feedback resistor values can be half the single ended configuration shown in figure 6.7. Another benefit of this configuration is the common-mode rejection of coupled noise. By feeding the output of the current-to-voltage conversion into a differential amplifier stage, noise will show up as a common mode signal on both inputs and have a canceling effect at the circuit output. Note however, that this configuration is not a total replacement for electrostatic shielding, but works well removing coupling that passes through shield imperfections. Several pulse oximeter manufacturers use this configuration in their pulse oximetry systems (Cheung *et al* 1993, Criticare 1990).

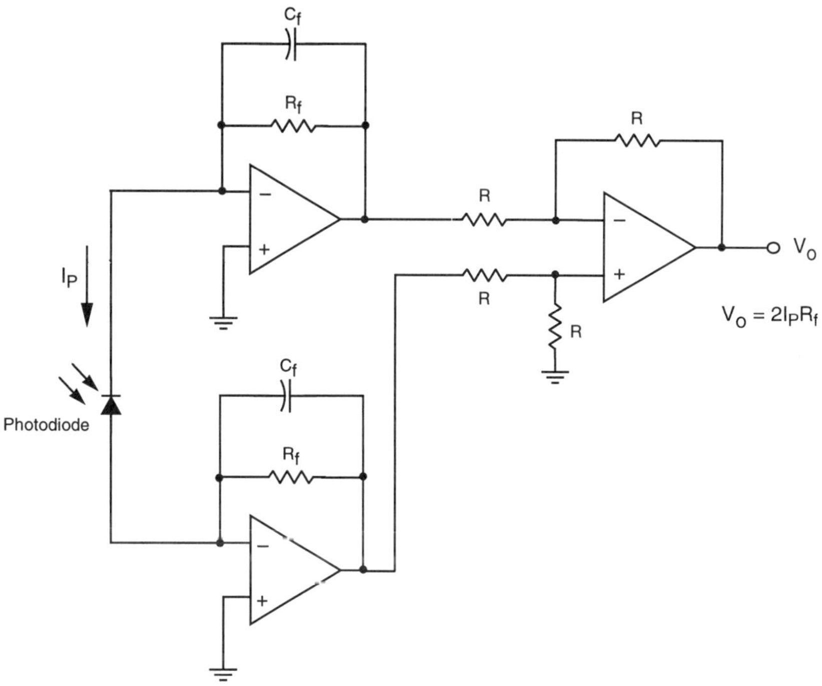

Figure 6.8 Differential current sensing transimpedance configuration.

6.4.3 Zeroing circuit

The purpose of the zeroing circuit shown in figure 6.9 is to remove the ambient light signal from the photodiode output signal. To do this, an FET switch is closed, so the *RC* network of the output is active. This allows the capacitor to charge up to the voltage equivalent of the ambient light level. The only critical factor in the selection of these components is to make sure the *RC* time constant allows for complete charging of the capacitor in the time period allowed (see chapter 8). When an LED is turned on, the FET switch opens, leaving the ambient light level voltage across the capacitor. This will have the net effect of subtracting the voltage across the capacitor from any LED output signal, thereby canceling out the ambient light level. The process repeats itself at the same rate at which the LEDs are pulsing so changes in light level are immediately accounted for. Potratz (1994) presents a similar but different implementation of this circuit with the same general functionality for use in pulse oximetry application.

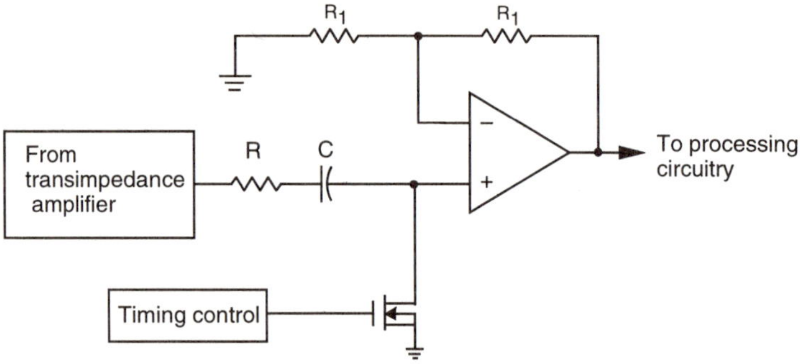

Figure 6.9 Typical zeroing circuit used in pulse oximetry applications to remove ambient light offset from the usable signal.

6.4.4 Future trends

At least one manufacturer (Protocol Systems 1992) provides for an electronic switching mechanism on the input to accommodate either a current input directly from a photodiode or a voltage input from an IC sensor. Burr-Brown (1994a,b,c) and Texas Instruments (1993) are two manufacturers that provide ICs that directly integrate the photodiode and transimpedance amplifier to convert a light intensity light directly to a voltage. As these devices drop in price, expect to see these ICs replace the photodiode as the photodetector of choice in future pulse oximetry applications.

REFERENCES

Burr-Brown 1994a OPT101 *Data Sheets: Monolithic Photodiode and Single-Supply Transimpedance Amplifier* (Tucson, AZ: Burr-Brown Corporation)

Burr-Brown 1994b *Application Bulletin AB-075. Photodiode Monitoring with Op Amps* (Tucson, AZ: Burr-Brown Corporation)

Burr-Brown 1994c *Application Bulletin AB-077. Designing Photodiode Amplifier Circuits with OPA128* (Tucson, AZ: Burr-Brown Corporation)

Cheung P W, Gauglitz K F, Hunsaker S W, Prosser S J, Wagner D O and Smith R E 1993 Apparatus for the automatic calibration of signals employed in oximetry *US patent 5,259,381*

Criticare 1990 *504/504-US Service Manual* (Waukesha, WI: Criticare Systems)

Cysewska-Sobusiak A 1995 Problems of processing reliability in noninvasive measurements of blood oxygen saturation *Optoelectronic and Electronic Sensors* ed R Jachowicz and Z Jankiewicz *Proc. SPIE* **2634** 163–71

Fraden J 1997 *Handbook of Modern Sensors, Physics, Designs and Applications 2nd edn* (Woodbury, NY: American Institute of Physics)

Graeme J 1992 Phase compensation optimizes photodiode bandwidth *EDN.* 7 May: 177–84

Graeme J 1994 *Applications Bulletin AB-094. Tame Photodiodes with Op Amp Bootstrap* (Tucson, AZ: Burr-Brown Corporation)

Hitachi 1992 *Opto Data Book* (Brisbane, CA: Hitachi America)

Kästle S, Noller F, Falk S, Bukta A, Mayer E and Miller D 1997 A new family of sensors for pulse oximetry *Hewlett-Packard J.* **48** (1) 39–53

Kirsten T R 1996 Increasing photodiode transimpedance bandwidth and SNR with a bootstrap buffer *Sensors* **13** 35–8

Marktech International 1993 *Optoelectronics data book* (Mendands, NY: Marktech International)

Nellcor 1993 *Controlling External Optical Interference in Pulse Oximetry. Reference Note: Pulse Oximetry Note Number 5* (Pleasanton, CA: Nellcor)

New W and Corenman E 1987 Calibrated pulse oximetry probe *US patent 4,700,708*

Pallas-Areny R and Webster J G 1991 *Sensors and Signal Conditioning* (New York: Wiley)
Potratz R S 1994 Condensed oximeter system with noise reduction software *US Patent 5,351,685*
Protocol Systems 1992 *Ultra-Portable Vital Signs Monitor Technical Reference Guide* (Beaverton, OR: Protocol Systems)
Schibli E G, Yee S S and Krishnan V M 1978 An electronic circuit for red/infrared oximeters *IEEE Trans. Biomed. Eng.* **BME-25** 94–6
Sharp 1988 *Optoelectronics Data Book* (Mahwah, NJ: Sharp Corporation)
Siemens 1993 *Optoelectronics Data Book* (Cupertino, CA: Siemens Electronics Corporation)
Sloan S R 1994 Photodetectors *Photonic Devices and Systems* ed R G Hunsperger (New York: Marcel Dekker)
Sprague Electric 1987 *Hall Effect and Opto Electronic Sensors* (Concord, NH: Sprague Electric Company)
Texas Instruments 1993 *Signal Conditioning 1993, Linear Design Seminars Reference Book* (Dallas, TX: Texas Instruments)
Vig R 1986 Light sensing using optical integrated circuits *Sensors* **3** 6–15
Wang T and Ehrman B 1994 *Application Bulletin AB-050. Compensate Transimpedance Amplifiers Intuitively* (Tucson, AZ: Burr-Brown Corporation)

INSTRUCTIONAL OBJECTIVES

6.1 Describe why photodiodes are used for light level detection in pulse oximeters.
6.2 Sketch the equivalent circuit of a photodiode and explain its operation.
6.3 Explain the difference between a *p–n* and a *p–i–n* diode.
6.4 Identify some of the most important characteristics to consider when selecting a photodiode.
6.5 Explain some of the techniques used to improve the signal-to-noise ratio when using photodiodes.
6.6 Describe some of the sources of optical interference in pulse oximeters.
6.7 Sketch a simple transimpedance amplifier configuration and explain its basic operation.
6.8 Given a light/current transfer curve of a photodiode, design a transimpedance amplifier.
6.9 Explain why the type of covering/filtering over a photodiode's exposed surface is important to the pulse oximeter designer.
6.10 Explain why photodiodes are normally configured to use current to indicate light level.
6.11 Explain the advantages of a differential amplifier configuration in a photodiode/transimpedance amplifier circuit.

CHAPTER 7

PROBES

Moola Venkata Subba Reddy

Light emitted by the light emitting diodes (LEDs) is partially reflected, transmitted, absorbed, and scattered by the skin and other tissues and the blood before it reaches the detector. The probe of a pulse oximeter consists of two LEDs of selected wavelengths and a detector. The wavelengths of the LEDs chosen are 660 nm and 940 nm (chapter 5) and the detector used is a photodiode (chapter 6). This assembly must be protected from the ambient light for the wavelengths to which the photodiode is sensitive.

The flexible cable connecting the probe and the pulse oximeter unit carries electric power to the LEDs and the signal from the photodiode. Depending on the design, the cable may also contain conductors for a temperature sensor, to detect the temperature of the probe and the underlying skin, and the coding resistor to compensate for the variation of the wavelengths of the emitted light from the LEDs.

7.1 TRANSMITTANCE PROBES

7.1.1 Principle

As the name suggests, a pulse oximeter with transmittance probes uses the light transmitted through the extremity to measure the arterial oxygen saturation of the blood. Figure 7.1 shows a general transmission probe.

The system employs two LEDs, with emission peak wavelengths at 660 nm in the red range and 940 nm in the infrared range. The LEDs are powered alternately so that light of one particular wavelength will pass through the tissue, and the transmitted light will be detected by the photodiode. The intensity of the light emerging from the tissue is attenuated by the amount of blood present in the tissue. This varies with the arterial pulse and is used as a measure to indicate the pulse rate. The absorption coefficient of oxyhemoglobin is different from that of deoxygenated hemoglobin for most wavelengths of light. For example, the infrared light is absorbed only by molecules made up of dissimilar atoms, because only such molecules (e.g., CO_2, CO, N_2O, H_2O and volatile anesthetic agents) possess an electric dipole moment with which the electromagnetic wave can interact. Symmetric molecules (e.g., O_2, N_2, H_2) do not have an electric dipole moment and therefore do not absorb infrared radiation (Primiano 1997). Thus differences in the amount of light absorbed by the blood at two different wavelengths can be used to indicate arterial oxygen saturation.

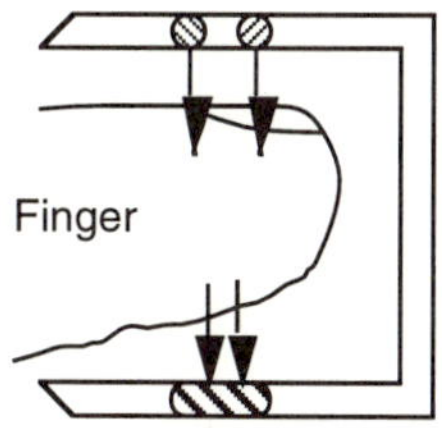

Figure 7.1 Probe using transmittance principle. Light from two LEDs passes alternately through the tissue of the finger and is detected by the photodiode.

7.1.2 Sensor placement

In transmission probes, as the photodiode has to detect the light transmitted through the tissue, the detector is placed in line with the LEDs so that the maximum amount of the transmitted light is detected. The photodiode should be placed as close as possible to the skin without exerting force on the tissue. The amount of force applied by reusable probes is much larger than the amount of force applied by disposable probes. The force applied also depends on the materials used to manufacture a particular probe and also on the company which produces the probes, e.g., Nellcor clip type probes exert less pressure than Ohmeda clip type probes. If the force exerted by the probe is significant, the blood under the tissue, where the probe is placed, may clot due to external pressure applied. And if we increase the distance between the LEDs and the photodiode (optical path length increases), the amount of detected light decreases as seen from Beer's law (section 4.1).

Thus we should place the LEDs and photodiode facing each other. Normally transmission probes are placed on the patient's finger, toe, ear or nose. In a clip type probe, the distance between the LEDs and the photodiode can be as much as 12 mm (without requiring much pressure).

7.2 REFLECTANCE PROBES

To measure arterial oxygen saturation, when pulse oximeters with transmission probes cannot be used, pulse oximeters with reflectance probes are used to monitor S_aO_2 based on the intensity of reflected light. The idea of using light reflection instead of light transmission in clinical oximetry was first described by Brinkman and Zijlstra (1949). They showed that S_aO_2 can be monitored by measuring the amount of light reflected (back scattered) from the tissue. The idea of using skin reflectance spectrophotometry marked a significant advancement in the noninvasive monitoring of S_aO_2 from virtually any point on the skin surface.

Even though it was a major advancement, difficulties in absolute calibration and limited accuracy were the major problems with early reflectance pulse oximeters.

7.2.1 Principle

The intensity of the back scattered light from the skin depends not only on the optical absorption spectrum of the blood but also on the structure and pigmentation of the skin.

S_aO_2 is measured by analyzing the pulsatile components of the detected red and infrared plethysmograms which make use of reflected light intensities. The light from the LEDs enters the tissue, is scattered by both the moving red blood cells and the nonmoving tissue, and a part of this back scattered light is detected by the photodiode. The output of the photodiode is processed by the pulse oximeter, and measures the S_aO_2 of the pulsatile blood.

7.2.2 Sensor placement

In reflectance pulse oximetry, the LEDs and the photodiode are placed on the same side of the skin surface as shown in figure 7.2. Normally the reflectance probe is placed on the forehead or temple, but is not restricted to only those two places. Reflectance probes can be used to measure arterial oxygen saturation at virtually any place on the human body where the probe can be placed.

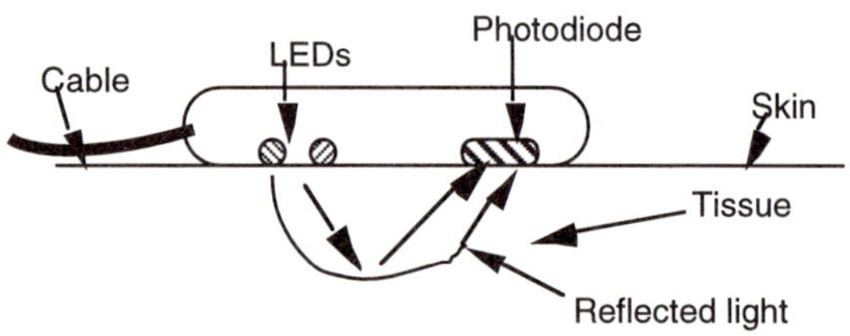

Figure 7.2 Reflectance probe. The light is transmitted into the tissue, travels through the tissue, and is detected at the photodiode.

7.2.2.1 Optimum distance between LEDs and photodiode. One of the major design considerations required in designing a reflectance pulse oximeter sensor is determining the optimum separation distance between the LEDs and the photodiode. This distance should be such that plethysmograms with both maximum and minimum pulsatile components can be detected. These pulsatile components depend not only on the amount of arterial blood in the illuminated tissue, but also on the systolic blood pulse in the peripheral vascular bed.

There are two techniques that can enhance the quality of the plethysmogram. One way is to use a large LED driving current, which determines the effective penetration depth of the incident light, which increases light intensity. So for a given LED/photodiode separation, using higher levels of incident light, we can illuminate a larger pulsatile vascular bed. As a result the reflected plethysmograms will contain a larger AC component. But, in practice, the LED driving current is limited by the manufacturer to a specified maximum power dissipation. The other way is to place the photodiode close to the LEDs. If we place the photodiode too close to the LEDs, the photodiode will be saturated as a result of the large DC component obtained by the multiple scattering of the incident photons by the blood-free stratum corneum and epidermal layers in the skin.

For a constant LED intensity the light intensity detected by the photodiode decreases roughly exponentially as the radial distance between the LEDs and the photodiode is increased and the same applies to the AC and DC components of the reflected plethysmograms as shown in figure 7.3. Figure 7.4 shows the effect of LED/photodiode separation on the relative pulse amplitude of the red and infrared plethysmograms. This is expected as the probability of the number of photons reaching the photodiode is decreased with the increase in separation.

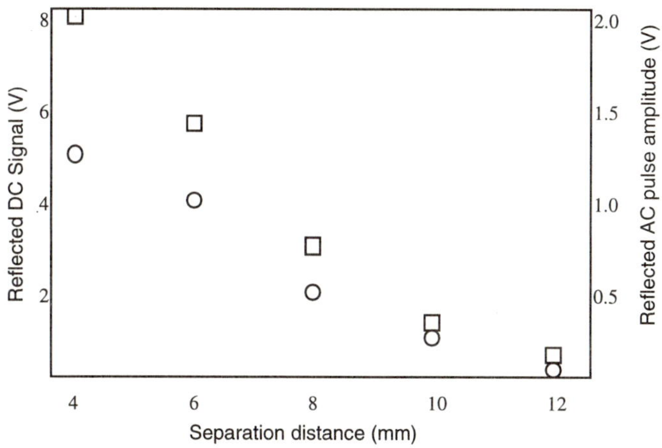

Figure 7.3 Effect of LED/photodiode separation on the DC (□) and AC (○) components of the reflected infrared plethysmograms. Measurements were performed at a skin temperature of 43 °C (adapted from Mendelson and Ochs 1988).

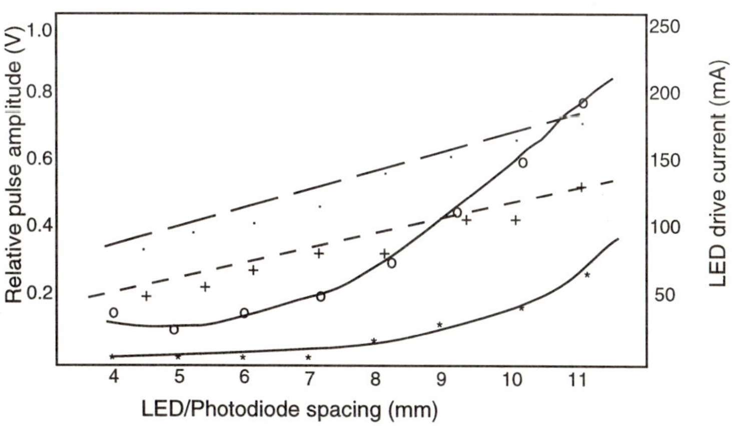

Figure 7.4 Effect of LED/photodiode separation on the relative pulse amplitude of the red (+) and infrared(·) plethysmograms. The driving currents of the red(o) and infrared(*) LEDs required to maintain a constant DC reflectance from the skin are shown for comparison (adapted from Mendelson and Ochs 1988).

Thus the selection of a particular separation distance involves a trade-off. We can achieve larger plethysmograms by placing the photodiode farther apart from the LEDs but we need higher LED driving currents to overcome absorption due to increased optical path length.

7.2.3 Effect of multiple photodiode arrangement

In a reflectance oximeter, the incident light emitted from the LEDs diffuses through the skin and the back scattered light forms a circular pattern around the LEDs. Thus if we use multiple photodiodes placed symmetrically with respect to the emitter instead of a single photodiode, a large fraction of back scattered light can be detected and therefore larger plethysmograms can be obtained.

To demonstrate this, Mendelson and Ochs (1988) used three photodiodes mounted symmetrically with respect to the red and infrared LEDs; this enabled them to triple the total active area of the photodiode and thus collect a greater fraction of the back scattered light from the skin. The same result can be obtained using a photodiode with three times the area.

7.2.4 Effect of skin temperature

Mendelson and Ochs (1988) studied the effect of skin temperature on the quality of signals detected by the photodiode. In their experiment, the LED/photodiode separation was kept constant and after attaching the reflectance sensor to the forearm, they increased the skin temperature to 45 °C in 1 °C step increments.

Figure 7.5 and figure 7.6 show that by increasing the skin temperature from 34 °C to 45 °C, they were able to obtain a five-fold increase in the pulse amplitude.

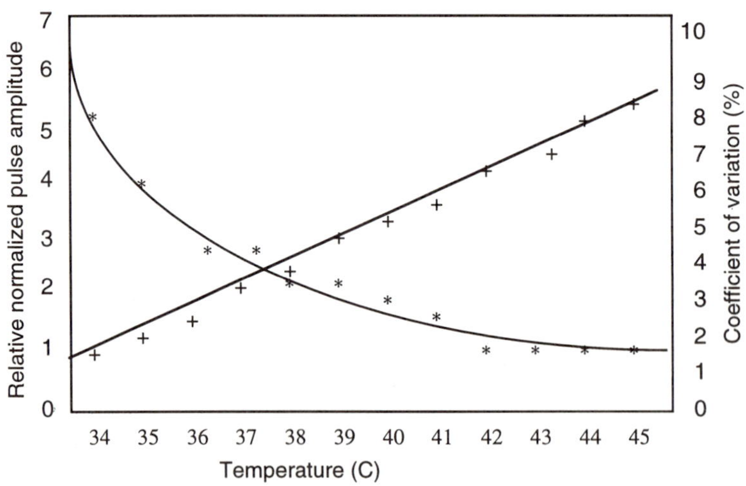

Figure 7.5 Effect of skin temperature on the mean pulse amplitude (+) and the corresponding decrease in the coefficient of variation (*) of the infrared plethysmograms. Each pulse amplitude was normalized to a constant separation of 4 mm (adapted from Mendelson and Ochs 1988).

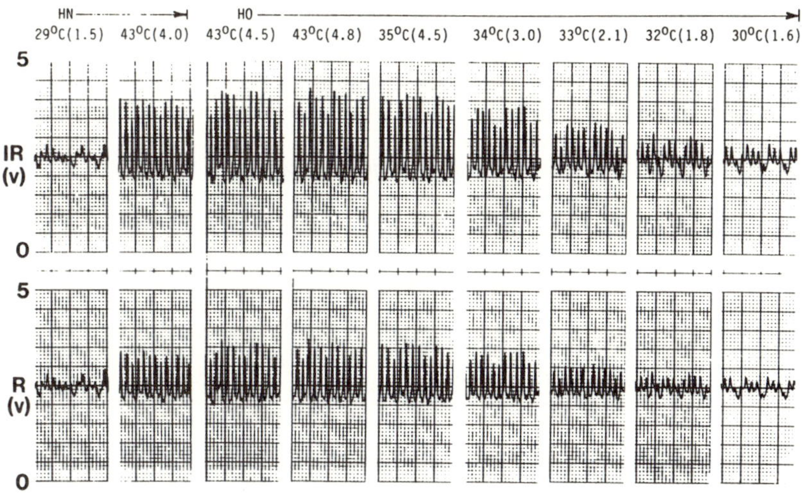

Figure 7.6 Simultaneous recording of the infrared and red plethysmograms from the forearm at different skin temperatures (from Mendelson and Ochs 1988).

7.2.5 Advantages and disadvantages of reflectance probes over transmittance probes

The basic advantage of transmittance probes over reflectance probes is the intensity of the light detected by the photodiode. As the amount of light passing through thin tissue is greater than the amount of light reflected and as the light passing through the tissue is concentrated in a particular area, the intensity of detected light is larger for transmittance probes. The major disadvantage of the transmittance probes is that the sensor application is limited to peripheral parts of the body such as the finger tips, toes, ear and nose in the adults or on the foot or palms in the infant. Reflectance probes can be placed on virtually any place on the body where we can expect light reflection due to tissue.

7.3 MRI PROBES

When a pulse oximeter with either transmission or reflection probes is used in the presence of magnetic resonance imaging (MRI), it may give erroneous readings. This is due to the very high magnetic field strengths involved in MRI which makes the use of conventional electronic monitoring equipment difficult. This is due to the radio frequency magnetic pulses generated in the magnetic field. Also if there is any metal connection to the skin of the patient, this could lead to burns.

In order to solve the problems involving MRI, the manufacturers have developed special pulse oximeters for use with MRI scanners. The MR-compatible sensor of Magnetic Resonance Equipment Corporation uses low attenuation optical filter bundles. The complete pulse oximeter unit is kept beyond the field of influence of the magnetic field from MRI and the light from the LEDs is transmitted through the optical fibers and the transmitted/reflected light is brought through the optical fibers to the photodiode. The LEDs, photodiode, and all the electronic equipment required are kept in a main unit which is kept far away (approximately 3 m) from the MRI equipment. The effect of the magnetic field is minimal on optical fibers when compared to the amplitude of plethysmograph and oximetry signals.

The probes are nearly the same, but instead of LEDs and a photodiode, the MRI probes use fiber optic cables. Figure 7.7 shows a typical clip type probe of Magnetic Resonance Equipment Corporation (MR Equipment 1995).

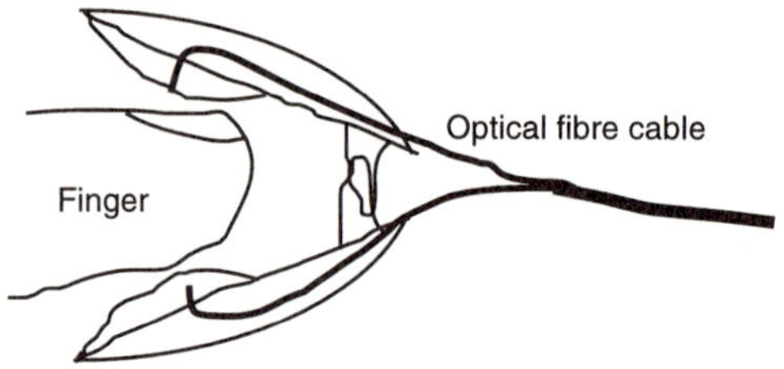

Figure 7.7 MRI compatible pulse oximeter probe using optical fibers.

7.4 PROBE CONNECTORS

The pulse oximeter may be connected to the subject through a disposable probe. The instrument is used on different subjects (adults, children and infants). The probe can be attached to the subjects by different means, for example the sensors can be attached to the subject's finger, foot, ear or forehead and these require a variety of probes. The hospital should stock a large number of different kinds of pulse oximeter disposable probes, which requires a large inventory.

In order to solve this problem, Goldberger *et al* (1995) used a probe connector, which connects the sensor elements and the cable section of the probe (which is connected to the pulse oximeter instrument). Here they used the fact that money is spent on the cable, so if we can save the cable for multiple use and still use disposable sensors, then we could save money.

The designing of the probe connector should be simple and reliable and should interconnect the sensor elements with the cable that connects the pulse oximeter instrument to the sensor element. The probe connector must be mechanically rugged and should prevent accidental disconnection between the sensor and the instrument. In order to minimize the cost, the electric contacts in the probe connector should be simple and should have low resistance. The conductors in both halves of the probe connector should be precisely aligned with each other in order to avoid susceptibility to electromagnetic interference. Figure 7.8 shows the probe connector used in Ohmeda pulse oximeter probes.

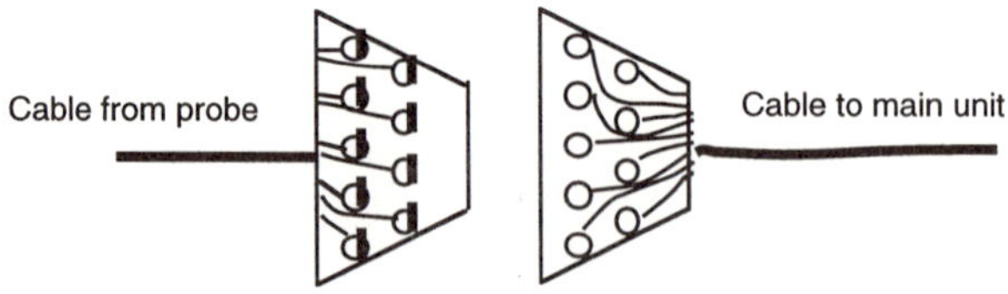

Figure 7.8 Probe connector used in Ohmeda pulse oximeter units. The cable from the probe has 9 male pins that mate with the 9 female sockets on the cable to the main unit.

7.5 REUSABLE PROBES

Probes which can be used more than once in monitoring S_aO_2 are called reusable probes. Generally all probes with nonadhesive or disposable adhesive sensors are reusable probes. Figure 7.9 shows the most common of them, which is is a clip (or clamp) type sensor used over the patient's finger. Figure 7.10 shows a reusable sensor with disposable adhesive wrap) and figure 7.11 shows a reusable reflectance sensor applied over the forehead with a disposable adhesive pad.

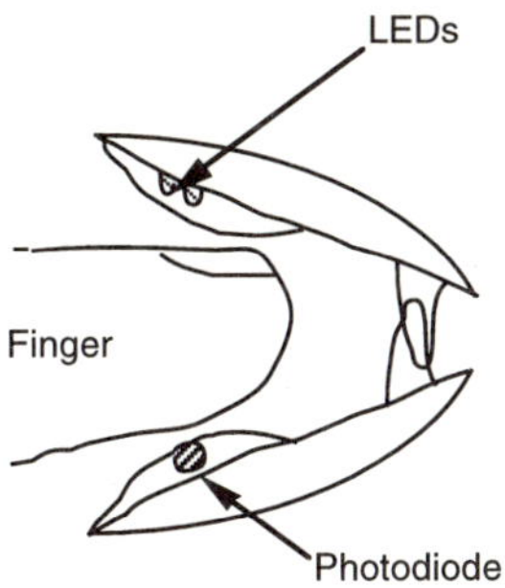

Figure 7.9 Clamp (or clip) type reusable probe.

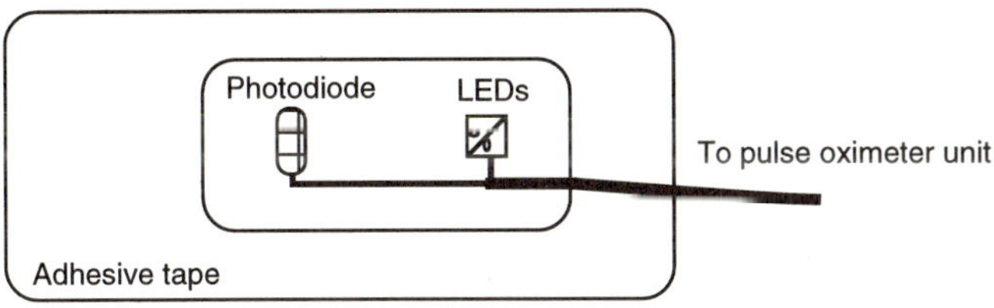

Figure 7.10 Reusable probe with disposable adhesive sensors.

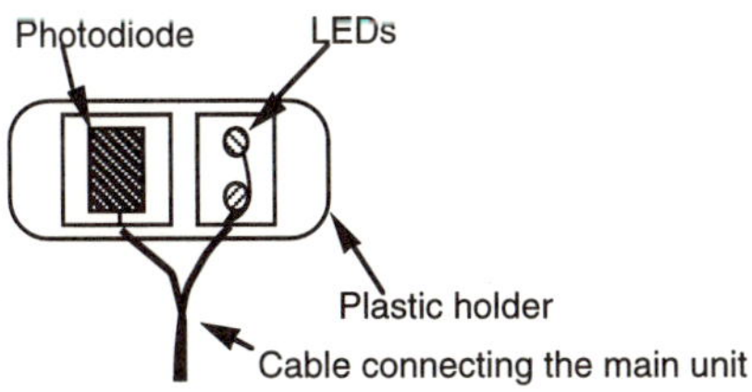

Figure 7.11 Reusable reflectance sensor.

The main advantage of the reusable probes is the low per use cost involved. By using the same probe over and over we reduce the total cost for the patient. However reusable sensors require cleaning between patients to minimize the risk of cross contamination. In the case of infected patients or patients with a high risk of infection (e.g., neonates and immunosuppressed patients) reusable probes are not recommended. Moreover, clip type sensors are more susceptible to signal-distorting motion artifacts.

Reusable sensors are commonly used for on the spot measurements or for short term monitoring (usually of less than four hours). Reusable sensors should be changed to another site at least every four hours.

Kästle *et al* (1997) describe design considerations for reusable probes that include functionality, performance and regulations. They used a thin, flexible probe cable to minimize movement artifacts. They designed watertight connectors to the heavier adapter cable to avoid leakage. Electrical shielding minimized electrical interference and a closed, opaque housing minimized optical interference.

7.6 DISPOSABLE PROBES

As the name indicates disposable probes are discarded after they have been used. Since disposable probes are used on a single patient, they eliminate the possibility of cross contamination. All adhesive sensors are disposable sensors. They decrease the effect of signal distortions as they secure the sensor in the proper position and the relative motion between the patient and the sensors is nearly zero. Adhesive sensors are most commonly used when there is a need for monitoring when the electromagnetic interference levels in the surroundings is high (or if the signal obtained is low), as the electromagnetic shielding around the sensor and cable protects the pulse oximetry signal.

Adhesive sensors are used for both short term and long term monitoring. Typically adhesive sensors are checked at least every eight hours. Figure 7.12 shows a typical disposable probe.

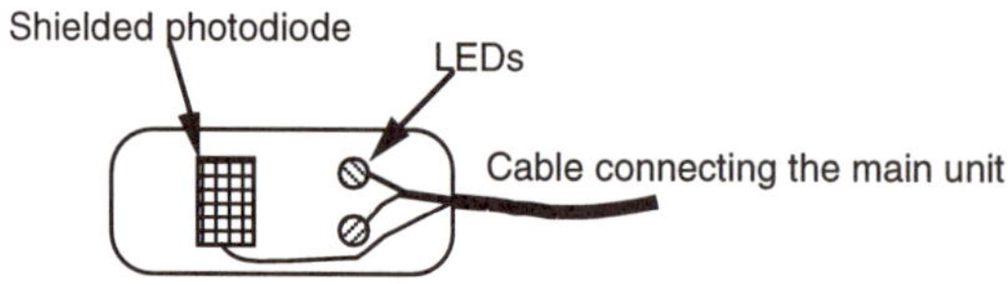

Figure 7.12 Disposable probe.

7.7 SOURCES OF ERRORS DUE TO PROBES AND PLACEMENT

7.7.1 Ambient light interference

Ambient light from sources such as sunlight, surgical lamps etc may cause errors in S_aO_2 readings. In order to prevent this, the simple solution is to cover the sensor site with opaque material which can prevent ambient light from reaching the photodiode.

7.7.2 Optical shunt

Optical shunting occurs when light from the sensor's LEDs reaches the photodiode without passing through the tissue. Optical shunting leads to erroneous readings in the value of S_aO_2 as the amount of light detected by the photodiode is greatly increased by optical shunting. This can be eliminated by choosing an appropriate sensor for the patient's size and by ensuring that the sensor remains securely in position.

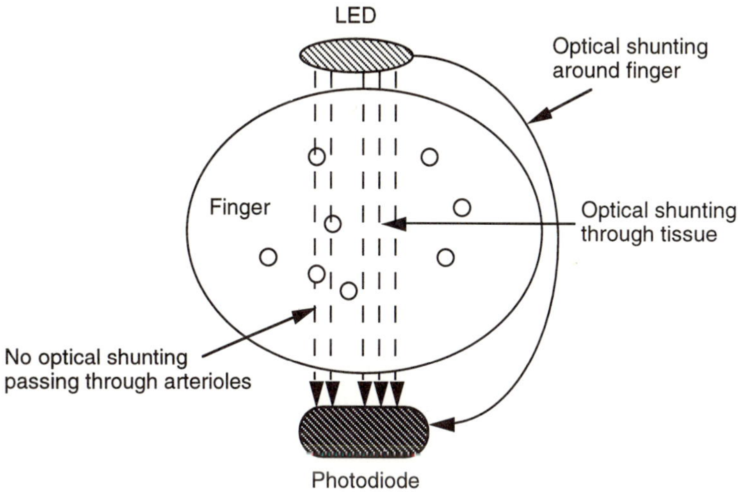

Figure 7.13 Light that does not pass through arterioles causes optical shunting.

7.7.3 Edema

Edema is defined as an abnormal accumulation of serous fluid in a connective tissue or in a serous cavity, in other words swelling in the body. When the probe is used over such a swelling, the resultant arterial oxygen saturation reading may not be accurate as the fluid in the swelling changes the absorbed and reflected light. This changes the intensity of light detected by the photodiode, resulting in an erroneous reading. By placing the probe on a nonedemous tissue, this error can be avoided.

7.7.4 Nail polish

Certain colors of nail polish, especially blues, greens, browns, and black, absorb so much light that the detected light is too small. Then the resultant S_aO_2 may be inaccurate. Thus nail polish of these colors should be removed.

REFERENCES

Brinkman R and Zijlstra 1949 Determination and continuous registration of the percentage of the percentage oxygen saturation in small amounts of blood *Arch. Chir. Neerl.* **1** 177–83

Cheung P W, Gauglitz K F, Hunsaker S W, Prosser S J, Wagner D O and Smith R E 1993 Apparatus for the automatic calibration of signals employed in oximetry *US patent 5,259,381*

Criticare 1995 *Product Catalog* (Waukesha, WI: Criticare Systems)

Delonzor R 1993 Disposable pulse oximeter sensor *US patent 5,246,003*

Goldberger D S, Turley T A and Weimer K L 1995 Pulse oximeter probe connector *US patent 5,387,122*

Kästle S, Noller F, Falk S, Bukta A, Mayer E and Miller D 1997 A new family of sensors for pulse oximetry *Hewlett-Packard J.* **48** (1) 39–53

Larsen V H, Hansen T and Nielsen S L 1993 Oxygen status determined by the photo-electric method – a circular finger probe constructed for detection of blood oxygen content, blood flow and vascular density *Lab Invest.* **53** Suppl. 214 75–81

Mannheimer P D, Chung C, Ritson C 1993 Multiple region pulse oximetry probe and oximeter *US patent 5,218,962*

Mendelson Y and Ochs B D 1988 Noninvasive pulse oximetry utilizing skin reflectance photoplethysmography *IEEE Trans. Biomed. Eng.* **35** 798–806

MR Equipment 1995 *Product Catalog* (Bay Shore, NY: Magnetic Resonance Equipment Corp)

Nellcor 1995 *Product Catalog* (Hayward, CA: Nellcor Incorporated)

Nelson D 1995 Molded pulse oximeter sensor *US patent 5,425,360*

O'Leary R J Jr, Landon M and Benumof J L 1992 Buccal pulse oximeter is more accurate than finger pulse oximeter in measuring oxygen saturation *Report* Department of Anesthesiology, University of California—San Diego

Ohmeda 1996 *Product Catalog* (Louisville, CO: Ohmeda)

Pedan C J, Daugherty M O and Zorab J S 1994 Fiberoptic pulse oximetry monitoring of anaesthetized patients during magnetic resonance imaging *Eur . J. Anaethesiol.* **11** 111–3

Pologe J A 1987 Pulse oximetry: technical aspects of machine design *Int. Anesthesiol. Clinics* **35** . 137–53

Primiano F P Jr 1998 Measurements of the respiratory system *Medical Instrumentation: Application and Design* 3rd edn J G Webster ed (New York: Wiley)

Santamaria T and Williams J S 1994 Pulse oximetry *Medical Device Research Report* **1** (2)

Sugiura K 1995 Pulse oximeter probe *US patent 5,413,101*

Young R L, Heinzelman B D and Lovejoy D A 1993 Noninvasive oximeter probe *US patent 5,217,012*

INSTRUCTIONAL OBJECTIVES

7.1 Explain how transmission probes work.
7.2 List the main constraints in the use of transmission probes.
7.3 Explain what we need to look at, when placing the emitter and detector of the pulse oximeter probe on the patient.
7.4 Explain how the reflection probes work.
7.5 Explain when we need to use reflectance probes.
7.6 Explain the effects of skin temperature over reflectance probes.
7.7 Explain the advantages of using multiple detectors in reflectance probes.
7.8 Explain why the need to use MRI probes arises.
7.9 List the precautions that should be taken in using MRI probes.
7.10 Compare reusable and disposable probes.
7.11 Explain the common sources of error in pulse oximeters due to probes and explain how we can prevent them.

CHAPTER 8

ELECTRONIC INSTRUMENT CONTROL

Ketan S Paranjape

The pulse oximeter consists of an optoelectronic sensor that is applied to the patient and a microprocessor-based system (MBS) that processes and displays the measurement. The optoelectronic sensor contains two low-voltage, high-intensity light-emitting diodes (LEDs) as light sources and one photodiode as a light receiver. One LED emits red light (approximately 660 nm) and the other emits infrared light (approximately 940 nm). The light from the LEDs is transmitted through the tissue at the sensor site. A portion of the light is absorbed by skin, tissue, bone, and blood. The photodiode in the sensor measures the transmitted light and this signal is used to determine how much light was absorbed. The amount of absorption remains essentially constant during the *diastolic* (nonpulsatile) phase and this measurement is analogous to the reference measurements of a spectrophotometer. The amount of light varies during the *systolic* (pulsatile) phase. This chapter describes the electronics that control the operation of the pulse oximeter. The heart of this unit is the MBS, which controls the operation of this device from the light input to the display output. The signal received by the photodiode is small and may contain noise, so the first step involves amplification and filtering. Then the signals are split into the infrared (IR) and the red (R) components. Synchronizing with the R wave of the ECG signal helps to minimize motion artifacts. This chapter describes the electronics for the optoelectronic sensors, MBS, analog signal processing, power requirements, display and finally the storage of data.

8.1 GENERAL THEORY OF OPERATION

Measurements of oxygen saturation require light of two different wavelengths, as explained in chapter 4 (Pologe 1987). Two LEDs (one IR and one R) emit light, which is passed through the tissue at the sensor site into a single photodiode. The LEDs are alternately illuminated using a four-state clock. The photodiode signal, representing light from both LEDs in sequence, is amplified and then separated by a two-channel *synchronous detector* (demodulator), one channel sensitive to the infrared light waveform and the other sensitive to the red light waveform. These signals are filtered to remove the LED switching frequency as well as electrical and ambient noise, and then digitized by an analog-to-digital converter (ADC). The digital signal is processed by the microprocessor to identify individual pulses and compute the oxygen saturation from the ratio of the signal at the red wavelength compared to the signal at the IR wavelength.

The pulse oximeter does not measure the functional oxygen saturation because along with oxygenated and deoxygenated hemoglobin other forms of hemoglobin also exist. It may produce measurements that differ from those instruments that measure fractional oxygen saturation. As the pulse oximeter uses two wavelengths it can estimate only the oxygenated and deoxygenated (i.e., functional) hemoglobin. It does not determine the significant amount of dysfunctional hemoglobin (MetHb or COHb). The oxygen saturation measured is not exactly the arterial oxygen saturation (S_aO_2), but is termed pulse oximeter measured oxygen saturation, S_pO_2.

8.1.1 Historic perspective

Various patents describe the electronics involved for saturation calculation. Nielsen (1983) describes a logarithmic amplifier to amplify the output current to produce a signal having AC and DC components and containing information about the intensity of light transmitted at both wavelengths. *Sample-and-hold* units demodulate the R and IR wavelengths signals. In the sample and hold circuits the common signal from the photodiode is split into the IR and R components by the control signals from the MBS. The mixed signal is fed into the sample-and-hold circuit, whose timings are controlled such that each circuit samples the signal input during the portion of the signal corresponding to the wavelength to which it responds. The DC components of the signals are then blocked by a series of bandpass filters and capacitors, eliminating the effect of fixed absorptive components from the signals. The resultant AC signal components are unaffected by fixed absorption components, such as hair, bone, tissue, and skin. An average of the peak-to-peak value of each AC signal is produced, and the ratio of the two averages is then used to determine the saturation from empirically determined values associated with the ratio. The AC components are also used to determine the pulse ratio. The pulse ratio is the ratio of the R ac signal and the IR ac signal.

Wilber (1985) describes a photodiode sensor used to produce a signal for each wavelength having a DC and AC component. A normalization circuit employs *feedback* to scale the two signals so that the nonpulsatile DC components of each are equal and the offset voltages are removed. *Decoders* separate the two signals into two channels, where the DC components are removed. The remaining AC components are amplified and multiplexed along with other analog signals prior to being sent into an ADC. The oxygen saturation is then determined using the MBS.

New (1987) describes each LED having a one-in-four *duty cycle*. A photodiode produces a signal in response that is then split into two channels. The one-in-four duty cycle allows negatively amplified noise signals to be integrated with positively amplified signals including the photodiode response and noise, thereby eliminating the effect of noise on the signal produced. The resultant signal has a large DC component along with the small AC component. To improve the accuracy of the ADC this DC component is first subtracted prior to conversion, and is subsequently added back by the MBS. A quotient of the AC to DC components is determined for each wavelength of transmitted light. The ratio of the two quotients is fitted to an empirical curve of independently derived oxygen saturation (CO-oximeter). See chapter 10 for further details. To compensate for the different transmission characteristics, an adjustable drive source for the LEDs is provided.

New (1987) uses a calibrated oximeter probe. This probe includes a coding resistor that is used to identify a particular combination of wavelengths of the two LEDs. The *coding resistor* value is sensed by the MBS, and in this manner the effect the different wavelengths have on the oxygen saturation is compensated for. The oxygen saturation is calculated using the empirical curve (New 1987).

8.2 MAIN BLOCK DIAGRAM

Figure 8.1 shows the block diagram of the pulse oximeter system. The probe houses the transmitting LEDs and the receiving photodiode. The patient module contains the ECG amplifier. The photodiode signal, the ECG signal and the coding resistance value are sent to the MBS unit via the patient cable.

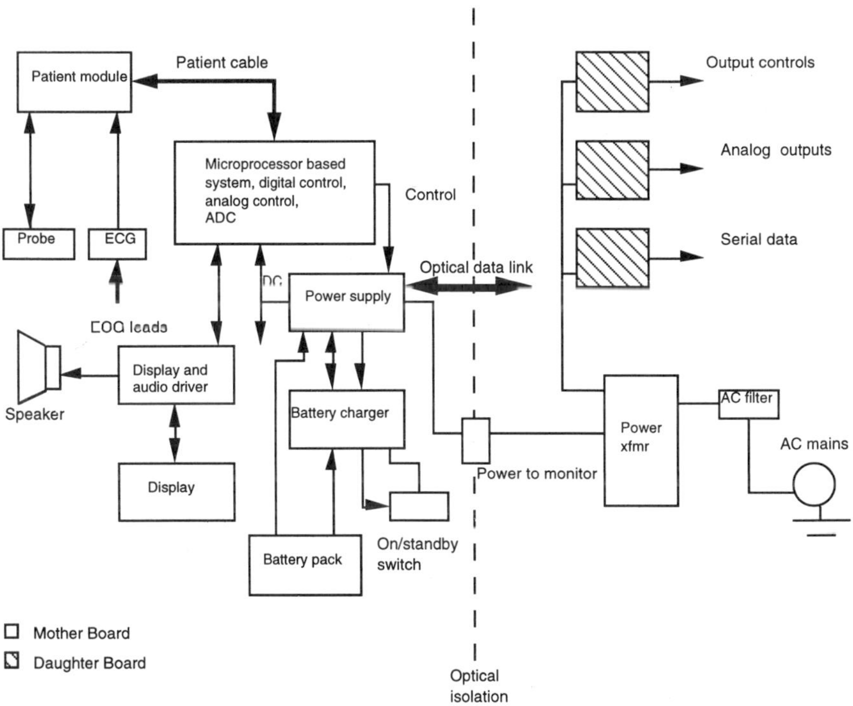

Figure 8.1 Main block diagram of a pulse oximeter system . Adapted from Nellcor N-200® (Nellcor 1989).

The MBS houses the digital and analog circuitry along with the ADC. The MBS is responsible for generating the various control signals of the system. The on board power supply is powered by a battery pack. A display driver drives the display section.

The section on the grounded side of the optical isolation consists of various cards such as the serial data communication card and certain analog and control outputs. The power transformer is located on this section. The main reason for the optical isolation is to prevent electric shock to the patient.

8.2.1 Input module

Figure 8.2 shows the input module or the patient module, which contains a preamplifier to generate the detector voltage (V_{det}) and electrocardiogram (ECG) signals used in ECG synchronization. Power for the circuitry is obtained from an on board power supply.

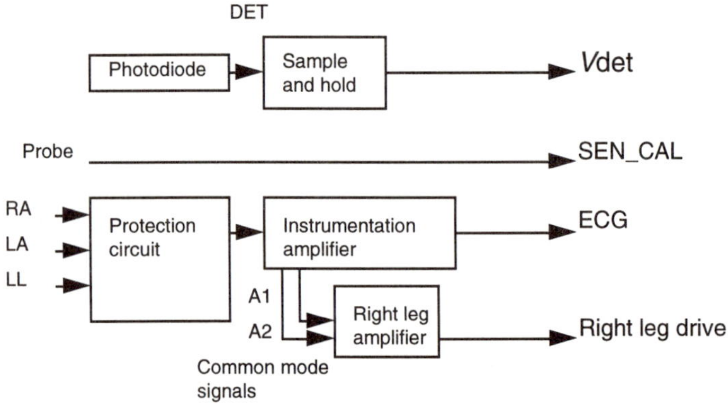

Figure 8.2 Input module or the analog front end module consisting of the detector, ECG unit, Protection unit and amplifiers. Adapted from Nellcor N-200® (Nellcor 1989).

The driver current for the pairs of probe LEDs is supplied from the LED driver circuit section. This waveform is a bipolar current drive which is passed through the input module to the back-to-back probe LEDs. A positive current pulse drives the IR LED and a negative current pulse drives the red LED. The drive current is controlled by a feedback loop in response to photodiode response. This feedback loop is controlled by the MBS.

The photodiode generates a current proportional to the amount of light received. The saturation preamplifier converts the photodiode current to a voltage. The conversion ratio is initially determined and fixed. Its units are mV/μA. A voltage regulator biases the preamplifier to a voltage output for zero current input. This bias helps to increase the swing of the current-to-voltage converter to its largest output. Some additional voltage margin is left in case of high ambient light conditions.

An *instrumentation amplifier* preamplifies the ECG signal used in ECG synchronization. The protection circuit consists of neon lamps to protect the instrumentation amplifier from potentially damaging high-voltage pulses which may result during defibrillation. Series resistors provide further isolation from high transient currents. Diodes shunt high-voltage transients to the low-impedance power supplies. Additional resistors pull the input signal lines to the power supply voltage levels when an ECG signal lead has become detached. By detecting a lead off, the pulse oximeter can indicate that the ECG synchronization is lost.

Common mode signals A1 and A2 from the instrumentation amplifier are summed, amplified, and inverted through the driven right leg amplifier. The output of this amplifier is fed back to the patient to drive the patient to a low common mode voltage by measuring the common mode voltage at the input sensing leads (driven right leg amplifier). In the *driven right leg* configuration, rather than the patient being grounded the right leg electrode is connected to the output of an auxiliary op amp. The body displacement current flows not to ground but rather to the op amp output circuit. This reduces the interference into the ECG amplifier and effectively grounds the patient. The ECG signal from the instrumentation amplifier goes directly to the ADC and finally to the MBS.

The probe connector contains a coding resistor that codes the wavelengths of the red and infrared LEDs mounted in the sensor (Sen_Cal). Because the wavelengths of the red and IR LEDs vary from one probe to another, an error would result in the computation for oxygen saturation if not corrected for by using the coding resistor. This coding resistor is measured and the value provided to the processing system. Since the probe is located near the patient, this coding resistor is sealed in epoxy to prevent damage from moisture and is nonrepairable. Therefore in case of any damage to the probe, or in the event of a failure the entire assembly has to be replaced.

8.3 DIGITAL PROCESSOR SYSTEM

8.3.1 Microprocessor subsection

The most important component of this system is the microprocessor. The microprocessor along with memory, input/output devices, communication circuits and additional peripheral devices constitutes the Microprocessor Based System (MBS). Depending on the application and the processing requirements, sometimes the microprocessor is replaced by a microcontroller. A *microcontroller* consists of a microprocessor, additional memory, ports and certain controls all built on the same chip. In portable pulse oximeters where power consumption and size are the main constraints microcontrollers may be used.

The processing power of a pulse oximeter lies in the microprocessor and how well it is configured along with memory to perform at a certain level. From the Intel line of microprocessors the 8085, the 8086, and the 8088 are the most commonly used devices, along with some other devices which may be designed for specified needs or dedicated for a certain kind of application. This chapter describes a standard Intel 8088, configured in the minimum mode, and used on Nellcor's N-200® series (Nellcor 1987).

8.3.1.1 Memory and memory mapping. The memory section usually consists of a mixture of *Random Access Memories* (RAMs) and *Read Only Memories* (ROMs). The memory has two purposes. The first is to store the binary codes for the sequences of instructions the subsystem is to carry out, such as determining the correct calibration curve depending on the probe used. The second is to store the binary-coded data with which the subsystem will work, such as the pulse rate or ECG data.

8.3.1.2 Input/output. This section allows the subsystem to take in data from the patient or send data out. Signals from the probe (photodiode output and ECG) are the input signals and the LED drive signals and display signals are the output signals. Ports are special devices used to interface the subsystem buses to the external system. The input port can receive signals from an ADC, and the output port sends signals to a printer or a digital-to-analog converter (DAC).

8.3.2 General block description

Figure 8.3 shows the most common configuration used for a microprocessor-based system. The main control signals may vary from make to make. If a microcontroller is used then there may be some reduction in the number of chips on board, thereby reducing the number of control lines on board.

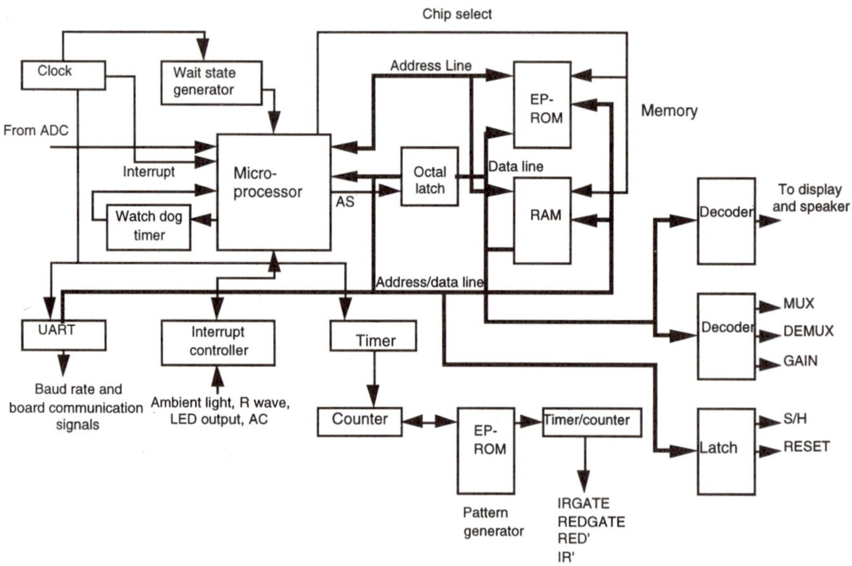

Figure 8.3 Generic microprocessor-based system. The ADC input to the microprocessor consists of the signal received from the photodiode, ECG signal for R-wave synchronization etc.

The microprocessor could be an Intel 8085 (Nellcor N-200)® or Zilog Z-80 (Ohmeda 3740(Ohmeda 1988))®. The clocking circuit consists of the clock generator and the *wait state generator*. For synchronized operation the clock rate must be constant and stable. For this reason a crystal oscillator is used. The wait state generator is used to slow down the microprocessor with respect to the input/output devices connected.

The communication section consists of the *Universal Asynchronous Receiver and Transmitter (UART)*, along with the *interrupt* control signals. This communication section makes use of the memory and control signals available on board. Timers and counters are employed to generate pulse trains required for the pattern generator section.

The pattern generator section is used to generate control sequences. This information is stored in the *EPROM* and is withdrawn using the address supplied by the counter that increments or decrements as desired. The IRGATE and the REDGATE are the two control signals used to *demodulate* the incoming photodiode output. The RED' and the IR' signals are used to synchronize the outputs of the two multiplexers so that they are combined before being fed into the voltage-to-current converter (figure 8.8), where a bipolar current output is generated to be fed into the two LEDs tied back-to-back that act as a source.

The memory on board consists of *latches*, buffers, decoders, RAMs, ROMs and EPROMs that are used to store the calibration curve data, digitized data from the photodiode that needs processing and storing data to generate the control signals. Oxygen saturation and pulse rate data can be stored in this memory. Octal transparent latches are needed to demultiplex the address and data bus information. EPROMs are erasable memory devices that store information such as the calibration curves, compensation requirements, etc., which may need occasional change. Therefore in such cases the technicians could reprogram or burn this new information into the chip. The set of instructions to be executed by the pulse oximeter is stored in the ROMs and RAMs. The code stored for example may be used for signal processing.

Finally decoders are used to decode the address and data information to generate the required control signals. The DEMUX/MUX signals are used to demultiplex the photodiode output into the individual IR and red signals, and the multiplexer is used to multiplex the IR and red signals, to drive the LEDs. Signals such as GAIN are used to adjust the gain requirements of offset amplifiers or *programmable gain amplifiers* used in the analog-to- digital conversion. RESET is generated in response to a *high* from the *watchdog timer*, which could mean temporarily shutting the system down.

8.3.3 Wait state generator

The pulse oximeter has a hardware–software interface that allows analog signals to be accepted, digitized, analyzed, processed, and finally converted back to analog to drive the LEDs. The rate at which the data enter the MBS or leave the MBS depends on the various components on the board and the communication/data transfer rate. These data may arrive at irregular intervals, and may need to be delayed before they are transferred to the output section. Therefore the MBS must generate some wait states to take care of these delays. The wait state generator is used to generate a wait state of one clock cycle. The microprocessor will insert the selected number of wait states in any machine cycle which accesses any device not addressed on the board, or any I/O device on the board. The purpose of inserting wait states is to give the addressed device more time to accept or output data. In this configuration, we use either a *shift register* or a *D flip flop*. A shift register or a D flip flop are digital devices which when controlled via clock signals can store and release data.

8.3.4 Clock generator, timer circuit and UART

The timing control on a microprocessor subsystem is of extreme importance. The rate at which the different components on the MBS receive data, analyze, and process it, is dependent on the clock rate. The clock controls the duty cycle. Duty cycle is defined as the fraction of time the output is high compared to the total time. When designing such subsystems we have to examine the duty cycle.

8.3.4.1 Clock generator and timer circuit. A 555 timer can be used to generate a *n*-minute timer, but isn't accurate enough for this kind of application. For more precise timing we usually use a signal derived from a crystal-controlled oscillator. This clock is stable but is too high in frequency to drive a processor interrupt input directly. Therefore, it is divided with an external counter device to an appropriate frequency for the interrupt input. Usually such a system contains counter devices such as the Intel 8253 or 8254, which can be programmed with instructions to divide an input frequency by any desired number.

The big advantage of using these devices is that you can load a count into them, and start them and stop them with instructions in a software program. Sometimes addition of a wait state may be needed along with this device to compensate for the delay due to the decoders and buffers on board.

We usually reset the circuit using simple resistors and capacitors, which are held low during power-on. This maintains the logic at a known state, while the crystal oscillator and the power supplies stabilize.

The timer circuit could control the following units on the subsystem:

1. Set the baud rate of the UART communication network.
2. Generate interrupts for controlling the display circuit, as these are usually multiplexed to avoid use of high current.
3. Audio frequency generator, for alarms.
4. Clock frequency for the notch filter, used to suppress the power line noise.
5. Synchronous circuit operation for pattern generator.

8.3.4.2 Watchdog timer circuit. This is a kind of fail-safe timer circuit, which turns the oximeter off if the microprocessor fails.

A counter controls the input to a D flip-flop, which is tied to a shutdown signal in the power supply. The counter is reset using a control signal from the microprocessor and a latch. Using some current-limiting protection, this signal is ac-coupled to the reset input of the counter. Therefore if the counter is not reset before the counter output goes high, the flip-flop gets set and the power supply is turned off.

8.3.4.3 UART. Within a MBS, data are transferred in parallel, because that is the fastest way to do it. Data are sent either *synchronously or asynchronously*. A UART (Universal Asynchronous Receiver Transmitter), is a device which can be programmed to do asynchronous communication.

8.3.5 Pattern generator

This is a multipurpose section which is primarily used to generate timing patterns used for synchronous detection gating, LED control, and for synchronizing the power supply. The heart of this system is the EPROM (Erasable Programmable ROM). Here preprogrammed bit patterns are stored and are cycled out through the counter, and are tapped off using certain address lines. Using the address the bit pattern is sent out and latched using an octal latch. There may be an additional latch used to deglitch the system, in which the last byte is held until the counter increments itself to the next address and the next pattern is obtained. Various patterns within the EPROM are used to select the sampling speeds of the LEDs or the synchronous detector pulse, the calibration patterns and diagnostic timing.

8.4 ANALOG PROCESSING SYSTEM (NELLCOR®)

8.4.1 Analog signal flow

Signals obtained are usually weak and may have electromagnetic interference. These signals must be filtered and then amplified. Usually a 50 or 60 Hz low-pass (for example may be a 2nd order Butterworth) filter is used. The signal is then ac coupled to stages of amplifiers and depending on the kind of response, variable gain circuits can be designed. The aim here is to maximize the signal before it enters the detector circuit, where the IR and red signal are separated, so the signal-to-noise ratio (SNR) is also kept as high as possible.

8.4.2 Coding resistor, temperature sensor, and prefiltering

Before examining the analog signal flow path, it is necessary to mention how the MBS decides what compensation to use for those LEDs which do not have their peak wavelength at the desired value. As LEDs are manufactured in bulk and tested in a random fashion, the probes may not always have the LEDs with the desired wavelength. Therefore the MBS generates some compensation, in order to solve this problem. Probes must be calibrated. Pulse oximeter systems have a coding resistor in every probe connector. A current is provided to the probe which allows the MBS to determine the resistance of the coding resistor by measuring the voltage drop across it. Thus the particular combination of LED wavelengths can be determined. Following this the MBS can then make necessary adjustments to determine the oxygen saturation.

As the wavelengths of the LEDs depend on the temperatures, for accurate measurements the effects of the temperatures must also be known, for adequate compensation (Cheung *et al* 1989). A temperature sensor may be employed, whose signal along with that of the coding resistor is used to select the calibration curves which are to be employed for compensation.

Despite efforts to minimize ambient light interference via covers over the probes and sometimes red optical filters, interfering light does reach the photodiode. Light from the sun and the incandescent lamp are continuous. The fluorescent light source emits ac light. This may overload the signal produced by the photodiode in response to the light received.

8.4.3 Preamplifier

The photodiode generates a current proportional to the light incident upon it. The signal from the photodiode is received by a preamplifier. Figure 8.4 shows that the preamplifier consists of a differential current-to-voltage amplifier and a single ended output amplifier. A gain determination resistor converts the current flowing through it into voltage. But along with the current-to-voltage conversion, external interference is also amplified, making the true signal difficult to extract from the resulting output. The differential amplifier produces positive and negative versions of the output. This dual signal is then passed via a single ended amplifier with unity gain, which results in a signal with twice the magnitude of that of the input. Due to the opposite signs of the outputs of the differential amplifiers, the external interference is canceled out. As the noise factor increases by a marginal factor the signal-to-noise ratio improves. The mixed signal is then fed into two sample-and-hold (S/H) circuits whose timings are controlled such that each circuit samples the signal input to the demodulator during the portion of the signal corresponding to the wavelength to which it responds.

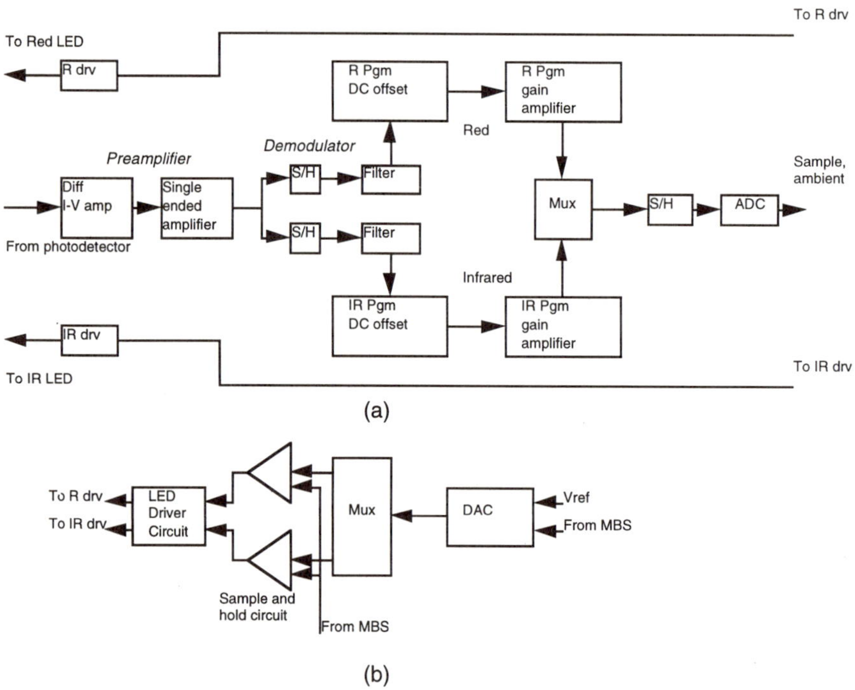

Figure 8.4 The analog signal flow path along with the signal demodulator and modulator circuit (from Cheung *et al* 1989).

8.4.4 Demodulator and filtering

This section splits the IR and the red signals from the mixed signal from the photodiode. The mixed signal is demultiplexed synchronously and steered depending on the type of signals present. The inputs to this circuit are the photodiode output and the timing or control signal from the MBS. The microprocessor along with the information stored in the EPROM calculates the time period each signal component is present in the photodiode output. Switching at the right time results in the two components getting separated. In order to eliminate the high-frequency switching noise, low-pass filters are provided. To optimize cost, size and accuracy, switched capacitor filters are used. These filters cause the two signals (red and infrared) to be identical in gain and phase frequency response. In order to filter out the noise generated by this switched capacitor a second filter follows in the cascade to filter out the switching frequency noise. This stage is a high roll-off stage, allowing the first stage to be the dominating one, resulting in higher accuracy. Then using programmable DC offset eliminators and programmable gain amplifiers, the two signals are multiplexed along with other analog signals prior to being fed into an ADC. Offset amplifiers offset the signals by a small positive level. This ensures that the offsets caused by the chain of amplifiers do not allow the signal to be negative as this is the input to the ADC, and the ADC only accepts inputs from 0 to 10 V. Also sometimes the gain of the red or the IR channel may be greater than the other, and therefore the offset must be compensated accordingly.

8.4.5 DC offset elimination

To exploit the entire dynamic range of the ADC the two signals (red and IR) have to be processed further. Before discussing how this processing is done let us examine why this is done.

We know that the mixed signal consists of a pulsatile and a nonpulsatile component. The nonpulsatile component approximates the intensity of the light received at the photodiode when only the absorptive nonpulsatile component is present at the site (finger, earlobe, etc). This component is relatively constant over short periods, but due to probe position variation and physiological changes this component may vary significantly over large intervals. But as we analyze these signals in small interval windows, this is not a major problem. Figure 8.5 shows that this nonpulsatile component may be S_LOW, with the difference between S_HIGH and S_LOW being the varying pulsatile component, due to the arterial pulsations at the site. This pulsatile component is very small compared to the nonpulsatile component. Therefore great care must be taken when determining and eventually analyzing these values, as we desire the pulsatile component.

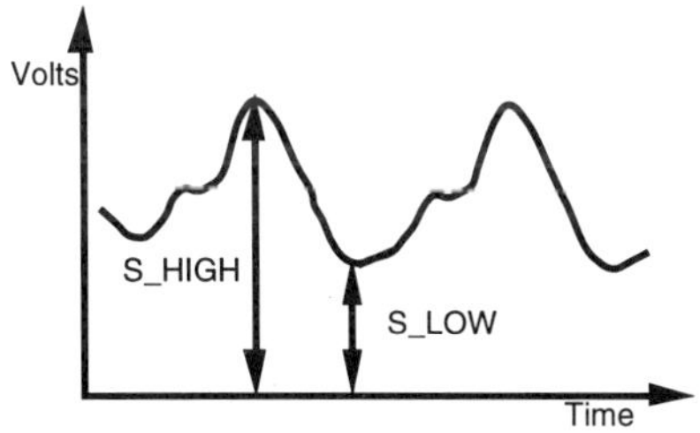

Figure 8.5 The nature of the signal transmission received by the photodiode circuit.

Amplifying and converting to digital form the substantial nonpulsatile component will use up most of the resolution of the ADC. Therefore in order to exploit the entire dynamic range we must eliminate this component, digitize it and later add it back to the pulsatile component.

For example, consider a ADC having an input range of 0 to 10 V. From figure 8.5 the AC may be 1% of the DC, let S_HIGH = 5.05 V and S_LOW = 5 V. For a 12-bit ADC, the resolution of this device is almost 2^{12}. This means that the total signal is discretized into 4096 levels. Therefore from the above value of the pulsatile component (S_HIGH – S_LOW), we see that only 20 levels are utilized. Therefore if the nonpulsatile component is removed we can use all the 4096 levels, improving the resolution of the ADC.

Cheung *et al* (1989) discuss this concept of nonpulsatile component elimination and addition. The photodiode output contains both the nonpulsatile and the pulsatile component. The programmable subtractors (offset amplifier) remove a substantial offset portion of the nonpulsatile component of each signal and the programmable gain amplifiers increase the gain of the remaining signal for conversion by the ADC. A digital reconstruction of the original signal is then produced by the MBS, which through the use of digital feedback information removes the gain and adds the offset voltages back to the signal.

Feedback from the MBS to the analog and the digital sections of the board is required for maintaining the values for the offset subtraction voltage, gain, and driver currents at levels appropriate for the ADC. Therefore for proper operation, the MBS must continuously analyze and respond to the offset subtraction voltage, gain, and driver currents.

Figure 8.6 shows that thresholds L1 and L2 are slightly below and above the maximum positive and negative excursions L3 and L4 allowable for the ADC input and are established and monitored by the MBS at the ADC. When the signal at the input of the ADC or at the output of the ADC exceeds either of the thresholds L1 or L2, the LED driver currents are readjusted to increase or decrease the intensity of light impinging upon the photodiode. In this manner the ADC is protected from overdrives and the margins between L3, L1, and L2, L4 helps ensure this even for rapidly varying signals. An operable voltage margin for the ADC exists outside the threshold, allowing the ADC to continue operating while the appropriate feedback does the required adjustments.

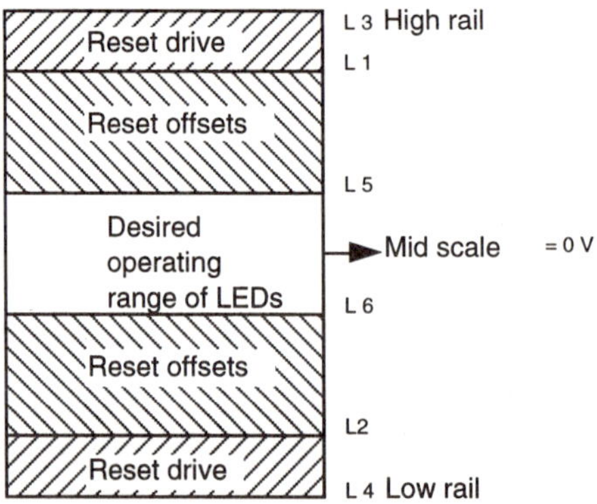

Figure 8.6 When the signal exceeds thresholds, the LED driver currents are readjusted to prevent overdriving the ADC (from Cheung *et al* 1989).

When the signal for the ADC exceeds the desired operating voltage threshold, L5 and L6, the MBS responds by signaling the programmable subtractor to increase or decrease the offset voltage being subtracted.

The instructions for the MBS program that controls this construction and reconstruction are stored in the erasable, programmable, read-only memory (EPROM).

8.4.6 Timing diagram (Nellcor®)

Figure 8.7 shows that the Nellcor pulse oximeter system uses a four state clock, or has a duty-cycle of 1/4, as compared to a Ohmeda system, where the duty-cycle is 1/3.

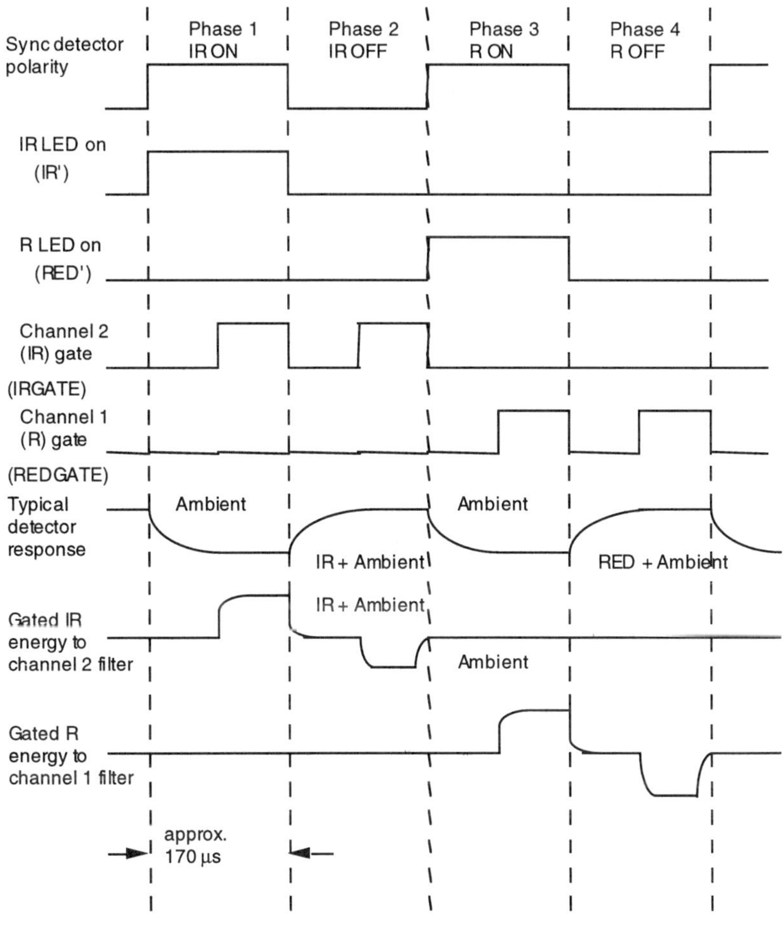

Figure 8.7 Timing diagram (reprinted with permission from Nellcor, Inc. ©Nellcor, Inc. 1989). Note that the typical detector response is inverted.

In the first quarter the IR LED is on and in the third quarter the R LED is on. In the second and the fourth quarters these LEDs are off. It is during this period that the ambient light measurements are done. The gate pulses are the sampling pulses applied to the input signal to separate out the R and the IR components from the input signal. The sample pulse during the OFF period of the respective LED is used to sample the ambient. The gradual rise or fall is due to the transients, which are smoothed out using low-pass filers. The ambient component is larger in the fourth quarter, compared to the value in the second.

Using suitable values for the gain in the programmable DC offset amplifiers we can eliminate this ambient component. The AC plus the DC components of the R and IR signals are digitized and sent to the MBS.

8.4.7 LED driver circuit

The need to drive both LEDs at different intensities requires analog switches that are used for gating the separate drive voltages. The factor that influences the amount of drive voltage necessary is the signal level from the photodiode and this value is set by the sample-and-hold section. The necessary control signals come from the pattern generator. The main purpose of this drive circuit is to convert this drive voltage to drive current.

Figure 8.8 shows the drive voltages VIR and VR and gating signals IR' and R'. These signals are used to multiplex the two signals back into one, and are fed into a voltage-to-current (*V–I*) converter such that the output of this *V–I* converter is a bipolar current signal, that is used to light up only one LED at a time. As the two LEDs are tied in a back to back configuration, this bipolar current drive ensures that only one LED is on at a time.

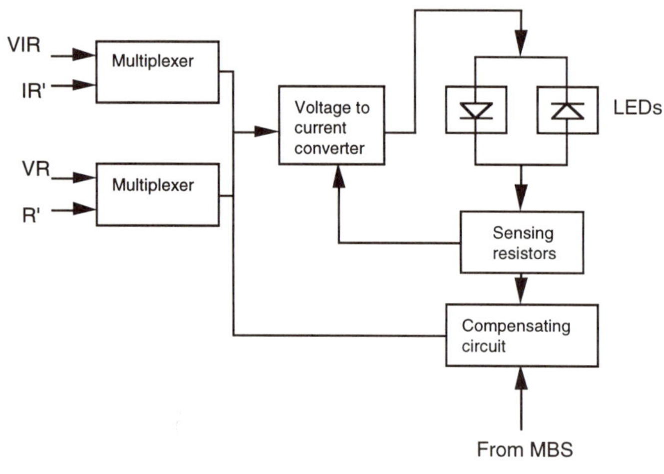

Figure 8.8 LED driver circuit with the sensor resistors to monitor and control the amount of current into the LED (adapted from Nellcor N-200® (Nellcor 1989)).

The drive current requires a control to convert the specified voltage to the proportional drive current. Within the voltage to current converter is an error amplifier that compares the voltage from the current sensing resistors with the specified voltage. There are two bridge networks with current boosters and drive and steering transistors which steer current around this conversion network. The drive output is connected to a pair of parallel back-to-back IR/R LEDs. The current through the LEDs is determined by a sensing resistor and fed back to the error amplifier to maintain a constant current proportional to the desired output voltage and to be independent of the other voltages present across the bridge circuit. Maximum LED current at 25% duty cycle is approximately 120 mA. The back-to-back configuration is such that when one LED is forward biased the other is not. Chapter 5 describes the LED driver circuit used in the Nellcor® system.

8.4.8 Analog processing system (Ohmeda®)

The main block diagram indicated how different signals on board a pulse oximeter system flow, and showed the signal transfer from one major block (for e.g. ECG, probe, MBS, power supply, etc) to the other. This section will elaborate on the analog signal flow from the photodiode output until the analog signal is ready to drive the LEDs to make another measurement. See figure 8.9.

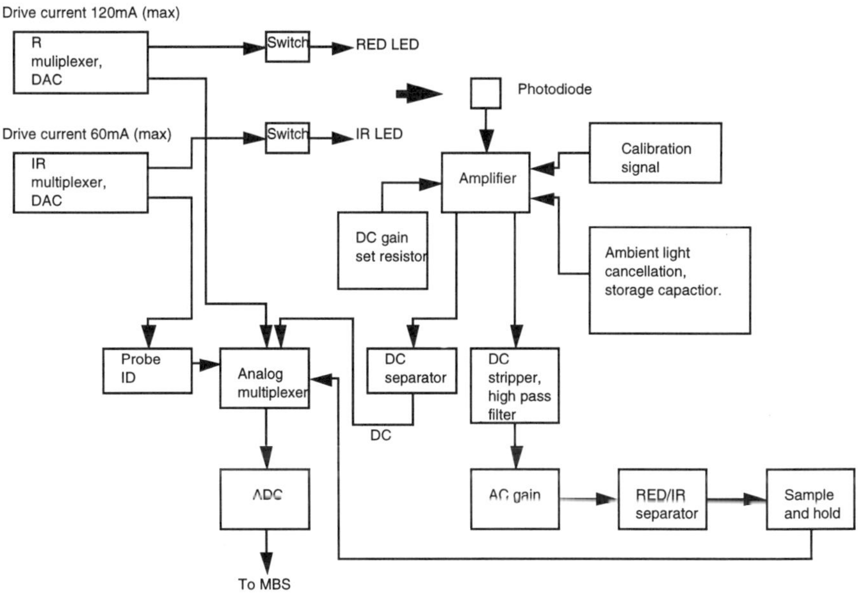

Figure 8.9 Functional block diagram of the pulse oximeter system showing all the main blocks involved in analog signal processing (adapted from Ohmeda 3740® (Ohmeda 1988)).

8.4.8.1 LED drive and monitor. The probe consists of the LEDs and the photodiode. The currents through the LEDs are controlled by a pair of multiplexers and switches and digital-to-analog converters (DACs). The maximum drive current is 120 mA through the R LED and 60 mA through the IR LED. The multiplexer and the switch turn the R and the IR LED drive on and off. The timing signal from the MBS controls the switches. The duty cycle of this timing signal is approximately 1/3 (Note that the duty cycle in devices from Nellcor® is 1/4). Therefore the subsequent hardware and analog and digital signal processing is different. The notable differences are in the multiplexers and sample and hold circuit. In the Ohmeda version, as the duty cycle is 1/3, first the R, then the IR, and finally the ambient component (measured when both the R and IR LEDs are off), are separated. In the Nellcor version as the duty cycle is 1/4 (see figure 8.7), the ambient component is measured twice.

The LED drive currents are monitored by switches and capacitors when both the R and IR LEDs are on individually and when both of them are off.

8.4.8.2 Calibration test signal. The signal received by the photodiode contains information on the AC and DC components of the pulsatile arterial blood flow measured by both the R and IR LEDs and also the ambient light component which is measured when both the R and IR LEDs are off. The calibration signal is a test signal injected into the signal path. The calibration signal is used to emulate the photodiode amplifier output which represents a known oxygen concentration and pulse rate of 150 to 210 beats per minute. The MBS checks the calibration of the oximeter by setting a test signal. This selects the calibration signal to be passed through the switch of the multiplexer in place of the photodiode amplifier output.

8.4.8.3 Ambient light cancellation. Ambient light cancellation is done to remove the effects of ambient light from the photodiode signal. A capacitor and a switch of a multiplexer are used to first charge up this capacitor to a voltage difference between the input signal and ground, when the input signal contains only the ambient component (R and IR LEDs are off). After this phase this voltage is subtracted from the input signal, now containing the R and IR components.

8.4.8.4 DC gain set resistor. The DC gain of the input signal is set under the control of the MBS. One resistor from a resistor bank is selected and along with another fixed resistor is used to set the gain of the amplifier.

8.4.8.5 DC separator. This block separates the DC components of the R and IR signals. This section consists of multiplexers and low-pass filters. The switches, controlled by the MBS, allow the red or the infrared component of the signal to pass through the low-pass filter. A set of amplifiers amplifies this DC before it is sent into the analog-to-digital (ADC) converter for conversion before being fed into the MBS. As a result of this stage we obtain the DC components of the R and the IR signals.

8.4.8.6 Low-pass filtering and DC stripping. A switched capacitor low-pass filter is used in this section. Since the DC components have been separated and measured previously, it is not necessary to filter during the ambient time. The R and IR components are low-pass filtered during the R and IR. time.

DC stripping is used to separate the pulsatile component from the signal. The low-passed signal is sent via a high-pass switching filter, and depending on the R and IR LED times, the pulsatile or the AC components of the R and IR signals are generated. This stage yields the AC components of the R and the IR signals.

8.4.8.7 Red/infrared separator. Multiplexers separate the red and infrared pulsatile signals into two independent channels. Low-pass filters are also used to smooth the separated signals. To compensate for the gain differences between the red and infrared signal paths, the gain of the infrared amplifier is adjustable by potentiometer.

8.4.8.8 Sample and hold circuits. Sample and hold circuits sample the red and infrared pulsatile signals simultaneously so that they can be measured by the ADC. An additional sampling signal controls the timing of the sampling of the pulsatile components at a rate synchronous to the power line frequency. This sampling frequency helps to suppress interference generated from sources connected to the line power.

8.4.8.9 Probe identification. This is the voltage generated by passing a known amount of current through the probe coding resistor to identify the wavelengths associated with the probe. This signal is digitized, compared to a lookup table in the MBS's memory, and the associated wavelength values are used for further processing.

An analog multiplexer is used to choose one of the many inputs and feed it to the ADC. The MBS for the Ohmeda system is similar to the one used in Nellcor, but uses Zilog's Z-8002. Motion artifact elimination using the R wave (ECG synchronization), as seen in Nellcor N-200 is not present in this system.

8.4.8.10 Timing diagram. In the Ohmeda Biox 3700® oximeter the LED on-off cycle is repeated at a rate of 480 Hz (figure 8.10). This cycling allows the oximeter to know which LED is on at any instant of time (Pologe 1987). The duty cycle in this system is 1/3. The red LED is on for the first 1/3 of the cycle, the infrared LED is on for the second 1/3 and both LEDs are off for the third 1/3, allowing for the ambient light measurements. This kind of measurement of ambient light is necessary so that it can be subtracted from the levels obtained when the LEDs are on.

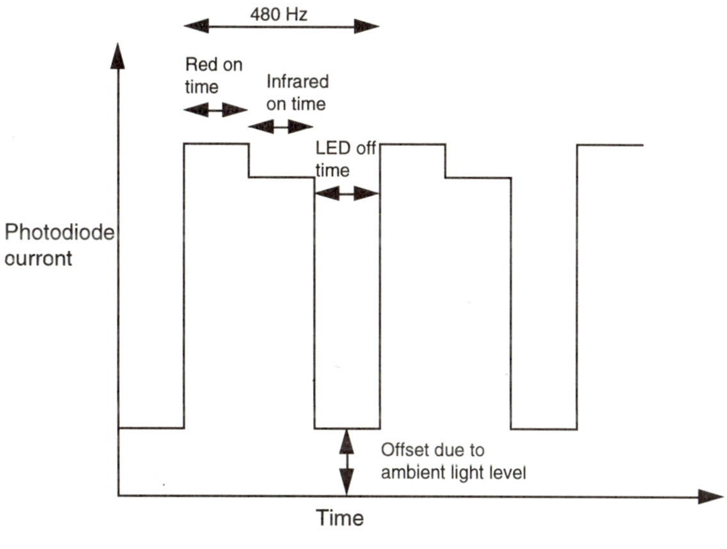

Figure 8.10 Output of the photodiode of the pulse oximeter system (adapted from Ohmeda 3700® (Ohmeda 1988)).

8.5 ECG SECTION

Pulse oximeters use the ECG to eliminate disturbances caused by motion artifacts and ambient light. There is a time delay between the electrical and the mechanical activity of the heart. When an ECG QRS electrical complex is detected, a mechanical pulse will be detected at the sensor after a transit delay of about 100 ms. This delay depends on factors such as the heart rate, the compliance of the arteries, and the distance of the probe from the heart. The pulse oximeter computes this delay and stores it, and an average delay is generated after a few pulses. This average delay is used to establish a time window, during which the pulse is expected at the probe site. So if a pulse is received within this time window, it is treated as real and is processed. Any pulse arriving outside this window is simply rejected. Note that the time averaging and the time window are constantly updated to account for the patient's physiological changes.

8.5.1 Active filters

Figure 8.11 shows that the ECG signal from the patient has to pass through a series of filters before it is used for processing. Usually these filter stages provide gain, as the signal level received is very small. The most commonly used filters are as follows.

1. A low-pass filter with a corner frequency of 40 Hz.
2. A switched capacitor notch filter at the power line frequency. The capacitor switching frequency is determined by the timer pulse, which is in turn set by the microprocessor. The microprocessor along with its associated circuitry determines the power line frequency, and accordingly sets the notch frequency.
3. A second 40 Hz low-pass filter may be used to filter out the transients generated by the capacitor switching.
4. A high-pass filter, with a corner frequency of 0.5 Hz, is used for the lower end of the range. This filter has a substantial gain and has a long time constant. The reset condition discharges this capacitor. The most common situations desiring a reset are the lead fall off condition, muscle contraction under the electrode, or a sudden shift in the baseline of the ECG, due to the already high combined gain due to the front end section and the filters preceding this stage.

When pulse oximeters are used in electromagnetic environments (MRI 1992), such as magnetic resonance imaging (MRI), special care has to be taken, as EMI interferences are quite disturbing for pulse oximeters. Probes and connectors are shielded, using faraday cages, and additional EMI elimination filters are incorporated in the pulse oximeter (see chapter 11).

8.5.2 Offset amplifiers

Analog-to-digital converters have a specified input dynamic range for obtaining the maximum digitized output. Usually these are in the positive range, from 0 to 5 or 0 to 10 V. Therefore an amplifier that can offset the analog signals to a value beyond 0 V and convert its peak value to 5 to 10 V is needed. For example if there is a signal from –0.7 V to +6.7 V and an ADC with dynamic range of 0 to 5 V, the offset amplifier will convert this range to 0 to 5 V. Then we can make use of the entire resolution of the ADC.

8.5.3 Detached lead indicator

ECG signals are sensed by the electrodes placed on the body and the signals are transferred from the site to the pulse oximeter via leads. If the electrode falls off from the surface of the body, the pulse oximeter's front end display must indicate this. The indicator system consists of a voltage comparator, absolute value circuit and a latching flip flop. This stage examines the ECG signal at the input to the switched capacitor notch filter (60 Hz), after it has passed through the low-pass stage preceding it. There is a biasing resistor network that drives the ECG signal to either ±15 V, if one or more ECG leads are detached from the patient's body. If the signal rails to –15 V, it is converted to positive voltage by a level shifter amplifier. Using this signal, the data are latched in a flip flop and the processor is notified that a lead has fallen off. After the processor recognizes this, it resets this latching flip flop so it is now ready to sense any other fall off.

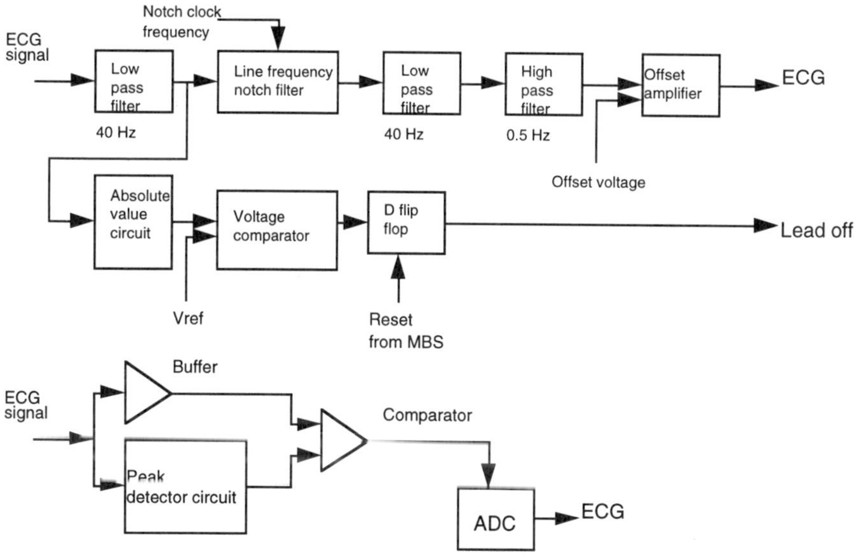

Figure 8.11 ECG signal processing section along with the lead fall off indicator and peak detector unit (adapted from Nellcor N-3000® (Nellcor 1991)).

8.5.4 Power line frequency sensing

Electric devices usually have the main power ac signal transformed to a root mean square value for power line analysis. A voltage comparator is used to generate a signal that interrupts the microprocessor at the frequency of the ac power line. This signal is then used to set the notch filter at the line frequency to eliminate the line frequencies. Most devices have provision for a 50 Hz or 60 Hz line.

8.5.5 ECG output

This section is used to generate pulses to synchronize the processor with the R-wave arrival. There is a peak follower circuit that stores the peak R-wave pulse in a slowly decaying fashion. This is employed to ensure that the capacitor doesn't discharge before the next R wave arrives. An adjustable threshold is provided for sensing each R-wave peak. This parameter is set by the processor, which in turn is influenced by many external parameters. A voltage comparator produces a high output whenever the ECG input exceeds the adjustable threshold determined by the previous R-wave peak.

8.6 SIGNAL CONVERSION

The signal conversion unit consists of an ADC or DAC. Signals have dc offsets subtracted and even amplified before processing. This enables us to extract signals having low modulation and riding on a high DC, and this helps improve the response time of the system.

8.6.1 Analog-to-digital conversion technique

Analog-to-digital conversion on the processor board is accomplished by using a sample-and-hold circuit, which holds a voltage until it is sampled by a routine written in the memory of the processor. Both software and hardware play a important role in the conversion.

Figure 8.12 shows that first the processor writes to an analog multiplexer to select one of its several analog inputs that desire digital conversion. These signals could be the demultiplexed and filtered IR or the red photodiode channel signal, the filtered ECG waveform, or filtered voltage from the coding resistor. The selected signal is first latched and the analog circuitry is notified of the amplitude level. This helps to set the gain of the programmable amplifiers, so that the voltages at the input of the ADC do not exceed the maximum range. This ensures that the entire range of the ADC is used. A sample is generated by the processor to trigger the sample-and-hold circuit. This causes the sample-and-hold chip to hold the current channel for conversion. The processor begins executing the appropriate software for conversion. Usually the conversion routine adopted is the successive approximation routine (SAR). In this system, the SAR performs a binary check, by setting up a voltage using the DAC, which is compared to the currently held voltage signal via a comparator. The result of this comparison is polled by the processor. This process continues till the least significant bit is converted. Usually a 12-bit conversion is done in approximately 100 μs.

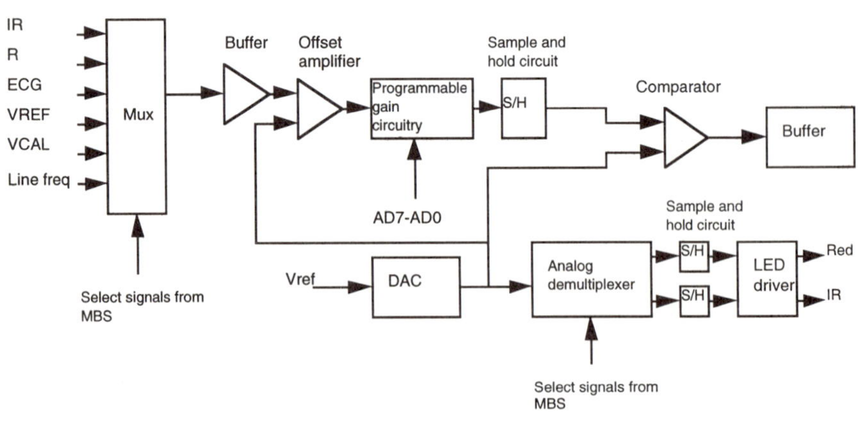

Figure 8.12 Generic analog-to-digital conversion circuit.

8.6.2 Digital-to-analog conversion

This is used to aid in digitizing the voltage at the ADC input, using the SAR. The DAC also has the following important application. The DACs have analog sample-and-hold circuits which are made using the analog demultiplexers and a series of variable gain amplifiers. Usually a DAC is used to update and store signals like the IR/red LED brightness control, or the speaker volume control. The analog signals are routed using the microprocessor. The processor puts out the analog voltage to the analog demultiplexer using the DAC. The processor selects which output will be written, using the address lines.

8.6.3 Sample-and-hold circuit

The analog conversion circuitry contains an *n*-bit DAC, a 1-to-8 bus-compatible analog multiplexer, switches for selecting the full scale analog output voltage range, and the analog sample-and-hold network. The DAC puts out an address of the task to be sampled and this information is decoded by the analog multiplexer and the desired sample-and-hold circuit is selected.

The sample-and-hold circuit is made up of storage capacitors and unity gain amplifiers. These amplifiers are usually the FET high-input-impedance devices. These circuits are protected via zener diodes that are needed to eliminate the short lived voltage transients. As we are driving high capacitive loads we need resistances to minimize these transients.

8.7 TIMING AND CONTROL

The time required to access a memory or an external device is as important as controling the various instruction executions within a microprocessor subsystem. The microprocessor adopts two techniques for the timing control. These are the polled processor I/O signal and the processor interrupts.

8.7.1 Polling and interrupt

In the polling technique the microprocessor has a signal that constantly polls or scans the various input waiting for a response. As soon as a valid signal is received at the polled input, the microprocessor starts the required task execution.

In the interrupt technique, the various chips and the inputs on the system are tied to the microprocessor via dedicated input lines. These lines are asserted high when a device requests help from the microprocessor. The microprocessor interrupts its current functionality and starts executing the interrupt routine. In the case of important activities these interrupt lines could be masked. Inputs may be provided with priority interrupt levels. When two or more interrupts are initiated at the same time, the higher priority interrupt performs first. Nested processing is done, in which within one interrupt execution another interrupt request can be answered.

For example, while the oxygen saturation is being measured along with the ECG signal, if there is a lead fall off situation, in which the device loses the ECG signal, a number of processes have to be initiated. First of all, the program in progress, calculating the oxygen saturation using the calibration tables has to be interrupted, as the R wave detected is no longer valid, and therefore the software used to eliminate motion artifacts will have to be terminated. An interrupt to the display/audio driver will start a routine to display the lead fall off information and generate some alarms. After this problem has been fixed, another interrupt would trigger the operation to resume. During interrupt routine execution the MBS stalls for a while until the process generating the interrupt has been serviced.

Also, if the physician wants to refer to the pulse rate of the patient recorded a few minutes back, the interrupt raised will cause the current routine to branch, retrieve the data from the memory and continue with the recording. Usually while user-triggered interrupts are generated, the main routine continues with the measurements and has this raised interrupt serviced in parallel.

8.8 POWER SUPPLY

The power supplies present on most boards are switched mode power supplies (SMPS). A SMPS-based power supply is either in the flyback or the flyforward converter mode. Figure 8.13 shows the power supply present on the Nellcor N-200®, which contains switching power supplies in flyback converter configuration. These power supplies are capable of generating low voltages at high currents. The supply is capable of providing 2 A at +5 V and 100 mA at ±18 V.

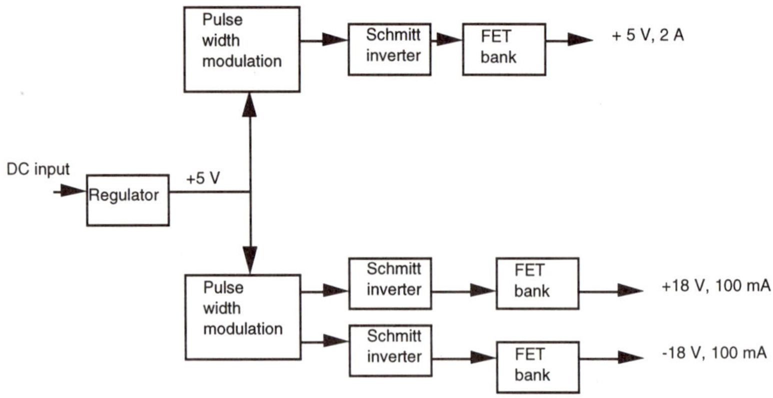

Figure 8.13 Basic power supply block diagram (adapted from Nellcor N-200® (Nellcor 1989)).

The essence of a SMPS supply is the pulse width modulator (PWM). In figure 8.13 the two PWMs control the 5 V and the ±18 V supply. The PWM senses the dc voltages at their inputs and controls the pulse width at the gates of switching FETs. A voltage regulator provides reference voltages for the two PWMs.

Field effect transistors are characterized by the rise and fall times of their drain currents. As the gates of the FETs are slightly capacitive, there is a need to minimize the rise and fall times of the drain current. Schmitt inverters are present to provide low impedance active current drive to these capacitive gates.

8.8.1 Recharging

Battery charger circuits are necessary to charge up a battery in case of a power line failure. In such an application when the main system is on line a battery charging circuit charges up a battery making use of the ac line voltage. In case of emergencies, because of a line failure, this system is set into the battery operated mode. However there is only a limited usage time available. Moreover the system becomes more bulky.

Figure 8.14 shows that ac power is taken from one of the secondaries of the transformers. It is rectified using a diode bridge arrangement (full wave rectifier) and filtered using a capacitor, to provide a positive voltage to the voltage regulator. Current limiting action is present via the use of a current-sensing resistor and a set of current-limiting transistors. Potentiometers are provided to trim the battery charging voltage. In order to avoid back discharge from the battery when the ac power is removed, a diode is present. Keeping in mind the efficiency of a power system, the expected voltage is 85% of the voltage provided by the battery charging circuit.

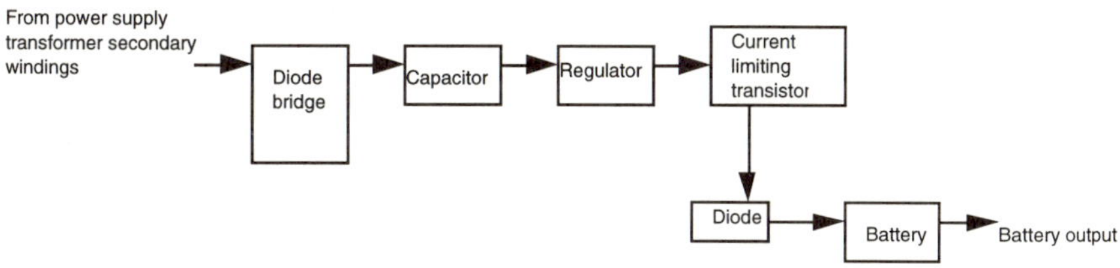

Figure 8.14 Battery charger block diagram (adapted from Nellcor N-200® (Nellcor 1989)).

8.9 ALARMS

When using pulse oximeters in critical applications, alarms are essential to give an indication to the physician that something is wrong. These alarms have to be in both audio and visual form. Comparators, power amplifiers, drivers and speakers constitute the audio alarm section. LCD bar graphs and blinkers are used for the visual section. Certain guidelines have been formulated by standardizing agencies such as the American Society for Testing and Materials (ASTM) regarding the color of the indicator, frequency of the indicator and the tone, audio level etc. Also the signals that need to be treated as emergency signal are classified (pulse rate, detached lead, etc).

8.10 STORAGE

Data concerning oxygen saturation and pulse rate can be collected and stored in memory. This may be used in the future to train the pulse oximeter system, using neural networks and artificial intelligence to generate control signals.

Memories are selected using address lines and data lines are used to load and unload data from them. In order to make the most efficient use of this storage mechanism, some care has to be taken while designing the memory system. When no power is applied to the memory system there is danger of losing data.

8.11 FRONT END DISPLAY

This section includes the display terminal on the front end of the pulse oximeter. Liquid crystal displays (LCD) or LED displays are used depending on the clarity, resolution, power consumption, and even aesthetics. Push buttons in the form of feather touch buttons or conventional switches are provided. Interface points, alarm indicators, and other important features are also displayed.

8.11.1 Front end driver circuit

Figure 8.15 shows that the driver circuit consists of two major driving techniques, one for the digit and bargraph and the other for lightbar and other front panel indicators.

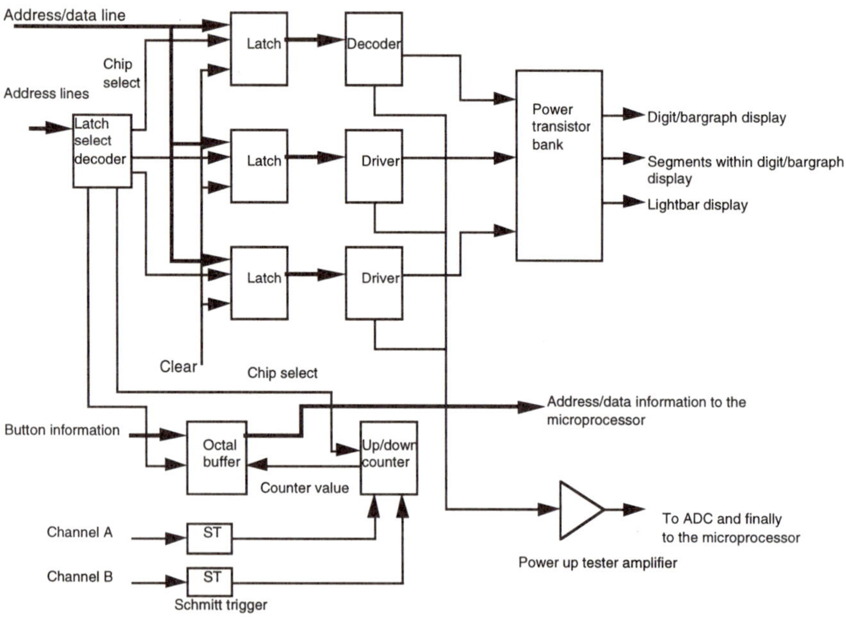

Figure 8.15 Generic display driver circuit.

In order to clear the front panel display, a reset circuit consisting of capacitors and resistors is used. During reset, these components generate a small duration reset pulse that is sent to all the latches on the driver board. This reset pulse clears all the front panel display elements.

Latches and decoders are used to generate the signals required to display information on the display elements. A set of power transistors are used to generate the drive current required to turn on the display elements. A chip select decoder is used to select the latch–decoder combination depending on the type of display needed. In the circuit layout, signals are required for the digit/bargraph display, to select particular segments within the digit/bargraph display and signals for light bar display. The latches generate the information to be displayed via address information that comes from the microprocessor. After the microprocessor-based system has decided what is to be displayed, address information is sent to Character Generator ROMs (CG-ROMs) or Dynamic Display RAMs (DD-RAMs) which generate the digit/display information. In these devices, bit information pertaining to a particular character is stored at a specific address location. Depending on the address at the input, the required character is generated.

8.11.2 Front panel control

The chip select decoder is used to select the octal buffer, which reads in inputs from the buttons on the front panel and the up/down counter which reads in the control knob rotation information which is relayed through it.

The control knob consists of a two-channel optical chopper, with the two channels mechanically 90 degrees out of phase with each other, and a dual channel optical slot detector. There are two Schmitt triggers, one per channel, to eliminate any transients present and to clean up the signal. The two channels are used to send control signals to the up/down counter. Depending on the direction in which the knob is turned, either the up or the down mode is selected. Channel A provides the clocking pulses for the counter and channel B provides the direction control, whether it is up or down. The processor reads the counter output to determine a change in the up/down mode of the counter. It then adds the count to the accumulated count. The processor then resets the counter.

8.11.3 Power up display tests

When we power up the system for the first time the system runs a few initialization tests. The software tests run are discussed in detail in chapter 9. The primary concern is to ensure that all the display elements are operational. We therefore have a power up tester amplifier which senses the power return line from the driver ICs by monitoring a voltage developed across a resistor. The driver ICs are used to boost the drive current into the segments of the digital displays. This is amplified and given to the ADC. The processor uses this signal during start up to check whether the display is faulty.

8.12 SPEAKERS

The speaker is an inductive load needing a positive and a negative signal. Figure 8.16 shows that currents to these two inputs are controlled by two different paths. Depending on the address/data information the demultiplexer generates many signals like the VRED, VIR and the volume control signal. A sample-and-hold circuit is used to hold this signal. This signal is then passed via a series of power transistors to boost the current flowing into the speaker.

A timer and counter chip generates a count using certain address/data information and temporarily saves it into a buffer. This tone signal is used to control a FET switch which alternately connects or disconnects the speakers negative input to ground. The frequency of the tone signal (determined by the timer/counter chip) determines the pitch of the sound produced. A capacitor is present to smoothen the sound. A diode is also present to suppress any transients from the inductive load.

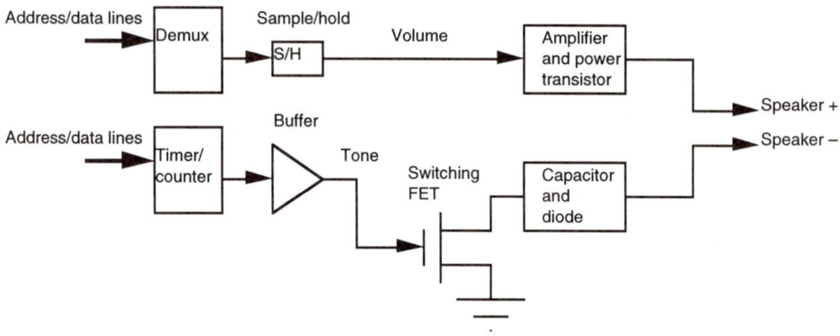

Figure 8.16 Speaker driver block diagram (adapted from Nellcor N-200® (Nellcor 1989)).

REFERENCES

Cheung P W, Gauglitz K F, Hunsaker S W, Prosser S J, Wagner D O and Smith R E 1989 Apparatus for the automatic calibration of signals employed in oximetry *US patent 5,259,381*

Corenman J E, Stone R T, Boross A, Briggs D A and Goodmann D E 1990 Method and apparatus for detecting optical pulses *US patent 4,934,372*

MRI 1992 *Service Manual, model 3500 MR-compatible oximeter* (Bay Shore, NY: MRI)

Nellcor 1989 *Service Manual, N-200 Pulse Oximeter* (Pleasanton, CA: Nellcor)

Nellcor 1991 *Service Manual, N-3000 Pulse Oximeter* (Pleasanton, CA: Nellcor)

New W Jr 1987 Pulse oximeter monitor *US patent 4,653,498*

Nielsen L L 1983 Multi-wavelength incremental absorbance oximeter *US patent 4,167,331*

Ohmeda 1988 *Service Manual, Model 3740 Pulse Oximeter* (Louisville, CO: Ohmeda)

Pologe J A 1987 Pulse oximetry: Technical aspects of machine design *Int. Anesth. Clinics* **25** (3) 137–53

Protocol 1991 *Service Manual* (Beaverton, OR: Protocol)

Sobusiak A C and Wiczynski G 1995 Specificity of SIF co-operating with optoelectronic sensor in pulse oximeter system *Proc. SPIE* **2634**

Wilber S A 1985 Blood constituent measuring device *US patent 4,407,290*

Yoshiya I, Shimada Y and Tanaka K 1980 Spectrophotometric monitoring of arterial oxygen saturation in the fingertip *Med. Biol. Eng. Comput.* **18** 27

INSTRUCTIONAL OBJECTIVES

8.1 Sketch the block diagram of the microprocessor subsystem, and highlight at least three main features that you think are vital for optimum operation of this system.
8.2 Explain the signal flow in the pulse oximeter from the photodiode to the front-end display.
8.3 Explain the kind of circuit protection associated with a patient module.
8.4 Explain how communication is established between the various chips on the microprocessor based system.
8.5 Describe the timing control involved in the microprocessor-based system.

8.6 Explain the operation of the synchronous detector and the demultiplexer in the pulse oximeter system.
8.7 Mention the need for active amplifiers and low-pass filters.
8.8 Explain the analog-to-digital conversion action involved in the pulse oximeter.
8.9 Explain the function of the pattern generator.
8.10 It is decided to improve the resolution of the ADC. List the steps you will take to improve the existing system. Explain how this will affect the system operation.
8.11 Explain the motivation for subtraction of the DC-level in the photodiode signal before the ADC.
8.12 Describe the components of an input module of a pulse oximeter.

CHAPTER 9

SIGNAL PROCESSING ALGORITHMS

Surekha Palreddy

Pulse oximeters measure and display the oxygen saturation of hemoglobin in arterial blood, volume of individual blood pulsations supplying the tissue, and the heart rate. These devices shine light through the tissue that is perfused with blood such as a finger, an ear, the nose or the scalp, and photoelectrically sense the transmittance of the light in the tissue. The amount of light that is transmitted is recorded as an electric signal. The signal is then processed using several signal processing algorithms to estimate the arterial oxygen saturation reliably in the presence of motion and other artifacts. Signal-processing algorithms implemented both in hardware and software play a major role in transforming the signals that are collected by the sensors and extracting useful information. In this chapter, the signal-processing to calculate S_aO_2 is discussed and ECG synchronization algorithms that enhance the reliability of S_aO_2 estimation and improve the signal-to-noise ratio are discussed. Commercial pulse oximeters use various algorithms for ECG synchronization. Some of these algorithms are discussed with reference to commercially available pulse oximeters such as from Nellcor® and Criticare®.

9.1 SOURCES OF ERRORS

The three general sources of errors dealt with by signal-processing algorithms are the *motion artifact*, *reduced saturation levels* (<80%) and *low perfusion levels* (Goodman and Corenman 1990). The motion artifact is a major problem that is usually due to the patient's muscle movement proximate to the oximeter probe inducing spurious pulses that are similar to arterial pulses. The spurious pulses when processed can produce erroneous results. This problem is particularly significant in active infants, and patients that do not remain still during monitoring. The quantity of motion required to disturb the signal is very small. Shivering and slight flexing of the fingers can make the signal erroneous.

Another significant problem occurs in circumstances where the patient's blood circulation is poor and the pulse strength is very weak. For example, poor circulation occurs in cases of insufficient blood pressure or reduced body temperature. In such conditions, it is difficult to separate the true pulsatile component from artifact pulses because of the low signal-to-noise ratio. Several time-domain and frequency-domain signal-processing algorithms are proposed to enhance the performance of pulse oximeters with improved rejection of noise, spurious pulses, motion artifact, and other undesirable aperiodic waveforms.

This chapter describes the algorithms required to estimate the arterial oxygen saturation based on the Beer–Lambert law.

9.2 BEER–LAMBERT LAW

Pulse oximetry measures the effect of arterial blood in tissue on the intensity of the transmitted light (Cheung *et al* 1989). The volume of blood in the tissue is a function of the arterial pulse, with a greater volume present at systole and a smaller volume present at diastole. Because blood absorbs most of the light passing through the tissue, the intensity of the light emerging from the tissue is inversely proportional to the volume of the blood present in the tissue. The emergent light intensity varies with the arterial pulse and can be used to indicate a patient's pulse rate. In addition, the absorbance coefficient of oxyhemoglobin is different from that of deoxygenated hemoglobin for most wavelengths of light. Differences in the amount of light absorbed by the blood at two different wavelengths can be used to indicate the hemoglobin oxygen saturation, which equals

$$\%S_aO_2 = [HbO_2]/([Hb] + [HbO_2]) \times 100\%. \tag{9.1}$$

The Beer–Lambert law governs the absorbance of light by homogeneous absorbing media. The incident light with an intensity I_0 impinges upon the absorptive medium of characteristic absorbance factor A that indicates the attenuating effect and a transmittance factor T that is the reciprocal of the absorbance factor ($1/A$). The intensity of the emerging light I_1 is less than the incident light I_0 and is expressed as the product TI_0. The emergent light intensity I_n transmitted through a medium divided into n identical components, each of unit thickness and the same transmittance factor T is equal to T^nI_0. I_n can be written in a more convenient base by equating T^n to $e^{-\alpha n}$, where α is the absorbance of medium per unit length and is frequently referred to as the relative extinction coefficient. The relative extinction coefficient α is related to the extinction coefficient ε (discussed in chapter 4) as $\alpha = \varepsilon C$, where C is the concentration of the absorptive material. The expression for the intensity of the light I_n emerging from a medium can be given by the following general equation called the Beer–Lambert law.

$$I_n = I_0 \, e^{-\alpha d} \tag{9.2}$$

where I_n is the emergent light intensity, I_0 is the incident light intensity, α is the absorbance coefficient of the medium per unit length, d is the thickness of the medium in unit lengths, and the exponential nature of the relationship has arbitrarily been expressed in terms of base e. Equation (9.2) is commonly referred to as the Beer–Lambert law of exponential light decay through a homogeneous absorbing medium (figure 9.1).

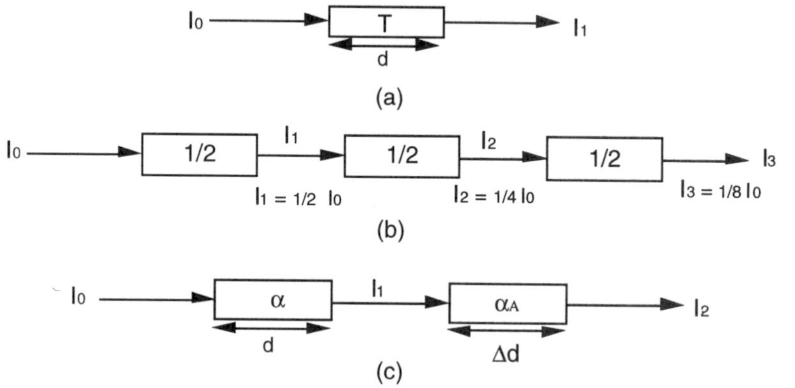

Figure 9.1. A block diagram illustrating the transmittance of light through a block model of the components of a finger. (a) Incident light having an intensity of I_0 impinges upon an absorptive medium with a characteristic transmittance factor T. (b) The effect of a medium divided into n identical components of unit thickness and same transmittance factor T on incident light intensity I_0. (c) For a finger model, the baseline component of the unchanging absorptive elements and the pulsating component of the changing absorptive portion are represented (Cheung *et al* 1989).

9.2.1 Estimation of oxygen saturation using the Beer–Lambert law

The absorbance coefficients of oxygenated and deoxygenated hemoglobin are different at most wavelengths, except at the *isosbestic* wavelength. If a finger is exposed to incident light and the emergent light intensity is measured, the difference between the two is the amount of light absorbed, which contains information relating to the oxygenated hemoglobin content of the blood in the finger. The volume of blood contained in the finger varies with the arterial pulse. The thickness of the finger also varies slightly with each pulse, changing the path length for the light that is transmitted through the finger. Also, the precise intensity of the incident light applied to the finger is not easily determined. Hence, it is desirable to eliminate the effects of intensity of the incident light and the thickness of the path length in estimating oxygen saturation. The Beer–Lambert law needs to be modified to eliminate the input light intensity and length of the path as variables.

9.2.1.1 Eliminating the input light intensity as a variable. The intensity of light transmitted through a finger is a function of the absorbance coefficient of both fixed components, such as bone, tissue, skin, and hair, as well as variable components, such as the volume of blood in the tissue. The intensity of light transmitted through the tissue, when expressed as a function of time is often said to include a baseline component, which varies slowly with time and represents the effect of the fixed components on the light, as well as a periodic pulsatile component, which varies more rapidly with time and represents the effect that changing tissue blood volume has on the light (Cheung *et al* 1989). The baseline component modeling the unchanging absorptive elements has a thickness d and an absorbance α. The pulsatile component representing the changing absorptive portion of the finger has a thickness of Δd and the relative absorbance of α_A representing the arterial blood absorbance (figure 9.1(c)).

The light emerging from the baseline component can be written as a function of the incident light intensity I_0 as follows

$$I_1 = I_0 \, e^{-\alpha d}. \tag{9.3}$$

Likewise, the intensity of light I_2 emerging from the pulsatile component is a function of its incident light intensity I_1 and can be written as follows

$$I_2 = I_1 e^{-\alpha_A \Delta d}. \tag{9.4}$$

Substituting the expression of I_1 in the expression for I_2, the light emerging from the finger as a function of the incident light intensity I_0 is as follows

$$I_2 = I_0 e^{-[\alpha d + \alpha_A \Delta d]}. \tag{9.5}$$

The effect of light produced by the arterial blood volume is given by the relationship between I_2 and I_1. Defining the change in transmittance produced by the arterial component as $T_{\Delta A}$, we have

$$T_{\Delta A} = I_2 / I_1. \tag{9.6}$$

Substituting the expressions for I_1 and I_2 in the above equation yields the following:

$$T_{\Delta A} = (I_0 e^{-[\alpha d + \alpha_A \Delta d]}) / (I_0 e^{-\alpha d}). \tag{9.7}$$

The term I_0 in the numerator and the denominator can be canceled by eliminating the input light intensity as a variable in the equation. Therefore, the change in arterial transmittance can be expressed as

$$T_{\Delta A} = e^{-\alpha_A \Delta d}. \tag{9.8}$$

A device employing this principle in operation is effectively self-calibrating, and is independent of the incident light intensity I_0.

9.2.1.2 Eliminating the thickness of the path as a variable. The changing thickness of the finger, Δd, produced by the changing arterial blood volume remains a variable in equation (9.8). To further simplify the equation, the logarithmic transformation is performed on the terms in equation (9.8) yielding the following

$$\ln T_{\Delta A} = \ln (e^{-\alpha_A \Delta d}) = -\alpha_A \Delta d. \tag{9.9}$$

The variable Δd can be eliminated by measuring arterial transmittance at two different wavelengths. The two measurements at two wavelengths provide two equations with two unknowns. The particular wavelengths selected are determined in part by consideration of a more complete expression of the arterial absorbance α_A

$$\alpha_A = (\alpha_{OA})(S_aO_2) - (\alpha_{DA})(1 - S_aO_2) \tag{9.10}$$

where α_{OA} is the oxygenated arterial absorbance, α_{DA} is the deoxygenated arterial absorbance, and S_aO_2 is the oxygen saturation of arterial Hb. α_{OA} and α_{DA} are substantially unequal at all light wavelengths in the red and near infrared wavelength regions except for the isosbestic wavelength of 805 nm. With an S_aO_2 of approximately 90%, the arterial absorbance α_A is 90% attributable to the oxygenated arterial absorbance α_{OA}, and 10% attributable to the deoxygenated arterial absorbance α_{DA}. At the isosbestic wavelength, the relative contribution of these two coefficients to the arterial absorbance α_A is of minimal significance in that both α_{OA} and α_{DA} are equal (figure 4.2).

Wavelengths selected are in a range away from the approximate isosbestic wavelength that is sufficient to allow the two signals to be easily distinguished. It is generally preferred that the two wavelengths selected fall within the red and infrared regions of the electromagnetic spectrum. The ratio of the transmittance produced by the arterial blood component at red and infrared wavelengths follows from equation (9.9).

$$\frac{\ln T_{\Delta AR}}{\ln T_{\Delta AIR}} = \frac{-\alpha_A(\lambda_R)\Delta d}{-\alpha_A(\lambda_{IR})\Delta d} \tag{9.11}$$

where $T_{\Delta AR}$ equals the change in arterial transmittance of light at the red wavelength λ_R and $T_{\Delta AIR}$ is the change in arterial transmittance at the infrared wavelength λ_{IR}. If the two sources are positioned at approximately the same location on the finger, the length of the light path through the finger is approximately the same for light emitted by each LED. Thus, the change in the light path resulting from arterial blood flow Δd is approximately the same for both the red and infrared wavelength sources. For this reason, the Δd term in the numerator and the denominator of the right side of equation (9.11) cancel, producing

$$\frac{\ln T_{\Delta AR}}{\ln T_{\Delta AIR}} = \frac{\alpha_A(\lambda_R)}{\alpha_A(\lambda_{IR})}. \tag{9.12}$$

Equation (9.12) is independent of the incident light intensity I_0 and the change in finger thickness Δd, attributable to arterial blood flow. Because of the complexity of the physiological process, the ratio indicated in equation (9.12) does not directly provide an accurate measurement of oxygen saturation. The correlation between the ratio of equation (9.12) and actual arterial blood gas measurement is therefore relied upon to produce an indication of the oxygen saturation. Thus, if the ratio of the arterial absorbance at the red and infrared wavelengths can be determined, the oxygen saturation of the arterial blood flow can be extracted from independently derived, empirical calibration curves in a manner dependent on I_0 and Δd. For simplicity, a measured ratio R_{OS} is defined from equation (9.12) as

$$\text{Ratio} = R_{OS} = \frac{\alpha_A(\lambda_R)}{\alpha_A(\lambda_{IR})}. \tag{9.13}$$

9.3 RATIO OF RATIOS

The Ratio of Ratios (R_{OS}) is a variable used in calculating the oxygen saturation level. It is typically calculated by taking the natural logarithm of the ratio of the peak value of the red signal divided by the valley measurement of the red signal. The ratio is then divided by the natural logarithm of the ratio of the peak value of the infrared signal divided by the valley measurement of the infrared signal (Cheung *et al* 1989).

9.3.1 Peak and valley method

A photodiode placed on the side of a finger opposite the red and infrared LEDs receives light at both wavelengths transmitted through the finger. The received red wavelength light intensity varies with each pulse and has high and low values R_H and R_L, respectively. R_L occurs during systole when arterial blood volume is at its greatest, while R_H occurs during diastole when the arterial blood volume is lowest (figure 9.2). Considering the exponential light decay through homogeneous media, it is observed that

$$R_L = I_0 e^{-[\alpha(\lambda_R)d + \alpha_A(\lambda_R)\Delta d]}. \tag{9.14}$$

Similarly,

$$R_H = I_0 e^{-\alpha(\lambda_R)d}. \tag{9.15}$$

Taking the ratio of equations (9.14) and (9.15) and simplifying, we have

$$\frac{R_L}{R_H} = e^{-\alpha_A(\lambda_R)\Delta d}. \tag{9.16}$$

Taking the logarithm of both sides of equation (9.16) yields

$$\ln\left(\frac{R_L}{R_H}\right) = -\alpha_A(\lambda_R)\Delta d. \tag{9.17}$$

Similar expressions can be produced for the infrared signal.

$$\ln\left(\frac{IR_L}{IR_H}\right) = -\alpha_A(\lambda_{IR})\Delta d. \tag{9.18}$$

The ratiometric combination of equations (9.17) and (9.18) yields

$$\frac{\ln\left(\frac{R_L}{R_H}\right)}{\ln\left(\frac{IR_L}{IR_H}\right)} = \frac{-\alpha_A(\lambda_R)\Delta d}{-\alpha_A(\lambda_{IR})\Delta d}. \tag{9.19}$$

Because the Δd terms in the numerator and denominator of the right side of the equation (9.19) cancel, as do the negative signs before each term, equation (9.19) when combined with equation (9.13) yields

$$\text{Ratio} = R_{OS} = \frac{\alpha_A(\lambda_R)}{\alpha_A(\lambda_{IR})} = \frac{\ln\left(\frac{R_L}{R_H}\right)}{\ln\left(\frac{IR_L}{IR_H}\right)}. \quad (9.20)$$

Thus, by measuring the minimum and the maximum emergent light intensities of both the red and infrared wavelengths (R_L, R_H, IR_L, IR_H), a value for the term R_{OS} can be computed. Empirically derived calibration curves are then used to determine the oxygen saturation based on R_{OS}.

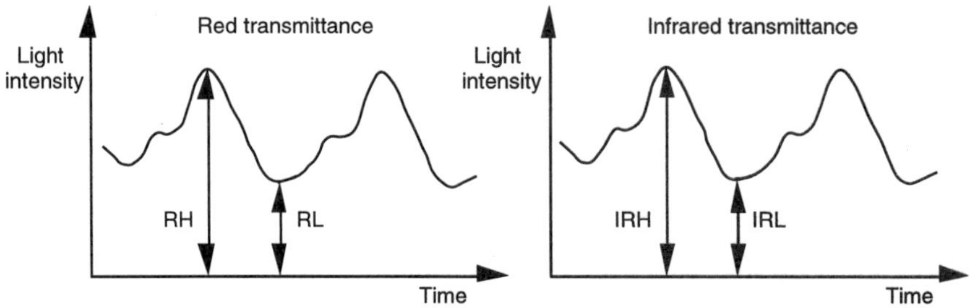

Figure 9.2. A graphical plot of transmitted light intensity converted into voltage. High (H) and low (L) signals are shown as a function of time of the transmittance of red (R) and infrared (IR) light through the finger.

9.3.2 Derivative method: noise reduction software

Yorkey (1996) derives the Ratio of Ratios by calculating using the separated AC and DC components of the measured signal. This mathematical derivation of the ratio of ratios is performed using the Beer–Lambert equation.

$$I_1 = I_0 \, e^{-\alpha L} \quad (9.21)$$

where I_1 is the emerging light intensity, I_0 is the incident light intensity, α is the relative extinction coefficient of the material and L is the path length. In this method, the Ratio of Ratios is determined using the derivatives. Assuming the change in path length is the same for both wavelengths during the same time interval between samples, the instantaneous change in path length (dL/dt) must also be the same for both wavelengths.

We can extend the general case of taking the derivative of e^u to our case

$$\frac{de^u}{dt} = e^u \frac{du}{dt} \quad (9.22)$$

$$\frac{dI_1}{dt} = I_0\, e^{-\alpha L}\left(-\alpha\frac{dL}{dt}\right) \tag{9.23}$$

Therefore,

$$\frac{(dI_1/dt)}{I_1} = -\alpha\frac{dL}{dt}. \tag{9.24}$$

Here, I_1 is equal to the combined AC and DC component of the waveform and dI_1/dt is equal to the derivative of the AC component of the waveform. Using two wavelengths we have

$$\text{R of R} = \frac{(dI_R/dt)/I_R}{(dI_{IR}/dt)/I_{IR}} = \frac{-\alpha(\lambda_R)}{-\alpha(\lambda_{IR})}. \tag{9.25}$$

Instead of using the previous method of calculating the Ratio of Ratios based on the natural logarithm of the peak and valley values of the red and infrared signals, the value of the R of R can be calculated based on the derivative value of the AC component of the waveform.

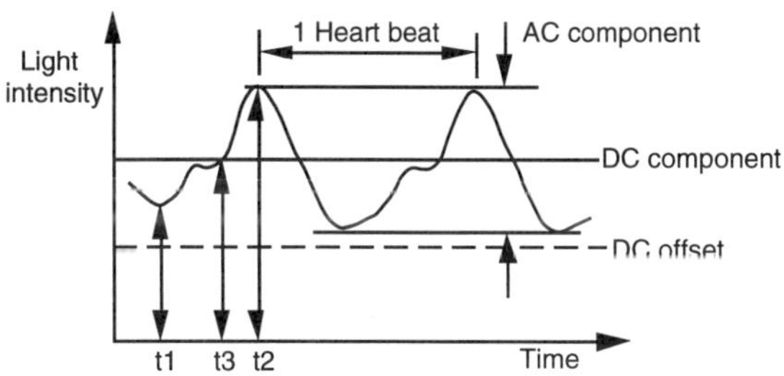

Figure 9.3. A waveform of the transmitted light intensity through a finger showing the AC component, the DC component and the DC offset.

Note in discrete time

$$\frac{dI_R(t)}{dt} \approx I_R(t_2) - I_R(t_1). \tag{9.26}$$

If we choose t_2 and t_1 to be the maximum and minimum of the waveform, we can refer to this difference as the AC value, and the denominator above evaluated at some point in time t_3 in between t_2 and t_1 as the DC value. So,

$$\frac{\dfrac{dI_R(t)/dt}{I_R}}{\dfrac{dI_{IR}(t)/dt}{I_{IR}}} \approx \frac{\dfrac{I_R(t_2)-I_R(t_1)}{I_R(t_3)}}{\dfrac{I_{IR}(t_2)-I_{IR}(t_1)}{I_{IR}(t_3)}} = \frac{\dfrac{AC_R}{DC_R}}{\dfrac{AC_{IR}}{DC_{IR}}} = R. \tag{9.27}$$

Potratz (1994) implemented another improved method for noise reduction called the derivative method of calculating the Ratio of Ratios. To calculate the Ratio of Ratios based on the derivative formula, a large number of sampled points along the waveform are used instead of merely the peak and valley measurements. A series of sample points from the digitized AC and AC + DC values for the infrared and red signals are used to form each data point. A digital FIR filtering step essentially averages these samples to give a data point. A large number of data points are determined in each period. The period is determined after the fact by noting where the peak and valley occur (figure 9.3).

From the AC signal, a derivative is then calculated for each pair of data points and used to determine the ratio of the derivatives for R and IR. A plot of these ratios over a period will ideally result in a straight line. Noise from the motion artifact and other sources will vary some values. But by doing the linear regression, a best line through a period can be determined, and used to calculate the Ratio of Ratios.

A problem with other systems was DC drift. Therefore, a linear extrapolation was performed between two consecutive negative peaks of the waveform. This adjusts the negative peak of the waveform as if the shift due to the system noise did not occur. A similar correction can be calculated using the derivative form of the waveform. In performing the correction of the DC component of the waveform, it is assumed that the drift caused by noise in the system is much slower than the waveform pulses and the drift is linear. The linear change on top of the waveform can be described by the function

$$g(t) = f(t) + mt + b \tag{9.28}$$

where m is equal to the slope of the waveform and b is equal to a constant.

The linear change added to the waveform does not affect the instantaneous DC component of the waveform. However, the derivative of the linear change will have an offset due to the slope of the interfering signal:

$$\mathrm{d}(f(t) + mt + b) / \mathrm{d}t = \mathrm{d}f(t) / \mathrm{d}t + m. \tag{9.29}$$

if we assume that the offset is constant over the period of time interval, then the Ratio of Ratios may be calculated by subtracting the offsets and dividing:

$$\mathrm{R\ of\ R} = \frac{Y}{X} = \frac{(y - m_y)}{(x - m_x)} \tag{9.30}$$

where y and x are the original values and m_x and m_y are the offsets.

Since the Ratio of Ratios is constant over this short time interval the above formula can be written as

$$\frac{(y - m_y)}{(x - m_x)} = R. \tag{9.31}$$

Therefore,

$$y = Rx - Rm_x + m_y. \tag{9.32}$$

Since it was assumed that m_1, m_2, and R are constant over the time interval, we have an equation in the form of $y = mx + b$ where m is the Ratio of Ratios. Thus, we do a large number of calculations of the Ratio of Ratios for each period, and then do the best fit calculation to the line $y = Rx + b$ to fit the optimum value of R for that period, taking into account the constant b which is caused by DC drift.

To determine the Ratio of Ratios exclusive of the DC offset we do a linear regression. It is preferred to take points along the curve having a large differential component, for example, from peak to valley. This will cause the mx term to dominate the constant b:

$$R = \frac{n\sum x_j y_j - \sum x_j \sum y_j}{n\sum x_j^2 - (\sum x_j)^2} \tag{9.33}$$

where n = # of samples, j = sample #, $x = I_R \, dI_{IR} / dt$, $y = I_{IR} \, dI_R / dt$.

Prior sampling methods typically calculate the Ratio of Ratios by sampling the combined AC and DC components of the waveform at the peak and valley measurements of the waveform. Sampling a large number of points on the waveform, using the derivative and performing a linear regression increases the accuracy of the Ratio of Ratios, since noise is averaged out. The derivative form eliminates the need to calculate the logarithm. Furthermore doing a linear regression over the sample points not only eliminates the noise caused by patient movement of the oximeter, it also decreases waveform noise caused by other sources.

9.4 GENERAL PROCESSING STEPS OF OXIMETRY SIGNALS

The determination of the Ratio of Ratios (R_{OS}) requires an accurate measure of both the baseline and pulsatile signal components (Frick *et al* 1989). The baseline component approximates the intensity of light received at the detector when only the fixed nonpulsatile absorptive component is present in the finger. This component of the signal is relatively constant over short intervals and does not vary with nonpulsatile physiological changes, such as movement of the probe. Over a relatively long time, this baseline component may vary significantly. The magnitude of the baseline component at a given point in time is approximately equal to the level identified as R_H (figure 9.2). However, for convenience, the baseline component may be thought of as the level indicated by R_L, with the pulsatile component varying between the values of R_H and R_L over a given pulse. Typically, the pulsatile component may be relatively small in comparison to the baseline component and is shown out of proportion in figure 9.3. Because the pulsatile components are smaller, greater care must be exercised with respect to the measurement of these components. If the entire signal, including the baseline and the pulsatile components, were amplified and converted to a digital format for use by microcomputer, a great deal of the accuracy of the conversion would be wasted because a substantial portion of the resolution would be used to measure the baseline component (Cheung *et al* 1989).

In this process, a substantial portion of the baseline component termed offset voltage V_{OS} is subtracted off the input signal V_1. The remaining pulsatile component is amplified and digitized using an ADC. A digital reconstruction is then produced by reversing the process, wherein the digitally provided information allows the gain to be removed and the offset voltage added back.

This step is necessary because the entire signal, including the baseline and pulsatile components is used in the oxygen saturation measurement process.

Feedback from the microcomputer is required to maintain the values for driver currents I_0, V_{os} and gain A at levels appropriate to produce optimal ADC resolution (figure 9.4). Threshold levels L1 and L2 slightly below and above the maximum positive and negative excursions L3 and L4 allowable for the ADC input are established and monitored by the microcomputer (figure 9.5). When the magnitude of the input to and output from the ADC exceeds either of the thresholds L1 or L2, the drive currents I_D are adjusted to increase or decrease the intensity of light impinging on the detector. This way, the ADC is not overdriven and the margin between L1 and L3 and between L2 and L4 helps assure this even for rapidly varying signals. An operable voltage margin for the ADC exists outside of the thresholds, allowing the ADC to continue operating while the appropriate feedback adjustments to A and V_{os} are made. When the output from the ADC exceeds the positive and negative thresholds L5 or L6, the microcomputer responds by signaling the programmable subtractor to increase or decrease the voltage V_{os} being subtracted. This is accomplished based on the level of the signal received from the ADC. Gain control is also established by the microcomputer in response to the output of the ADC (Cheung *et al* 1989).

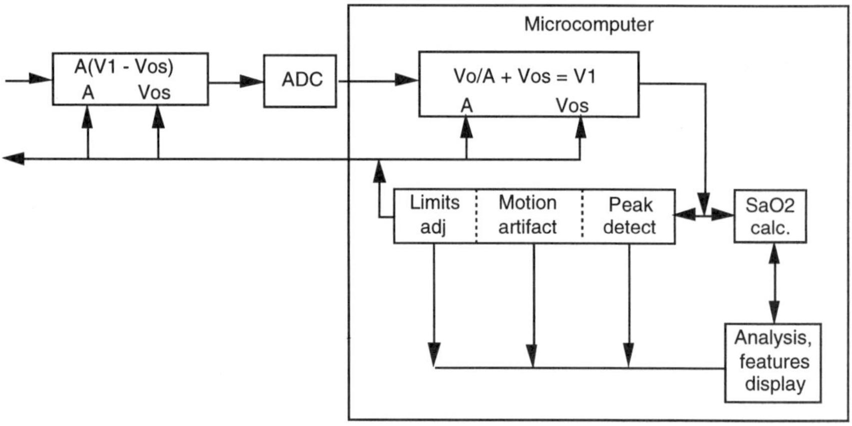

Figure 9.4. A functional block diagram of the microcomputer feedback illustrating the basic operation of the feedback control system. The DC value of the signal is subtracted before digitizing the waveform to increase the dynamic range of conversion. The removed DC value is later added to the digitized values for further signal processing (Cheung *et al* 1989).

A program of instructions executed by the Central Processing Unit of the microcomputer defines the manner in which the microcomputer provides servosensor control as well as produces measurements for display. The first segment of the software is the interrupt level routine.

9.4.1 Start up software.

The interrupt level routine employs a number of subroutines controlling various portions of the oximeter. At the start up, calibration of the oximeter is performed. After calibration, period zero subroutine is executed which includes five states, zero through four (figure 9.6).

Period zero subroutine is responsible for normal sampling

State 0: Initialize parameters
State 1: Set drive current
State 2: Set offsets
State 3: Set gains
State 4: Normal data acquisition state.

Probe set-up operations are performed during the states zero to three of this subroutine. During these states probe parameters including the amplifier gain A and offset voltage V_{OS} are initialized, provided that a finger is present in the probe. State 4 of the interrupt period zero subroutine is the normal data acquisition state. The signals produced in response to light at each wavelength are then compared with the desired operating ranges to determine whether modifications of the driver currents and voltage offsets are required. Finally state 4 of the period zero subroutine updates the displays of the oximeter. Sequential processing returns to state 0 whenever the conditions required for a particular state are violated (Cheung *et al* 1989).

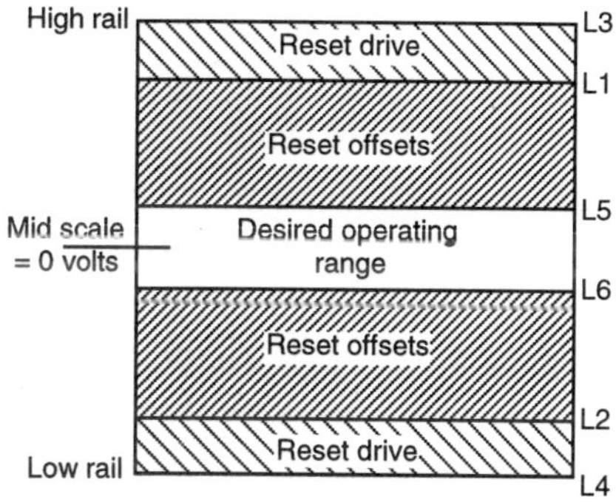

Figure 9.5 A graphical representation of the possible ranges of digitized signal, showing the desired response of the I/O circuit and microcomputer at each of the various possible ranges (Cheung *et al* 1989).

9.5 TRANSIENT CONDITIONS

The relative oxygen content of a patient's arterial pulses and the average background absorbance remain about the same from pulse to pulse. Therefore, the red and infrared light that is transmitted through the pulsatile flow produces a regularly modulated waveform with periodic pulses of comparable shape and amplitude and a steady state background transmittance. This regularity in shape helps in accurate determination of the oxygen saturation of the blood based on the maximum and minimum transmittance of the red and infrared light.

Changes in a patient's local blood volume at the probe site due to motion artifact or ventilatory artifact affect the absorbance of light. These localized changes often introduce artificial pulses into the blood flow causing the periodic pulses ride on a background intensity component of transmittance that varies as blood volume changes. This background intensity component variation, which is not necessarily related to changes in saturation, affects the pulse to pulse uniformity of shape, amplitude and expected ratio of the maximum to minimum transmittance, and can affect the reliability and accuracy of oxygen saturation determination (Stone and Briggs 1992).

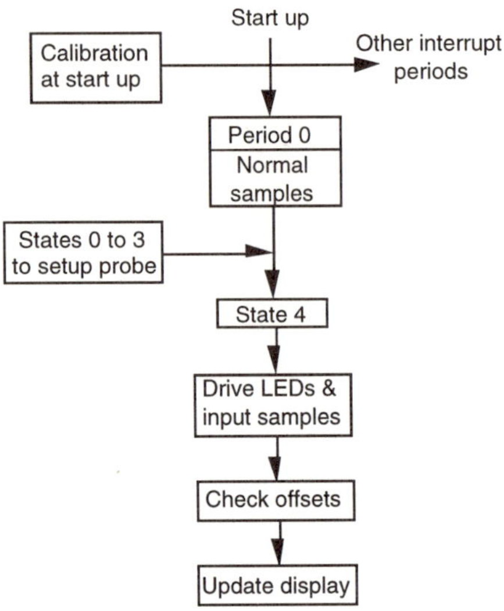

Figure 9.6. Flow chart of a portion of an interrupt level software routine included in the microcomputer (Cheung *et al* 1989).

In addition, there are times when the patient's background level of oxygen saturation undergoes transient changes, for example, when the patient loses or requires oxygen exchange in the lungs while under gaseous anesthesia. The transient waveform distorts the pulse shape, amplitude, and the expected ratio of the pulses, which in turn affects the reliability and accuracy of the oxygen saturation determination.

With changes in the background intensity absorbance component due to artifacts from changes in blood volume or transient saturation changes, the determined saturation value is not accurate and it would not become accurate again until the average absorbance level stabilizes.

The saturation calculations based upon transient signals provide an overestimation or underestimation of the actual saturation value, depending upon the trend. The transmittance of red light increases as oxygen saturation increases resulting in a signal value having a smaller pulse, and the transmittance of the infrared light decreases as saturation increases resulting in the infrared pulsatile amplitude increasing. For these wavelengths, the transmittance changes with saturation are linear in the range of clinical interest, i.e., oxygen saturation between 50% and 100%. The accuracy of the estimation is of particular concern during rapid desaturation. In such a case, the determined saturation based on the detected signals indicates a greater drop than the actual value. This underestimation of oxygen saturation may actuate low limit saturation alarms that can result in inappropriate clinical decisions.

The pulsatile amplitude is usually quite small, typically less than 5% of the overall intensity change and any small change in overall or background transmittance, such as slight changes in average blood saturation, can have a relatively large effect on the difference in maximum and minimum intensity of the light levels. Because the change in transmittance with changing oxygen saturation is opposite in direction for the red and infrared, this can result in overestimation of the pulsatile ratio during periods when saturation is decreasing, and underestimation during periods when saturation is increasing. It is therefore essential to compensate for the effects of transient conditions and localized blood volume changes on the actual signal, thereby providing a more accurate estimation of the actual oxygen saturation value.

This can be achieved by using a determined rate of change from pulse to pulse, using interpolation techniques and by using the low frequency characteristics of the detected signal values.

The transient error is corrected by linear interpolation where the determined maxima and minima for a first and second optical pulses are obtained, the second pulse following the first. The respective rates of change in the transmittance due to the transient are determined from the maximum transmittance point of the first detected pulse to the second detected pulse (Stone and Briggs 1992). The determined rates of change are then used to compensate any distortion in the detected transmittance of the first detected pulse introduced by the transient in accordance with the following algorithm

$$V_{max}(n)^* = V_{max}(n) + [V_{max}(n) - V_{max}(n+1)] \times \frac{[t_{max}(n) - t_{min}(n)]}{[t_{max}(n+1) - t_{max}(n)]} \qquad (9.34)$$

where $t_{max}(n)$ is the time of occurrence of the detected maximum transmittance at the n maximum, $t_{min}(n)$ is the time of occurrence of the detected minimum transmittance of the wavelength at the n minimum, $V_{max}(n)$ is the detected optical signal maximum value at the maximum transmittance of the wavelength at the n maximum $V_{max}(n)^*$ is the corrected value, for n being the first optical pulse, and $n + 1$ being the second optical pulse of that wavelength.

By application of the foregoing linear interpolation routine, the detected maximum transmittance value at $t_{max}(n)$ can be corrected, using the values $t_{max}(n+1)$, detected at the next coming pulse, to correspond to the transmittance value that would be detected as if the pulse were at steady state conditions. The corrected maximum value and the detected (uncorrected) minimum value thus provide an adjusted optical pulse maximum and minimum that correspond more closely to the actual oxygen saturation in the patient's blood at that time, not withstanding the transient condition. Thus, using the adjusted pulse values in place of the detected pulse values in the modulation ratio for calculating oxygen saturation provides a more accurate measure of oxygen saturation than would otherwise be obtained during transient operation.

Similarly, the respective rates of change in the transmittance are determined from the minimum transmittance point of the first detected pulse to the minimum of the second detected pulse. The determined rates of change are then used to compensate for any distortion in the detected minimum transmittance of the second detected pulse introduced by the transient in accordance with the following algorithm

$$V_{min}(n)^* = V_{min}(n-1) + [V_{min}(n) - V_{min}(n-1)] \times \frac{[t_{max}(n) - t_{min}(n-1)]}{[t_{min}(n) - t_{min}(n-1)]} \quad (9.35)$$

where $t_{max}(n)$ is the time of occurrence of the detected maximum transmittance at the n maximum; $t_{min}(n)$ is the time of occurrence of the detected minimum transmittance of the wavelength at the n minimum; $V_{min}(n)$ is the detected optical signal minimum value at the minimum transmittance of the wavelength at the n minimum; $V_{min}(n)^*$ is the corrected value, for n being the second optical pulse, and $n - 1$ being the first optical pulse of that wavelength.

By application of the foregoing linear interpolation routine, the detected minimum transmittance value at $t = n$ can be compensated using the detected values at the preceding pulse $t = n - 1$, to correspond to the transmittance value that would be detected as if the pulse were detected at steady state conditions. The compensated minimum value and the detected (uncompensated) maximum value thus provide an adjusted optical pulse maximum and minimum that correspond more closely to the actual oxygen saturation in the patient's blood at that time, notwithstanding the transient condition. Thus, using the adjusted pulse values in place of the detected pulse values in the modulation ratio for calculating oxygen saturation provides a more accurate measure of oxygen saturation than would otherwise be obtained during transient operation.

As is apparent from the algorithms, during steady state conditions the compensated value is equal to the detected value. Therefore, the linear interpolation routine may be applied to the detected signal at all times, rather than only when transient conditions are detected. Also, the algorithm may be applied to compensate the detected minimum or maximum transmittance values by appropriate adjustment of the algorithm terms. The amount of oxygen saturation can then be determined from this adjusted optical pulse signal by determining the relative maxima and minima as compensated for the respective wavelengths and using that information in determining the modulation ratios of the known Lambert–Beer equation.

The Nellcor® N-200 oximeter is designed to determine the oxygen saturation in one of the two modes. In the unintegrated mode the oxygen saturation determination is made on the basis of optical pulses in accordance with conventional pulse detection techniques. In the ECG synchronization mode the determination is based on enhanced periodic data obtained by processing the detected optical signal and the ECG waveform of the patient.

The calculation of saturation is based on detecting maximum and minimum transmittance of two or more wavelengths whether the determination is made pulse by pulse (the unintegrated mode) or based on an averaged pulse that is updated with the occurrence of additional pulses to reflect the patient's actual condition (the ECG synchronized mode).

Interrupt programs control the collection and digitization of incoming optical signal data. As particular events occur, various software flags are raised which transfer operation to various routines that are called from a main loop processing routine.

The detected optical signal waveform is sampled at a rate of 57 samples per second. When the digitized red and infrared signals for a given portion of detected optical signals are obtained, they are stored in a buffer called DATBUF and a software flag indicating the presence of data is set. This set flag calls a routine called MUNCH, which processes each new digitized optical signal waveform sample to identify pairs of maximum and minimum amplitudes corresponding to a pulse. The MUNCH routine first queries whether or not there is ECG synchronization, then the MUNCH routine obtains the enhanced composite pulse data in the ECG synchronization mode. Otherwise, MUNCH obtains the red and infrared optical signal sample stored in DATBUF, in the unintegrated mode. The determined maximum and minimum pairs are then sent to a processing routine for processing the pairs. Preferably, conventional techniques are used for evaluating whether a detected pulse pair is acceptable for processing as an arterial pulse and performing the saturation calculation, whether the pulse pair is obtained from the DATBUF or from the enhanced composite pulse data.

The MUNCH routine takes the first incoming pulse data and determines the maximum and minimum transmittance for each of the red and infrared detected optical signals, and then takes the second incoming pulse data, and determines the relative maximum and minimum transmittance. The routine for processing the pairs applies the aforementioned algorithm to the first and second pulse data of each wavelength. Then the oxygen saturation can be determined using the corrected minimum and detected maximum transmittance for the second pulses of the red and infrared optical signals. Some of the examples demonstrate the above application.

Example 1

Figure 9.7(*a*) shows the representative plethysmographic waveforms in a steady state condition for the red and infrared detected signals. V_{max}R(1) equals 1.01 V, and V_{min}R(1) equals 1.00 V, for n = 1, 2 and 3 pulses. V_{min}R(n) is the detected optical signal minimum value at the minimum transmittance at the n pulse minimum. The modulation ratio for the maxima and minima red signal is:

$$\frac{V_{max}\mathrm{R}(n)}{V_{min}\mathrm{R}(n)} = \frac{1.01v}{1.00v} = 1.01.$$

For the infrared wavelength, V_{max}IR(n) equals 1.01 V and V_{min}IR(n) equals 1.00 V and the determined modulation ratio is 1.01.

Using these determined modulation ratios in the formula for calculating the ratio R provides:

$$R = \frac{\ln[V_{max}\mathrm{R}(n) / V_{min}\mathrm{R}(n)]}{\ln[V_{max}\mathrm{IR}(n) / V_{min}\mathrm{IR}(n)]} = \frac{0.01}{0.01} = 1.00.$$

A calculated R = 1 corresponds to an actual saturation value of about 81% when incorporated into the saturation equation. A saturation of 81% corresponds to a healthy patient experiencing a degree of hypoxia for which some corrective action would be taken.

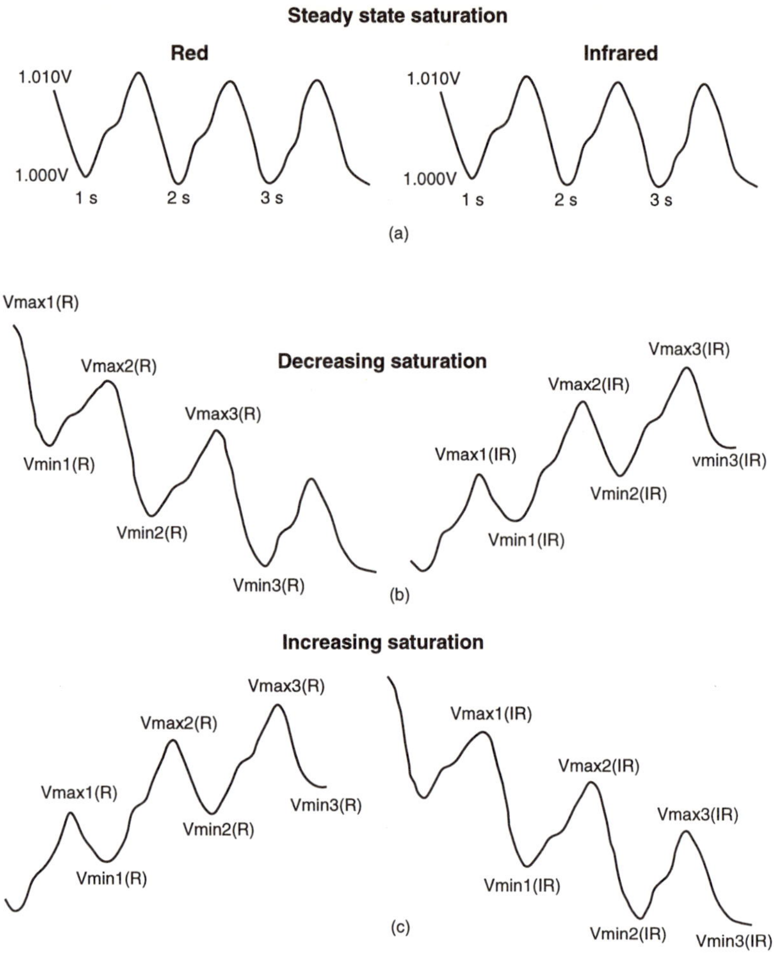

Figure 9.7. Graphical representation of detected optical signals during the steady state and transient conditions (Stone and Briggs 1992).

Example 2

Figure 9.7(*b*) shows the representative plethysmographic waveforms for a patient during desaturation or decreasing saturation transient conditions for the red and infrared detected signals having optical pulses n = 1, 2, and 3. However, in this transient example, it is known at n = 1, that the actual saturation of the patient is very close to that during the steady state conditions in example 1. In this transient example, the detected values are as follows for both the red and infrared signals:

$t_{max}(1) = 1.0$ s	$V_{max}R(1) = 1.012$ V	$V_{max}IR(1) = 1.008$ V
$t_{min}(1) = 1.2$ s	$V_{min}R(1) = 1.000$ V	$V_{min}IR(1) = 1.000$ V
$t_{max}(2) = 2.0$ s	$V_{max}R(2) = 1.002$ V	$V_{max}IR(2) = 1.018$ V
$t_{min}(2) = 2.2$ s	$V_{min}R(2) = 0.990$ V	$V_{min}IR(2) = 1.010$ V
$t_{max}(3) = 3.0$ s	$V_{max}R(3) = 0.992$ V	$V_{max}IR(3) = 1.028$ V
$t_{min}(3) = 3.2$ s	$V_{min}R(3) = 0.980$ V	$V_{min}IR(3) = 1.020$ V

Calculating the oxygen saturation ratio R at $n = 1$, using the detected optical signal provides the following

$$\begin{aligned} R &= \frac{\ln[V_{max}R(1)/V_{min}R(1)]}{\ln[V_{max}IR(1)/V_{min}IR(1)]} \\ &= \ln[1.012/1.000]/\ln[1.008/1.000] \\ &= \ln[1.012]/\ln[1.008] \\ &= 0.012/0.008 \quad = 1.5. \end{aligned}$$

The calculated saturation ratio of 1.5 based on the detected transmittance corresponds to a calculated oxygen saturation of about 65 for the patient, which corresponds to severe hypoxia in an otherwise healthy patient. This contrasts with the known saturation of about 81% and demonstrates the magnitude of the underestimation of the oxygen saturation (overestimation of desaturation) due to the distortion in transmittance of the red and infrared light caused by transient conditions.

Applying the correction algorithm to correct the distorted maximum transmittance point of the detected red signal during the transient condition:

$$\begin{aligned} V_{max}R(1)^* &= V_{max}R(1) - [V_{max}R(1) - V_{max}R(2)] \times \frac{[t_{max}(1) - t_{min}(1)]}{[t_{max}(2) - t_{max}(1)]} \\ &= 1.012 - [1.012 - 1.002] \times [1.0 - 1.2]/[1.0 - 2.0] \\ &= 1.010. \end{aligned}$$

and correspondingly for the maximum transmittance of the detected infrared signal

$$\begin{aligned} V_{max}IR(1)^* &= 1.008 - [1.008 - 1.018] \times [1.0 - 1.2]/[1.0 - 2.0] \\ &= 1.010 \end{aligned}$$

Thus, by replacing $V_{max}R(n)$ with $V_{max}R(n)^*$ and replacing $V_{max}IR(n)$ with $V_{max}IR(n)^*$ in the calculations for determining the oxygen saturation ratio R, we have

$$\begin{aligned} R &= \frac{\ln[V_{max}R(1)^* / V_{min}R(1)]}{\ln[V_{max}IR(1)^* / V_{min}IR(1)]} \\ &= \ln[1.010/1.00]/\ln[1.010/1.00] \\ &= 0.01/0.01 \\ &= 1.0. \end{aligned}$$

Thus, basing the saturation calculations on the corrected maximum transmittance values and the detected minimum transmittance values, the corrected R value corresponds to the same R for the steady state conditions and the actual oxygen saturation of the patient.

Example 3

Figure 9.7(c) shows the representative plethysmographic waveforms for a patient during desaturation or decreasing saturation transient conditions for the red and infrared detected signals having optical pulses n = 1, 2 and 3. However, in this transient example, it is known that at n = 2, the actual saturation of the patient is very close to that during the steady state conditions in example 1. In this transient example, the detected values are as follows for both the red and infrared signals:

t_{max} (1) = 1.0 s	V_{max}R(1) = 1.022 V	V_{max}IR(1) = 1.002 V
t_{min} (1) = 1.2 s	V_{min}R(1) = 1.008 V	V_{min}IR(1) = 0.992 V
t_{max} (2) = 2.0 s	V_{max}R(2) = 1.012 V	V_{max}IR(2) = 1.012 V
t_{min} (2) = 2.2 s	V_{min}R(2) = 0.998 V	V_{min}IR(2) = 1.002 V
t_{max} (3) = 3.0 s	V_{max}R(3) = 1.002 V	V_{max}IR(3) = 1.022 V
t_{min}(3) = 3.2 s	V_{min}R(3) = 0.988 V	V_{min}IR(3) = 1.012 V

Calculating the oxygen saturation ratio R at n = 2, using the detected optical signal provides the following

$$R = \frac{\ln[V_{max}\,\mathrm{R}(2)/V_{min}\,\mathrm{R}(2)]}{\ln[V_{max}\,\mathrm{IR}(2)/V_{min}\,\mathrm{IR}(2)]}$$
$$= \ln[1.012/0.998]/\ln[1.012/1.002]$$
$$= 0.01393/0.0099 \quad = 1.4.$$

Thus, the calculated saturation ratio of 1.4 based on the detected transmittance corresponds to a calculated oxygen saturation of about 51% for the patient, which corresponds to severe hypoxia in an otherwise healthy patient. This contrasts with the known saturation of about 81% and demonstrates the magnitude of the underestimation of the oxygen saturation (overestimation of desaturation) due to the distortion in transmittance of the red and infrared light caused by transient conditions.

Applying the correction algorithm to correct the distorted minimum transmittance point of the detected red signal during the transient condition, we find the following:

$$V_{min}\,\mathrm{R}(2)^* = V_{min}\,\mathrm{R}(2) - [V_{min}\,\mathrm{R}(2) - V_{min}\,\mathrm{R}(1)] \times \frac{[t_{max}(2) - t_{min}(1)]}{[t_{min}(2) - t_{max}(1)]}$$
$$= 1.008 - [0.998 - 1.008] \times [2.0 - 1.2]/[2.2 - 1.2]$$
$$= 1.0$$

and correspondingly for the minimum transmittance of the detected infrared optical signal we have:

$$V_{min}\text{IR}(2)^* = 0.992 - [1.002 - 0.992] \times 0.8$$
$$= 1.0.$$

Thus, by replacing $V_{min}\text{R}(n)$ with $V_{min}\text{R}(n)^*$ and replacing $V_{min}\text{IR}(n)$ with $V_{min}\text{IR}(n)^*$ in the calculations for determining oxygen saturation ratio R we have:

$$R = \frac{\ln[V_{max}\text{R}(2) / V_{min}\text{R}(2)^*]}{\ln[V_{max}\text{IR}(2) / V_{min}\text{IR}(2)^*]}$$
$$= \ln[1.012 / 1.0] / \ln[1.012 / 1.0]$$
$$= 1.0.$$

Thus, basing the saturation calculations on the corrected minimum transmittance values and the detected maximum transmittance values, the corrected R value corresponds to the same R for the steady state conditions and the actual oxygen saturation of the patient.

9.6 ECG SYNCHRONIZATION ALGORITHMS

Electrical heart activity occurs simultaneously with the heartbeat and can be monitored externally and characterized by the electrocardiogram waveform. The ECG waveform comprises a complex waveform having several components that correspond to electrical heart activity of which the QRS component relates to ventricular heart contraction. The R wave portion of the QRS component is typically the steepest wave therein having the largest amplitude and slope, and may be used for indicating the onset of cardiac activity. The arterial blood pulse flows mechanically and its appearance in any part of the body typically follows the R wave of the electrical heart activity by a determinable period of time. This fact is utilized in commercially available pulse oximeters to enhance their performance. Another advantage of recording ECG is that it provides a redundancy in calculating the heart rate from both the ECG signal and the optical signal to continuously monitor the patient even if one of the signals is lost (figure 9.8).

With ECG synchronization, the pulse oximeter uses the electrocardiographic (ECG) QRS complex as a timing indicator that the optical pulse will soon appear at the probe site. The R portion of the ECG signal is detected and the time delay by which an arterial pulse follows the R wave is determined to establish a time window an arterial pulse is to be expected. By using the QRS complex to time the oximeter's analysis of the optical pulse signal, ECG processing synchronizes the analysis of oxygen saturation and pulse rate data. The established time window provides the oximeter with a parameter enabling the oximeter to analyze the blood flow only when it is likely to have a pulse present for analysis. This method of signal processing passes those components of the signal that are coupled to the ECG (i.e., the peripheral pulse), while attenuating those components that are random with respect to the ECG (e.g., motion artifact or other noise in the signal).

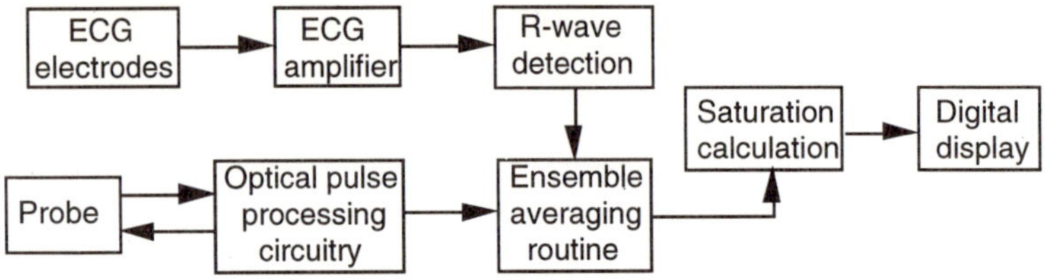

Figure 9.8. Block diagram illustrating the ECG processing components, its subcomponents and their relationship in an oximeter.

9.6.1 Nellcor® system

C-LOCK ECG synchronization enhances the signal-processing capabilities of Nellcor® systems such as the N-200 pulse oximeter and the N-1000 multifunction monitor. This improves the quality of the optical signal in certain clinical settings in which the performance of a conventional pulse oximeter may deteriorate, e.g. when a patient is moving or has poor peripheral pulses. Consequently, C-LOCK signal processing extends the range of clinical situations in which pulse oximetry may be used. Patient movement and poor peripheral pulses present similar problems for a conventional pulse oximeter: performance may deteriorate because the oximeter is unable to distinguish between the true optical pulse signal and background noise. C-LOCK ECG synchronization improves signal quality in these difficult signal-detection settings (Goodman and Corenman 1990).

The digital optical signal is processed by the microprocessor of the Nellcor N-1000 Pulse Oximeter in order to identify individual optical pulses and to compute the oxygen saturation from the ratio of maximum and minimum pulse levels as seen by the red wavelength compared to the pulse seen by the infrared wavelength.

Noninvasive pulse oximeters process optical signals which are prone to motion artifacts caused by the muscle movement proximate to the probe site. The spurious pulses induced in the optical signals may cause the pulse oximeter to process the artifact waveform and provide erroneous data. This problem is particularly significant with infants, fetuses, or patients that do not remain still during monitoring. Another problem exists in circumstances where the patient is in poor condition and the pulse strength is very weak. In continuously processing the optical data, it can be difficult to separate the true pulsatile component from the artifact pulses and noise because of low signal to noise ratio. Inability to reliably detect the pulsatile component in the pulsatile signal may result in a lack of the information needed to calculate oxygen blood saturation.

By incorporating the patient's heart activity into the pulse oximeter, problems due to motion artifact and low signal-to-noise ratio can be solved. Processing of the signals that occur during a period of time when the optical pulses are expected to be found, increases the likelihood that the oximeter will process only optical waveforms that contain the pulsatile component of arterial blood, and will not process spurious signals.

The software incorporated into the microprocessor for processing the ECG signals and displaying the calculated ECG pulse rate receives the digitized version of diagnostic ECG signal (DECG) and filtered ECG signals (FECG). The microprocessor calculates the amplitude of the ECG waveform and controls the AGC (automatic gain control) amplifier, so that DECG and FECG will fall within the voltage range limits of the electronic circuitry used to process these signals.

The microprocessor regularly searches a status input latch at a rate of 57 cycles per second. The output of detected R wave (DRW) sets the latch to a logical 1 when the R wave is detected. Depending on the status, the microprocessor selects the next operation and resets the DRW latch to 0. At this first level, the microprocessor counts the time interval beginning from the detection of an R wave pulse until the occurrence of the next logical 1 at the status input latch. Based on this time interval, the pulse oximeter displays the pulse rate. After averaging several time intervals and establishing a regular ECG pulse rate, the microprocessor will change to the second level of processing.

After the detection of an R wave pulse, the microprocessor separately analyzes the digital optical signal and correlates the period of time by which an optical pulse follows the detected R wave pulse to establish the time window during which the optical pulse is likely to occur. During this second level, the pulse oximeter just calculates and displays the time period or pulse rate between DRW pulses.

The third level of processing starts after a time window has been established. On detecting an R wave pulse, the microprocessor activates the time window so that only optical signals detected within the time window following the occurrence of an R wave pulse will be evaluated for acceptance or rejection and for use in calculating and displaying vital measurements such as oxygen saturation, pulse flow, and pulse rate. The evaluation of a detected pulse is made in conjunction with a preselected confidence factor that is associated with the quality of the optical signals. The higher the optical signal quality, the better the correlation between the recorded pulse history and the detected pulse, and the higher the confidence level. The confidence level may be set automatically by the microprocessor, or it may be adjusted by the operator. The microprocessor will reject any detected pulses occurring outside the time window. A typical time window for an adult male using a fingertip oximeter probe may be about 50 ms ±10 ms after the occurrence of an R wave. The oximeter will also reject any additional pulses detected after an optical pulse is detected within the same time window, even though the time window has not expired.

However, if the optical pulse is not found within an opened time window, the microprocessor will continue to search for optical pulses using the degraded criteria during the time window period for about three successive detected R wave (DRW) pulses, after which it continues to search with degraded criteria. After a specific interval, e.g. 10 s, without detecting an optical pulse, the microprocessor will revert to independent or nonintegrated processing of the optical and ECG signals, returning the pulse oximeter to startup conditions. Therefore, if the oximeter cannot establish or maintain a reliable correlation between the R wave and the optical pulse, the waveforms will be processed independently. The display will indicate whether the pulse oximeter is operating in integrated or nonintegrated mode. After attaining the third level of processing, losing either the ECG or optical pulse signals will activate an alarm and return the program to the startup condition.

9.6.1.1 R-wave determination routine. The R-wave determination routine begins with electric signals received from the ECG leads and calculating the R–R period RRPER between the last detected R wave and the present R wave (figure 9.9). The average period HISTORY from the previous R waves and the present R wave is calculated and the determined RRPER is compared to the average period HISTORY (Goodman and Corenman 1990). If RRPER does not correspond to HISTORY, the R wave ECG flag is reset and the routine is exited to await another R wave. If RRPER does correspond to HISTORY, a timer is activated to measure the interval from the occurrence of the R wave to the occurrence of the optical pulse. Output HR (ECG heart rate) is calculated based on successive R waves. The system determines whether a series of R–R periods have been synchronized (ECG synchronization). If not synchronized, then the system checks for alarms by comparing output HR to a preselected heart rate and generates an alarm if the output HR is too low. If the ECG is synchronized but the optical pulse to optical pulse is not synchronized, the output HR is sent to the display and then checked for alarms. If the optical signal is synchronized, then the system just checks for alarms. Only if the ECG is synchronized, the optical pulse is not synchronized, and the R wave looks like a valid R wave by comparison with HISTORY, then HISTORY is updated using the new R wave. After updating HISTORY, the system itself is updated (TIME OUT) to maintain synchronization. If TIME OUT is not updated for a period of five seconds, then ECG synchronization is lost and the routine must begin building a new history.

9.6.1.2 The systems routine. The system routine for processing digital optical pulse information for optical pulses to send to LEVEL 3 is flow charted (figure 9.10). The system begins by continuously evaluating the data from the detected digital optical signal (Goodman and Corenman 1990). The data are first evaluated for compatibility with signal processing. If the data are over or undervalued electronically, i.e., beyond the voltage range of the circuitry, then the system exits the routine, and the LED intensities are adjusted to correct the electrical values accordingly. When the data are compatible, they are next evaluated for a maximum signal. A relative maximum is determined and saved. The next value is compared to the saved value, and if it is a new maximum, it is saved instead. When the value found is not a new maximum, then a MAX FLAG is set. Thereafter, the system evaluates the following data received, by passing the maximum value section, to find the maximum slope, again by successive comparisons. When the largest slope value is found, it is saved and the SLOPE FLAG is set. Thereafter, the following data are evaluated, by passing the maximum and slope calculations, to find the minimum value corresponding to the end of the pulse. When the smallest minimum is found, it is saved and the slope value that was saved is compared with a pre-established minimum threshold to determine whether it is large enough to be a possible optical pulse. If it is not large enough, then the pulse is rejected, the flags are reset, and the routine begins processing the next possible pulse. If the slope is large enough, then the pulse parameters, maximum, minimum, and slope, are saved in memory for use by LEVEL 3 processing in evaluating the possible pulse. Then, the time delay from the R wave to the possible pulse is calculated. Thereafter, the DATA FLAG is set indicating to LEVEL 3 that there is a possible pulse to be evaluated, the MAX and SLOPE FLAGs are reset, and the routine begins again to process the following data, looking for new maximum values corresponding to possible pulses.

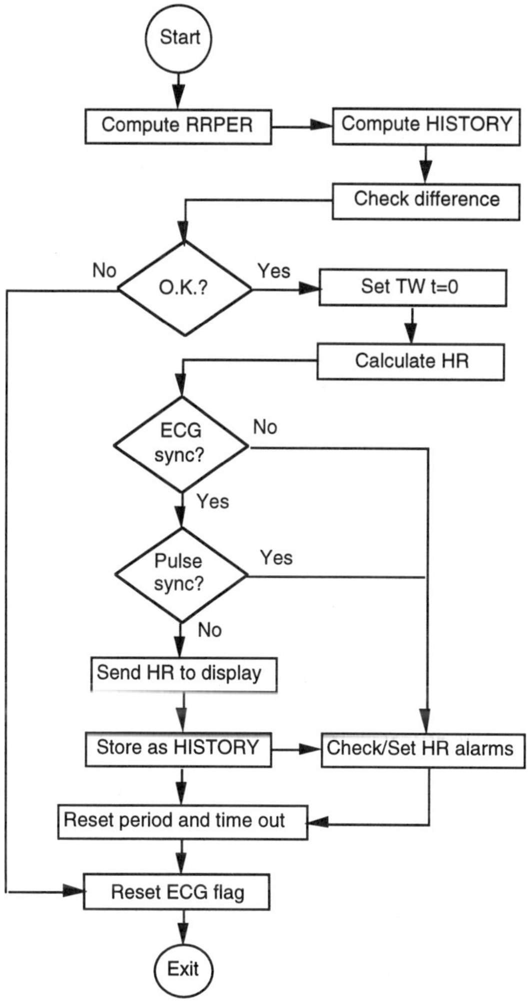

Figure 9.9. The R wave determination routine calculates RRPER, compares it with the average period HISTORY. If RRPER corresponds to HISTORY, the interval between the occurrence of R wave and occurrence of pulse is measured. The algorithm checks for ECG synchronization, alarms and displays heart rate (HR) (Goodman and Corenman 1990).

9.6.1.3 LEVEL 3 software. Figure 9.11 shows LEVEL 3 of software for computing the saturation measurements (Goodman and Corenman 1990). The system starts by acquiring a potential optical pulse after a DATA FLAG has been set and inquiring whether there is ECG synchronization i.e., a regular ECG period has been established. If a DATA FLAG has not been set, then the system exits the routine. If there has not been ECG synchronization, then the microprocessor processes the optical pulse signals independent of the ECG.

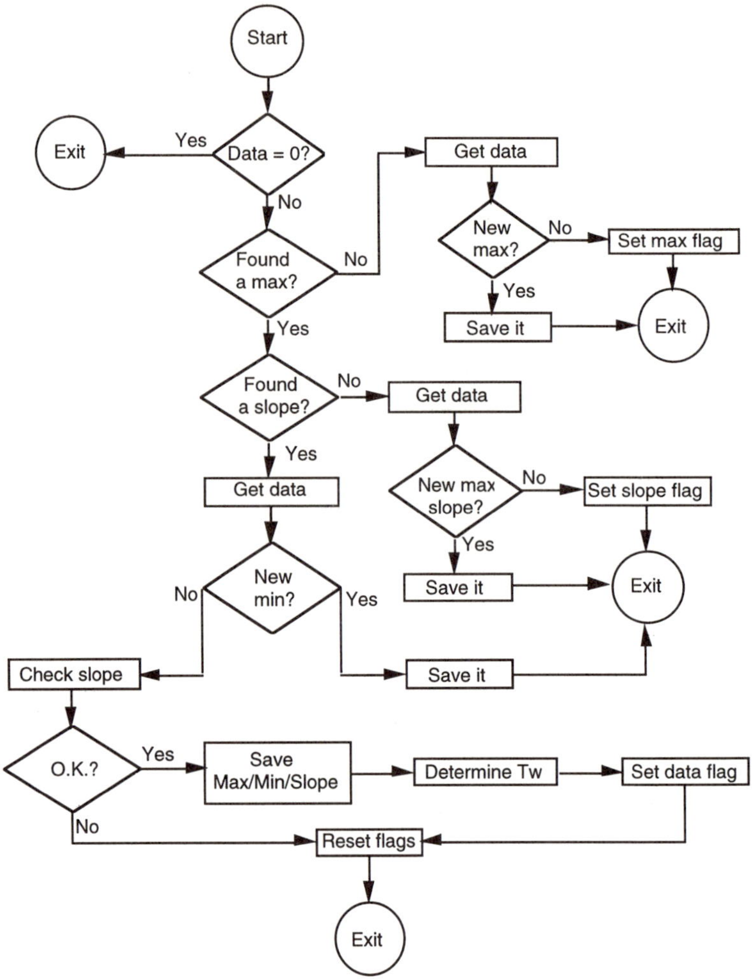

Figure 9.10. The system routine measures the maximum and minimum values in the data presented and calculates the largest slope. The slope value is compared with the normal expected values to determine whether it is a possible optical pulse (Goodman and Corenman 1990).

If there is ECG synchronization, but no R wave has occurred, then the system exits and the pulse is not processed. If there is ECG synchronization and a R wave has occurred, then the microprocessor processes the pulse. The LED intensity is evaluated to see if adjustment is necessary. The reset system gain, based on minimum LED intensity required for adequate signal strength, is checked to see if adjustment is required to the optical pulse historic period, amplitude and ratio. The system then inquires whether the ECG apparatus is operating between an R wave and the following optical pulses for the previous four pulses is computed to give the TIME WINDOW (TW). Then the pulse waveform is analyzed to see if it is a dicrotic notch rather than a real optical pulse. The downward slope of a dicrotic notch or other artifact can be misinterpreted as an optical pulse, but typically the pulse amplitude is less than half the amplitude of an actual pulse. If the pulse is determined to be a notch or artifact, then the system exits and the next pulse presented will be processed. If not determined to be a notch, then it is analyzed to determine if it is a pulse.

Assuming the ECG is synchronized, then the system determines if two criteria are met. The first is whether the time delay falls within the above-computed TIME WINDOW. If it does not, then the microprocessor rejects the pulse. The second criterion tested is whether or not the ratio is within acceptable limits. Only if the pulse satisfies both criteria is the pulse accepted and a saturation calculation made.

If the ECG is not synchronized then the pulse must pass any two of three criteria regarding (1) pulse period, (2) amplitude, and (3) ratio, to be accepted, e.g., pulse and period, period and amplitude, pulse and amplitude, or all three. If the pulse is accepted, then the oxygenation saturation is calculated.

After the system is turned on (POWER UP) after a TIME OUT alarm (a 10 s period with no valid optical pulse found) a series of consistent pulses must be found to generate an optical pulse history before the oxygenation saturation will be sent to the display. Thus, if there is no optical pulse synchronization, there will be no saturation display. All optical pulses, those accepted and those not accepted, excluding pulses rejected as artifacts, enter the calculation routine section. If the ECG is not synchronized then a pulse-to-pulse period and either an amplitude or a ratio must exist for the optical heart rate (OHR) calculation to be made. If either the ECG or the optical pulse is synchronized, then the HR calculation made will be displayed. If there is no synchronization, then the OHR is not displayed. The system is evaluating the status for pulse evaluation, i.e., whether signals should continue to be processed after a TIME WINDOW period has expired then TIME WINDOW is closed until opened by the detection of the next R wave. The blood oxygen saturation is calculated using the Ratio of Ratios.

9.6.2 Criticare® systems

The patient wears three standard ECG electrodes which provide the pulse oximeter with an ECG signal which if present is used to enhance the quality of the optical waveforms. The oximeter computes oxygen saturation from the enhanced waveform and displays it on a screen (Conlon *et al* 1990).

An ECG amplifier and an R-wave detection algorithm routine process the ECG signal provided by the electrodes and determine the timing for an ensemble averaging algorithm routine. An oxygen saturation value is calculated by a microcomputer in a calculation algorithm routine using the ensemble averaged waveform as input, and is then displayed digitally on a screen.

If an ECG signal is not present, the absence is detected by the R wave detection algorithm routine which causes the ensemble averaging routine to be bypassed and the unenhanced optical pulse to be input into the calculation algorithm routine. The microcomputer executes the software comprising the R wave detection, ensemble averaging, calculation, and display algorithm routines.

The three lead ECG signal is amplified by a differential amplifier. This amplifier amplifies the differential component of the signal, which is the desired ECG waveform, while rejecting a large portion of the common-mode voltage. The output of this amplifier is AC-coupled by a capacitor to an amplifier which provides further gain. The gain provided by the amplifier is adjustable and can be set to 1/2 or 2 by the microprocessor. The amplifier can also accept an additional high level input which is intended to be connected to the output of an external ECG monitoring device, thus obviating the need for an additional set of ECG electrodes on the patient. The output of the amplifier is processed by a low-pass filter to remove the unwanted artifact such as 60 Hz and electrosurgery induced noise, and is converted to a serial, digital signal by an ADC. The digitized signal then passes through an optoisolator to a serial port which resides on the bus of the microcomputer. The optoisolator serves to isolate the patient ECG leads from the external power supply and is incorporated for reasons of patient safety.

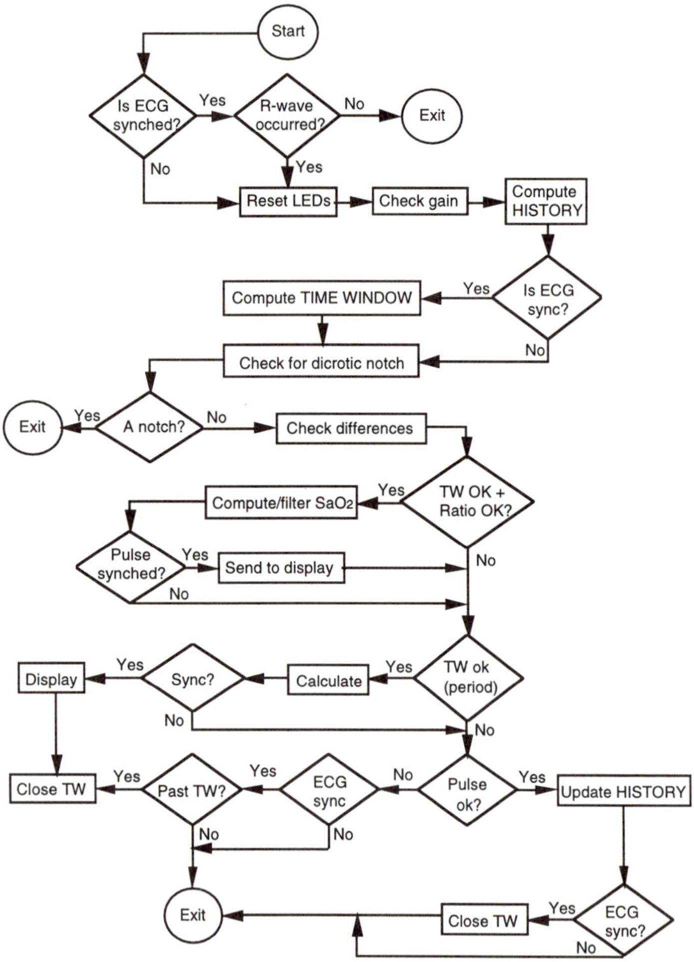

Figure 9.11. The LEVEL 3 software checks for ECG synchronization and processes the data appropriately to calculate the oxygen saturation (Goodman and Corenman 1990).

The oximeter is software driven and the operation of the software involves the process of removing motion artifact and enhancing waveform quality in low perfusion situations.

ECG synchronization is used to provide a reliable time frame upon which to base ensemble averaging, and a robust and accurate R wave detection algorithm is an integral part of the system. The R wave detection process involves three stages of processing: a low-pass digital filter, a peak excursion finding algorithm and a peak discrimination algorithm. The ECG input signal from the ADC is sampled at a rate of 240 Hz. The resulting digital waveform is low-pass filtered, with a corner frequency of 12 Hz, to remove artifact such as 60 Hz and muscle noise.

9.6.2.1 Peak excursion finding algorithm. The filtered ECG waveform then undergoes transformation by the *peak excursion finding algorithm*. The purpose of this transformation is to amplify those characteristics of the ECG waveform which are inherent in QRS complexes while inhibiting those which are not (Conlon *et al* 1990). This algorithm continually matches the ECG waveform to one of the two templates as shown in figure 9.12. The algorithm maintains a queue buffer of length N, which is searched in order to determine the parameters P1, P2, P3, and P4. The algorithm routine is called for $N = 8$, 12, 16, 20, and 24, and the individual excursion values are summed so as to give a total transformation value. More weight is placed on lower values of N in order to emphasize narrower spikes over wider ones. The newest sample is added to the buffer at each instant and the oldest sample is removed from the buffer. The maximum and the minimum values and their positions are searched in the buffer and depending on their relative positions, the matched template is chosen. The parameters P2 and P3 are assigned the appropriate maximum and minimum values accordingly. The parameters P1 and P4 are then found based on the template. For example, if the buffer matches template (a), the maximum value after P2 is assigned to P1, and the minimum before P3 is assigned to P4. Finally, the peak closed excursion on the interval N is computed as (P3 – P2 – (P4 – P1)) if the buffer matches template (a) or (P2 – P3 – (P1 – P4)) if the buffer matches template (b).

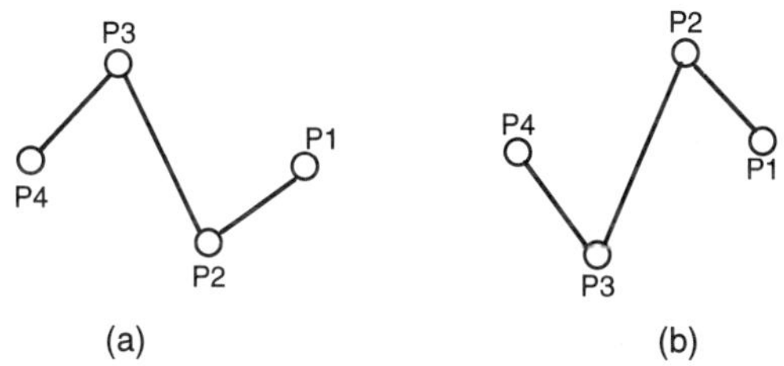

Figure 9.12. Two ECG waveform templates utilized in R-wave detection.

9.6.2.2 Peak discrimination algorithm. After transformation of the ECG waveform, the peak discrimination algorithm classifies the spikes found in the transformed waveform as either QRS complexes or artifact. The peak discrimination algorithm is a state machine with three states: peak, valley, and noise peak. The thresholds are set based upon the past history of the ECG waveform.

The algorithm enters the peak state if the algorithm is in the valley or noise state and exceeds a set threshold (threshold 2). The algorithm exits the peak state and enters the valley state when the waveform drops below one fourth of the maximum value attained in the peak state. The algorithm in valley state enters the noise state whenever the waveform climbs above four times the minimum value attained during the valley state. The algorithm in noise state enters the valley state when the waveform drops half the distance between the maximum value attained during the noise state and the minimum during the previous valley state. The detection of QRS spike is signaled upon the transition into the peak state. The conditions for state changes are summarized in the table 9.1. The algorithm maintains an average of the last eight QRS peaks in order to set the threshold for detecting the next peak in the waveform. An average of the noise peak levels found between the last four QRS peaks, is also maintained to aid the rejection of artifact while accepting valid QRS spikes. The averages are updated whenever there is a transition between the peak and valley states.

Table 9.1 A summary of conditions for state changes.

Present state	Condition	Next state
Valley or Noise states	Exceeds a set threshold	Peak state
Peak state	< 1/4 max in peak state	Valley state
Valley state	> 4* min in valley state	Noise state
Noise state	< 1/2 (max in noise – min in previous valley)	Valley state

Additional rejection of artifact is gained by examining the length of time which has elapsed between a new peak and the last accepted peak (interval figure 9.15). If it is less than 5/8 of the previous R–R interval, the spike is assumed to be noise and is not counted as a QRS spike. If it is greater than 7/8 of the previous R–R interval, it is accepted unconditionally. If it is greater than 5/8, but less than 7/8 of the previous R–R interval, the spike is accepted on probation as long as it exceeds a second threshold (threshold 1) which is set based on the noise peaks encountered during the last four beats. It is counted as a valid QRS spike but the previous state information is also saved in order to undo acceptance of the spike if a better candidate is found. The probation interval is equal to 9/8 of the previous R–R interval minus the length of time which has elapsed since the last accepted peak. During this interval any spike which meets the threshold requirements overrides the acceptance of the spike in question.

When the algorithm is found to be in the peak state, the maximum value encountered in this state is noted. If the waveform is not a local maximum, the routine checks to see if the waveform has fallen to one fourth of the last local maximum. If it has, the routine determines whether the current peak is a noise peak or a QRS peak. If it was a noise peak, the average of the noise levels over the last four beats is calculated. If it was a QRS peak, the average of the last eight QRS peaks is updated using the local maximum. The threshold values needed to detect the next QRS peak are then determined. Threshold 1 is halfway between the current eight-beat peak average and the current four-beat noise average. Threshold 2 is one-half of the current eight beat peak average (figure 9.13).

Before exiting the peak discrimination algorithm, parameters reflecting the quality of the ECG waveform are tested. If the time elapsed since the spike was accepted exceeds four times the R–R interval and/or the baseline of the transformed signal exceeds one-half the peak value, the ECG waveform is assumed to be lost and the routine disengages the synchronization.

9.6.2.3 Ensemble averaging algorithm. The ensemble averaging algorithm makes use of the output of the R–wave peak discrimination algorithm to enhance that part of the red and infrared plethysmographic waveforms which are correlated with the ECG, while diminishing all which is unrelated, to yield a signal with an improved signal to noise ratio (Conlon *et al* 1990).

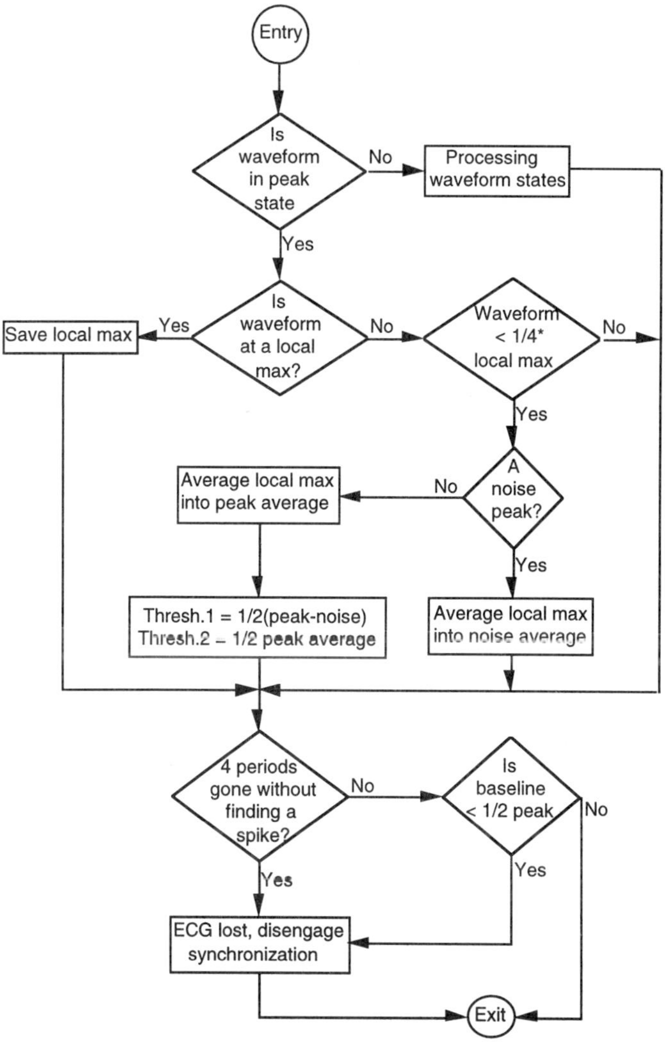

Figure 9.13. The R wave peak discrimination algorithm (Conlon *et al* 1990).

The algorithm relies on the assumption that instances of moderate to severe motion, and of low perfusion, can be detected as the plethysmographic waveforms are being sampled (figure 9.14). To do this, it was found to be advantageous to buffer these waveforms while they are being sampled, and to delay the actual averaging until the R peak is detected. The averaging weight of the current waveform cycle can then be adjusted, depending on whether the plethysmographic waveform just acquired is weak or exhibits the influence of excessive motion artifact. An additional benefit of this buffering stage is that the oximeter is able to discard waveform pulses during which optical pulse processing circuitry has saturated and distorted the waveform. Yet another benefit of this buffering stage is that it allows a level of error tolerance in the R wave detection process whereby the peak_discrimination routine can accept certain marginal QRS spikes on probation while maintaining the flexibility to correct the error if a better candidate is subsequently detected (figure 9.15).

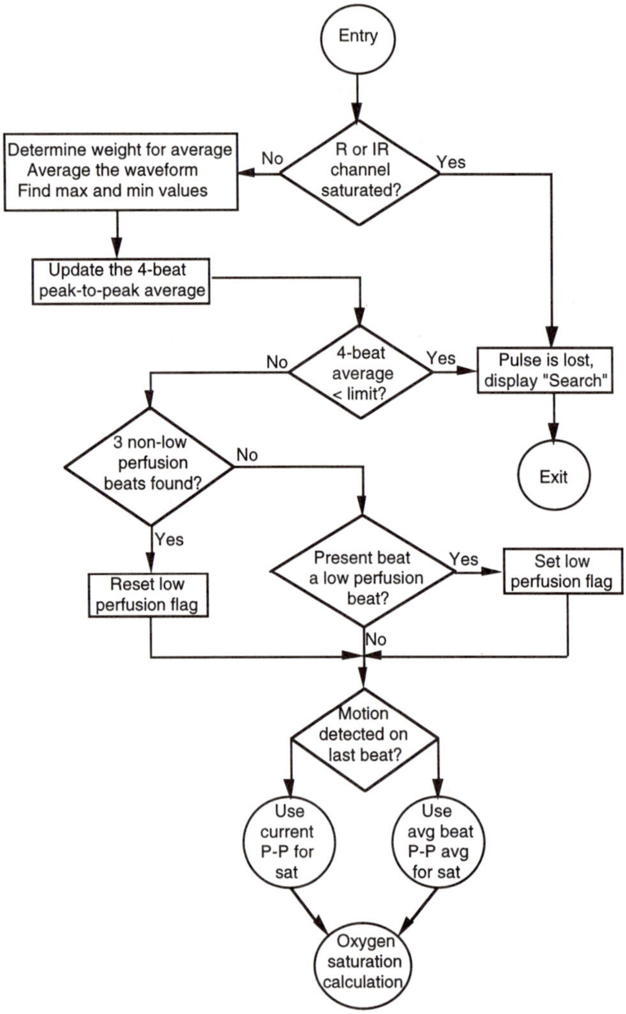

Figure 9.14. The ensemble averaging algorithm (Conlon *et al* 1990).

9.6.2.4 Motion determination algorithm. In order to give less weight to waveform pulses which are distorted by motion artifact, a criterion by which motion can be measured is established. This routine assumes that a plethysmograph unaffected by motion varies only slightly between one pulse and the next. In addition, a change in the amplitude, not shape, of the pulse comprises the majority of the observed difference between one pulse and the next. A plethysmograph containing artifact, however, differs greatly from the previous signal. A point-by-point subtraction of the latest pulse from the one preceding it yields a signal with an average amplitude less than that of the signal. The value of the difference signal is more or less constant while the signal itself changes rapidly. The integration of the difference signal yields a good indication of the amount of motion present in the pulse. A noisy signal yields a large value on integration compared to a clean signal. This routine checks for the occurrence of an R-wave spike which would be detected by the R-wave detection algorithm. If a spike was detected, the routine saves the integrated value as an indication of the level of motion present in the pulse, and initializes the variables to prepare for the next pulse (Conlon *et al* 1990).

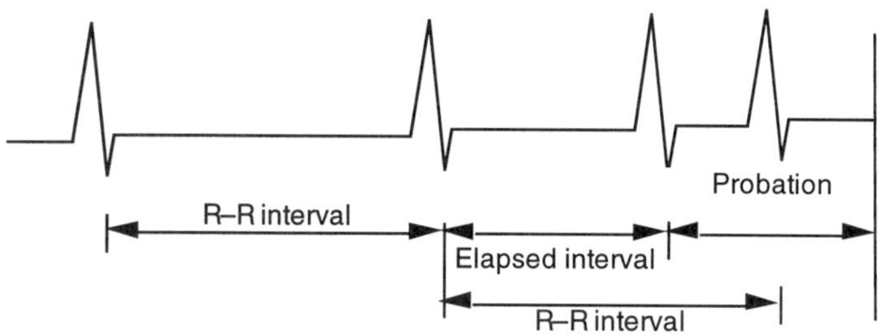

Figure 9.15. R wave artifact rejection timing subroutine (Conlon *et al* 1990).

9.6.2.5 Pulsatile waveform weight determination algorithm. It is generally known that ensemble averaging with a set of N waveforms increases signal-to-noise ratio by a factor of the $\sqrt{N}$ for uncorrelated, random noise. Thus, ensemble averaging will decrease the influence of the uncorrelated motion artifact and will enhance a low perfusion signal (which may be buried in noise) at the expense of response time. At the same time, a maximum limit on response time is set in order to ensure that the displayed saturation value is reasonably current (Conlon et al 1990).

The variable weight average is used in order to provide flexibility over a broad spectrum of pulsatile waveforms. It attempts to give a large weight to waves which are largely motion-free, while diminishing the weight given to those which have motion. Additionally, if a low perfusion situation is detected, less weight is given to all pulses until several strong pulses are found. Furthermore, the algorithm takes into account the pulse rate when determining the averaging weight. Since the averaging occurs each time a beat is detected, more averaging can be used on a patient with a fast pulse rate than one with a slow pulse rate while maintaining a constant response time. More averaging is needed in cases of motion artifact and low perfusion because the signal-to-noise ratio of these pulses is less than normal pulses (figure 9.16).

The weight determination algorithm uses two empirically determined thresholds to determine whether the motion is significant. One of these thresholds applies during the normal perfusion, while the other is used in cases of low perfusion. The algorithm decides which of the two thresholds to use by checking for the low perfusion state. If the low perfusion has not been detected, the algorithm checks the motion against the high motion threshold. If significant motion is not found, the algorithm checks whether the heart rate is above 120 bpm. If it is, the beat is assigned an average weight of 1/8, otherwise an average weight of 1/4. If significant motion is found, the algorithm checks for the heart rate and if it is above 120 bpm, assigns an average weight of 1/16. Further, if the heart rate is below 60 bpm, the algorithm assigns an average weight of 1/8.

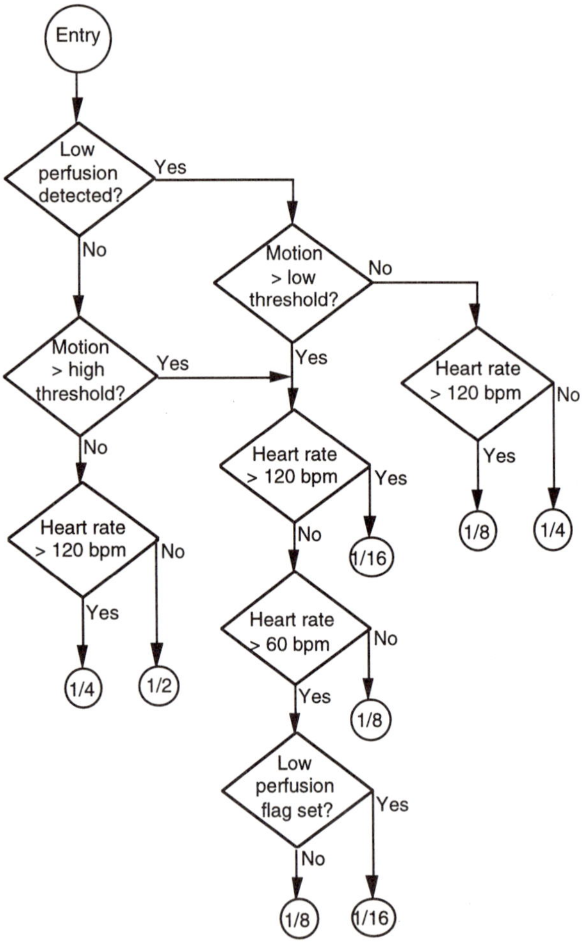

Figure 9.16. The weight determination algorithm (Conlon *et al* 1990).

If the heart rate is above 60 bpm and less than 120 bpm, the software has to differentiate between low perfusion with motion and motion alone. The algorithm checks the low perfusion flag and if set, assigns an average weight of 1/16, otherwise it assigns a weight of 1/8. The ensemble_averaging routine employs the weight determination algorithm to find the average weight of the waveform and averages the buffered waveform with the composite averaged waveform stored in the microcomputer memory using a tail-weight average of the form (W × NEW) + (1 – W) × COMPOSITE, with W being the averaging weight. Because the averaged pulses are of varying duration, some pulses will overlay more points of the averaged waveform than others. Thus, the tail of the averaged waveform may not accurately reflect the most recent plethysmographic information. Hence the minimum and maximum of the averaged waveform were found only up to the minimum length of the last eight pulses. After determining the minimum and maximum values, the four beat average of peak-to-peak values are updated.

The algorithm then checks to ensure that the average has not fallen below the minimum low perfusion threshold. If it has, the pulse is considered lost. Then the algorithm checks if three non-low perfusion beats have been found. If so, it resets the low perfusion flag. If not, it checks if the current beat is a low perfusion beat, setting the perfusion flag appropriately. The algorithm then checks for motion in the last beat. If there is motion, it sends the four-beat, peak-to-peak average to the saturation_calculation algorithm routine. Otherwise, the last peak-to-peak value of the routine is sent to the saturation_calculation algorithm which calculates the oxygen saturation and displays it.

9.7 SPECTRAL METHODS OF ESTIMATING S_pO_2

Arterial oxyhemoglobin saturation (S_pO_2) values are currently computed using weighted moving average (WMA) techniques (Rusch *et al* 1994). These methods process the time domain signals and give a precision of no better than ± 2% (± one standard deviation). Researchers have explored other digital signal processing algorithms for improved estimation of S_pO_2. The fast Fourier transform (FFT) and discrete cosine transform (DCT) were identified as potentially superior algorithms (Rusch *et al* 1994) and useful to optimize the portability of pulse oximetry systems. Preliminary studies indicate that a 64-point FFT, with a 15 Hz sample rate, over a data collection period of 4.3 s was found to be the optimal combination for pulse oximetry applications, minimizing hardware expense, footprint, and power consumption. S_pO_2 values were calculated from a transform size of 64 points using

$$S_pO_2 = 110 - 25 \times R \tag{9.36}$$

where R is the ratio of the red and infrared normalized transmitted light intensity. The R value is

$$R = \frac{\mathrm{AC_R/DC_R}}{\mathrm{AC_{IR}/DC_{IR}}}. \tag{9.37}$$

The AC component is the signal variation at the cardiac frequency and the DC component is the average overall transmitted light intensity. The AC component is selected as the highest spectral line in the cardiac frequency band.

REFERENCES

Cheung P W, Gauglitz K, Mason L R, Prosser S J, Smith R E, Wagner D O and Hunsaker S W 1989 Feedback-controlled method and apparatus for processing signals used in oximetry *US patent 4,819,646*

Cheung P W, Gauglitz K, Mason L R, Prosser S J, Smith R E, Wagner D O and Hunsaker S W 1990 Method and apparatus for offsetting baseline portion of oximeter signal *US patent 4,892,101*

Conlon B, Devine J A and Dittmar J A 1990 ECG synchronized pulse oximeter *US patent 4,960,126*

Corenman J E, Stone R T, Boross A, Briggs D A and Goodman D E 1990 Method and apparatus for detecting optical signals *US patent 4,934,372*

Frick G, McCarthy R and Pawlowski M 1989 Waveform filter pulse detector and method for modulated signal *US patent 4,867,571*

Goodman D E and Corenman J E 1990 Method and apparatus for detecting optical signals *US patent 4,928,692*

Jaeb J P and Branstetter R L 1992 Composite signal implementation for acquiring oximetry signals *US patent 5,094,239*

Pologe J A 1987 Pulse oximetry: technical aspects of machine-design *Int. Anesthesiol. Clinics* **25** 137–53

Potratz R S 1994 Condensed oximeter system with noise reduction software *US patent 5,351,685*

Scharf J E and Rusch T L 1993 Optimization of portable pulse oximetry through fourier analysis *Proc. IEEE Twelfth Southern Biomedical Engineering Conf.* Tulane University pp 233–5

Smith R E 1989 Method and apparatus for processing signals used in oximetry *US patent 4,800,495*

Stone R T and Briggs D A 1992 Method and apparatus for calculating arterial oxygen saturation based plethysmographs including transients *US patent 5,078,136*

Yorkey T J 1996 Two 'rat rat' derivation *Personal communication* (Hayward, CA: Nellcor Inc)

INSTRUCTIONAL OBJECTIVES

9.1. Name the general sources of error that could be corrected with signal processing algorithms.
9.2. Explain the process of eliminating incident light intensity and thickness of the path as variables from Beer–Lambert law.
9.3. How is R_{OS} (Ratio of Ratios) estimated from the red and infrared optical signals?
9.4. Discuss the advantages of estimating R_{OS} using the derivative method over the peak and valley method. Explain how noise reduction is achieved using the derivative method.
9.5. Discuss the role of the construction–reconstruction process in improving the accuracy of S_aO_2 estimation.
9.6. Explain the function of the start-up interrupts.
9.7. Discuss the function of the five different states in the period zero subroutine.
9.8. Discuss the C-Lock ECG synchronization algorithm used in Nellcor®.
9.9. Explain the motion detection algorithm used in Criticare.
9.10. Name the advantages of using spectral methods in estimating oxygen saturation.
9.11. Explain the advantages of using ECG synchronization.

CHAPTER 10

CALIBRATION

Jeffrey S Schowalter

The calibration curves of *R* (Ratio of Ratios) values used to calculate oxygen saturation levels are critical to the accuracy of the entire pulse oximeter system. Without an accurate table of appropriate *R* values, the pulse oximeter has no way of determining oxygen saturation levels. As such, it is important to understand how the pulse oximeter calibration curve data are acquired. In addition, it is important to understand some of the past and present simulation techniques used to test the accuracy and functionality of pulse oximeters.

10.1 CALIBRATION METHODS

Chapter 4 states that Beer's law does not apply for a pulse oximetry system due to the scattering effects of blood. Therefore, pulse oximeter manufacturers are currently forced to use an empirical method of determining the percentage of arterial oxygen saturation for a given *R* ratio.

10.1.1 Traditional in vivo calibration

The traditional method of pulse oximeter calibration involves comparison of oximeter *R* value to the oxygen saturation ratio obtained from *in vivo* samples using human test subjects. In fact, this was the only method used to calibrate these devices up until 1993 (Moyle 1994). Although this method requires a variety of laboratory instrumentation and is typically done in a hospital setting, this data collection process is only required during the design and development of the device.

10.1.1.1 Procedure. In general, the calibration procedure is fairly straightforward. Test subjects are fitted with an indwelling arterial cannula, which is placed in the radial artery. A sample of blood is taken and analyzed with a CO-oximeter (see chapter 3) to determine the subject's levels of COHb and MetHb. In most cases, samples are taken over a broad population. Typically, data come from nonsmokers with background carboxyhemoglobin levels between 1% and 2%. Wukitsch *et al* (1988) mentions that subjects used for the Ohmeda Biox 3700 calibration had an average COHb level of 1.6% and a MetHb level of 0.4%.

Once a low level of COHb and MetHb are verified, the subject is also fitted with one or more pulse oximeter probes. The test begins by first ensuring that the subject is at the 100% oxygen saturation level. The subject breathes an oxygen/air mix so as to bring the arterial oxygen saturation level to 100% (as determined from arterial blood analyzed with the CO-oximeter). Oxygen saturation level is incrementally decreased by breathing gas mixtures of progressively less oxygen and more nitrogen. At each level where the pulse oximeter indicates a stable reading, an arterial blood sample is immediately taken and analyzed with the CO-oximeter. Corresponding readings are recorded and the data are then plotted with oxygen saturation percentage (as determined by the CO-oximeter) on the *y*-axis and *R* ratio (as determined by the pulse oximeter under test) on the *x*-axis yielding a traditional *R* curve as shown in figure 4.7. Typical values for the *R* ratio vary from 0.4 to 3.4 (Pologe 1989). A best fit calibration equation is then calculated from the data. If the pulse oximeter manufacturer has selected LEDs for their probes that have relatively narrow bands of center wavelength (as discussed in chapter 5), then only one curve is required. If they have a number of probes with differing red and infrared center wavelengths, then each probe with unique LED combination must be tested to obtain its unique curve characteristics. Some manufacturers have as many as 30 different probes.

10.1.1.2 Problems. One of the problems with this traditional method is the limited range of oxygen saturation that can be acquired. Ethical issues prevent intentional desaturation of healthy subjects below a certain point due to risk of hypoxic brain damage. As a result, saturation levels can only be reduced to around 60%. This leaves a large range of values on the curve that need to be calculated by extrapolation. This has the potential to induce errors and in fact, Severinghaus *et al* (1989) tested 14 pulse oximeter models and showed that most pulse oximeters performed poorly under relatively low levels of saturation (see chapter 11). Another problem of this calibration method is that it does not address the spacing and number of data points needed to build a curve. Moyle (1994) states that well spaced data points over the entire range from 100% down to 80% is more accurate than having many data points clustered between 95% and 100%.

There has been a great deal of debate over the years as to what the pulse oximeter is actually measuring and as such, a unique term has been created to specify an oxygen saturation reading as determined by a pulse oximeter. The problem is that the pulse oximeter uses two wavelengths to measure oxygen saturation. However, there are four common species of hemoglobin (Hb, HbO_2, COHb, and MetHb). Since there are routinely four light absorbing substances in a sample in a system which is assuming it is measuring only two substances, much discussion and misconception arise as to what the pulse oximeter is actually measuring (Pologe 1989). Equation (4.5) shows that functional S_aO_2 is the ratio of oxygenated hemoglobin to the sum of oxygenated and reduced hemoglobin. If a person were found that had no COHb or MetHb, this is what the pulse oximeter would measure. However, since some COHb and MetHb are typically present in everyone's blood, and these terms show up in the fractional S_aO_2 formula, it is easy to assume that the pulse oximeter is measuring fractional S_aO_2. However, this is not the case either. Moyle (1994) state that the conventional two-wavelength oximeter measures what should be defined as 'oxygen saturation as measured by a pulse oximeter', or S_pO_2.

Payne and Severinghaus (1986, p 47) state that the pulse oximeter reports

$$\frac{HbO_2 + COHb + MetHb}{HbO_2 + COHb + MetHb + Hb} \times 100\% \qquad (10.1)$$

and subtracting this quantity from 100% gives the percentage of reduced hemoglobin or Hb%. He suggests that to eliminate confusion pulse oximeters should display this value instead. If this were done, however, conventional thinking would have to change because readings would increase from zero as opposed to S_pO_2 readings which decrease currently from 100. The bottom line is that COHb and MetHb do have an effect on the accuracy of pulse oximeter readings (Reynolds *et al* 1993a,b) so they cannot be ignored as part of the calibration process.

10.1.1.3 Effects of COHb and MetHb. The effects of COHb and MetHb are typically handled in one of two ways. Some manufacturers subtract 2% for these factors so they are displaying fractional saturation (assuming a patient with nominal levels of COHb and MetHb) and others do not subtract this factor so they are displaying functional saturation (Ackerman and Weith 1995).

In a sense, the pulse oximeter will measure what it has been calibrated to measure based on the test subject profile. Tremper (Payne and Severinghaus 1986) states that Nellcor calibration data were originally based on five Olympic athletes in virtually perfect physical condition. These individuals probably had as low levels of COHb and MetHb as are found in humans. As such, anyone being tested with higher (normal) levels of COHb and MetHb yielded inaccurate readings. Today, by using a more representative subject to build the calibration curve, pulse oximeter manufacturers account for some of this during the calibration process. However, individuals with relatively high levels of COHb and MetHb will have inaccurate S_pO_2 readings.

10.1.1.4 Field calibration. Another issue of concern is field calibration. Using this technique, once the *R* curves are established, the transmitting wavelengths of the LEDs and corresponding *R* curve are provided via a coding resistor (see chapter 5) and as such only a two-point check to verify the correctly selected calibration curve is required. Typically this check will identify a problem due to a malfunctioning LED or photodiode or an incorrect coding resistor. However, other than this cursory check there is no type of field calibration done on the pulse oximeter. Cheung *et al* (1993) have proposed a system for compensating for the effects of temperature variations on the LEDs. Since the pulse oximeter photodiode cannot detect a shift in LED wavelength, the proposed system provides the capability for the temperature of the probe LEDs to be measured and thus an alternative calibration curve, as shown in figure 10.1, can be used for the new set of LED wavelengths. This system seems to be of limited usefulness however, since Reynolds *et al* (1991) have shown that the peak wavelength of a red LED will only shift by 5.5 nm and an infrared LED will shift by 7.8 nm with a temperature shift from 0 °C to 50 °C. Applying this information to a theoretical computer model based on Beer's law, causes negligible changes in accuracy of the pulse oximeter.

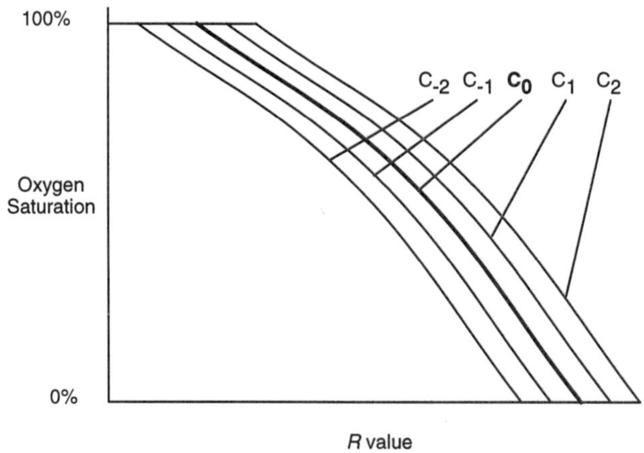

Figure 10.1 Temperature compensation curves as proposed by Cheung *et al* (1993). C values indicate different curves for use with different sensed temperatures.

10.1.2 In vitro calibration using blood

Figure 10.2 shows an *in vitro* test system that requires whole blood (Reynolds *et al* 1992). Blood is pumped through a cuvette acting as a model finger. The pulse oximeter probe is then attached to the model finger. Blood is pulsed within the system using a computer controlled peristaltic pump head capable of generating almost any shape of pulsatile waveform. Blood is oxygenated by passing through a membrane oxygenator using a gas mixture of O_2, N_2, and CO_2. The composition of the gas mixture passing through the membrane oxygenator is controlled with a gas mixing pump. A variety of model fingers were tried with the final model finger consisting of a cuvette made of two thin (0.5 mm) silicone rubber membranes and a rigid Plexiglas central section. When using whole blood, the model finger is covered with a diffuser made from translucent paper. Blood enters one end of the cuvette and flows in a thin (1 mm) layer through the cuvette over the fingertip end and back along the bottom side. Both inlet and outlet are tapered to prevent flow separation. The silicone rubber membrane is flexible enough such that pulsating blood produced volume changes in the tubes giving an AC/DC ratio in the physiological range. Readings from the pulse oximeter are recorded and a simultaneous sample taken from the sample port and analyzed by the CO-oximeter in a similar fashion to the procedure described in section 10.1.1. This system yields calibration values that are accurate to 50% and lower. Most pulse oximeters have no specified accuracy below 50%. One problem is that the system is sensitive to blood flow rate, due to changes in blood cell orientation with flow. This was verified using a hemoglobin solution instead of whole blood in the test system.

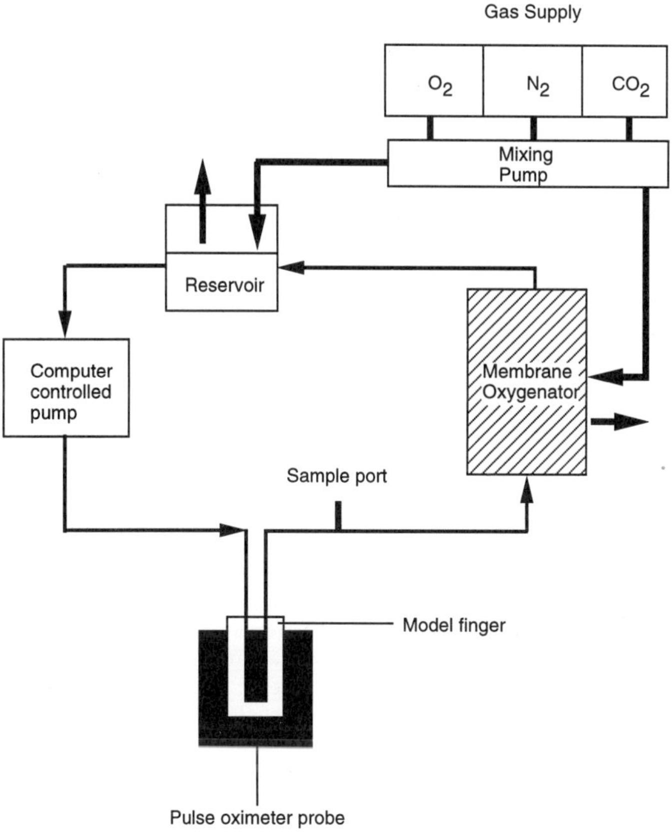

Figure 10.2 Block diagram of *in vitro* test system developed by Reynolds *et al* (1992).

10.2 TESTING SIMULATORS

Devices which check the functionality of pulse oximeters are becoming increasingly popular. Many of these devices use some type of artificial finger to verify that the pulse oximeter is functioning correctly. When pulse oximeters first came out, the only way technicians had to verify the functionality of the pulse oximeter was to use their own fingers. This, however, only indicates basic functionality at best with no way to control any parameters and with nothing with which to compare. Several devices have been developed that simulate the optical properties of the human finger and its pulsatile blood flow. In addition, optoelectronic systems, which simulate the human finger electronically, have also been developed. Finally, pulse oximeter manufacturers themselves have developed simple simulators that essentially simulate a probe's electron signals.

10.2.1 Simulators using blood

Several simulators have been proposed that need whole blood to test the functionality of the pulse oximeter. These simulators are all based on the concept of being able to simulate the absorbance of human tissue (normally the finger) between the LEDs and the photodiode of the pulse oximeter under test. Since few substances have been found that simulate the optical properties of blood, these types of systems typically provide the most accurate simulation.

10.2.1.1 Reynolds system. The system described in section 10.1.2 functions equally well as a simulator to test the functionality of pulse oximeters. In fact this system has been used to compare ten commercially available oximeters (Reynolds *et al* 1992), and has been used to evaluate the effects of dyshemoglobins on pulse oximeter accuracy (Reynolds *et al* 1993a,b). However, this *in vitro* test system is not practical in a hospital setting where most pulse oximeters are used. The system requires a laboratory setting, is not portable, uses oxygenated whole blood and needs a CO-oximeter for comparison. However, this instrument is generally considered the *gold standard* for calibrating and testing a pulse oximeter over its complete range.

10.2.1.2 Vegfors system. The Vegfors system is similar to the Reynolds system but with a focus on the artificial finger or 'finger phantom' used. Vegfors *et al* (1993) describe a system where their artificial fingers consist of silicone rubber tubes inserted in plastic Delrin cubes. The tubing system chosen was based on its characteristics of tubing diameter, wall elasticity, and blood flow velocity to simulate normal physiological characteristics of blood in motion. Delrin was used because it has similar optical scattering properties to human tissue. Figure 10.3 shows three models. Two different finger models, one with one tube and another with five tubes were tested along with a third artificial finger consisting of 15 silicone rubber tubes mounted in silicone rubber in the form of glue. The object was to develop an optical model which simulated the arterial bed of the human finger containing blood vessels and surrounding tissue. The results of these different finger configurations determined that physical dimensions of the artificial bed are of minor significance for pulse oximeter readings.

10.2.1.3 Single wedge system. Several other less complicated simulators using whole blood have been proposed. In one system, proposed by Yount (1989), a light-absorbing wedge shaped vessel containing blood of known oxygen saturation level is placed in the pulse oximeter's optical path. If the wedge (figure 10.4) is moved repetitively back and forth perpendicular to this optical path, either manually or with the aid of a mechanical device, both the pulse rate and shape of the pulse can be altered. Pulse rate can be simulated by changing the frequency at which the wedge is moved across the optical path. The shape of the pulse can be changed by altering the speed at which the wedge is moving.

10.2.1.4 Dual wedge system. In another arrangement of the system, two wedges are used. One is filled with 100% oxygen saturated blood and the other with completely unsaturated blood. The wedges are placed as shown in figure 10.5 and by varying the position along the optical path of this arrangement, virtually any saturation level can be obtained. Note however that with this second arrangement, an additional external device is needed to obtain a pulsatile variation in the simulator. figure 10.6 shows the polarization filter system proposed by Yount (1989) to achieve this pulsatile variation needed. A pair of polarizing disks simulate the changes in transmittance expected by the pulse oximeter. A stepper motor controls the motion of one disk. This changes the angle of polarization between the two disks, and therefore the amount of light transmitted. By varying the rate of angle change, this system can simulate both the shape and pulse rate seen by the pulse oximeter. This particular system also has several glass windows. This allows for multiple samples to be loaded on the same disk so different oxygen saturation levels can be simulated by rotating the appropriate sample into the probe.

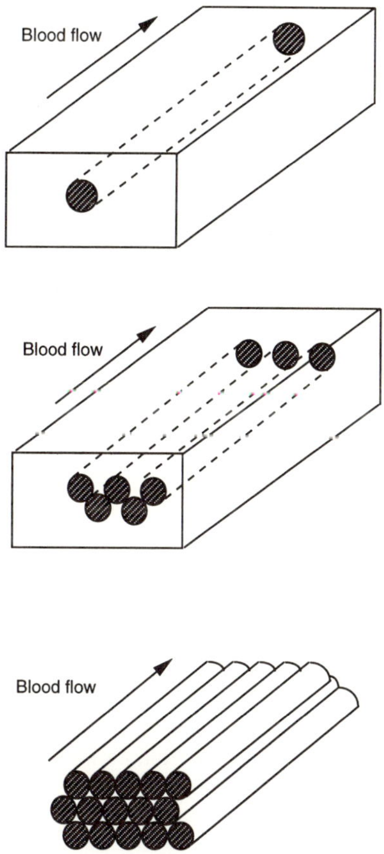

Figure 10.3 Block diagram of various artificial fingers as proposed by Vegfors *et al* (1993).

One limitation of these wedge systems is that if blood is used as the medium in the wedge, the samples either need to be prepared shortly before use or steps need to be taken to stabilize the blood.

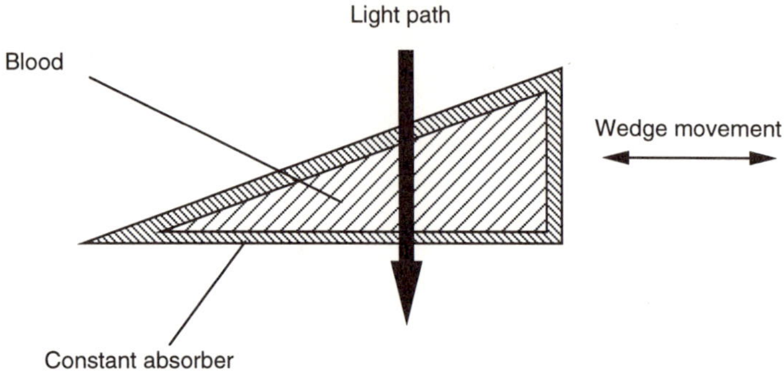

Figure 10.4 Block diagram of a wedge system as proposed by Yount (1989).

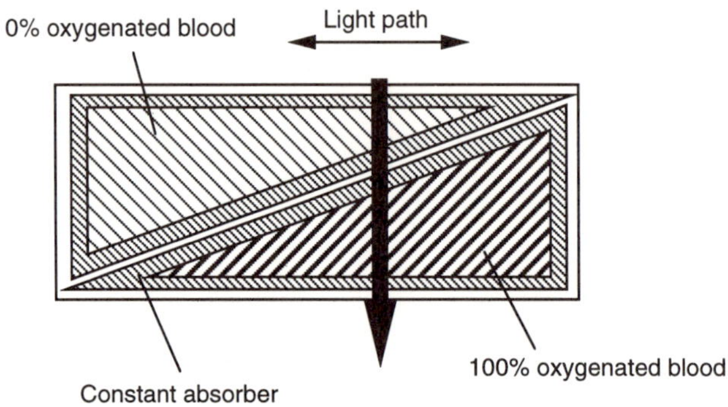

Figure 10.5 Block diagram of the dual wedge system as proposed by Yount (1989).

10.2.1.5 Bulb device. Volgyesi (1989) proposed a simple mechanical design to simulate a pulsing finger. Figure 10.7 shows the tube and bulb type device. It requires a 0.5 to 1 mL blood sample for each saturation level to be tested. A piece of silicone rubber tubing is placed inside a disposable plastic test tube which contains a blood specimen. The operator then manually squeezes the bulb at regular intervals which causes the silicone rubber tubing and the blood in the annular space between the silicone rubber tubing and the test tube to deform or *pulse*. Samples of heparinized blood are externally altered to different saturation levels so different levels of oxygen saturation can be tested. With a variety of oxygen saturation level samples prepared in individual test tubes, the pulse oximeter can be applied to the device. After the operator is able to rhythmically squeeze the bulb for a consistent plethysmograph (rate and amplitude), a reading is recorded from the pulse oximeter and the sample is sent to a CO-oximeter for a comparison reading. The main advantage of this system is its simple implementation. The disadvantage is that the pulsatile nature of the system is operator dependent and samples of known oxygen saturation levels of blood need to be prepared.

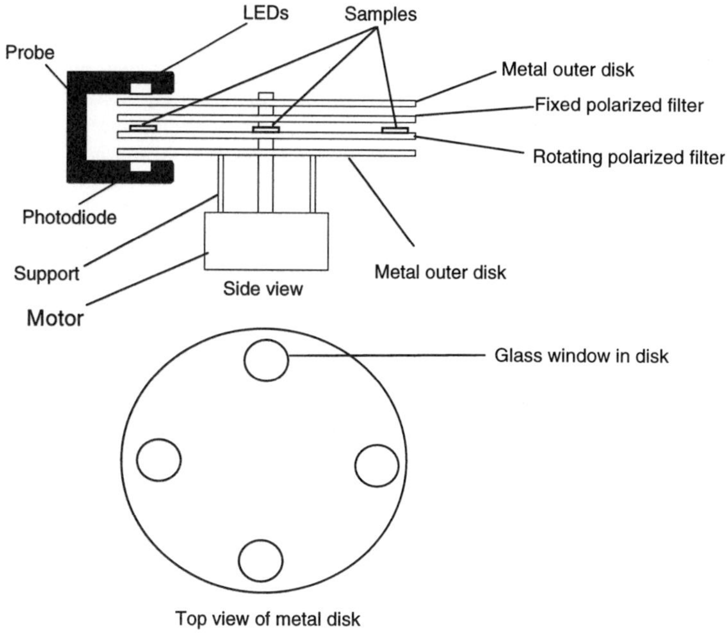

Figure 10.6 Schematic diagram of polarization system (adapted from Yount 1989).

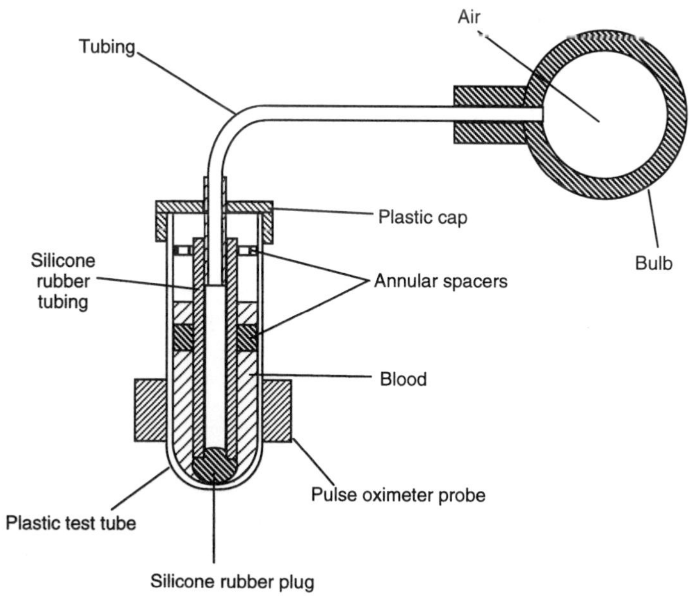

Figure 10.7 Block diagram of tube and bulb device.

10.2.2 Nonblood simulators

Nonblood simulators, like simulators that use blood, are also based on the concept of being able to simulate the absorbance of human tissue (normally the finger) between the LEDs and the photodiode of the pulse oximeter under test. These devices use colored materials to simulate blood. These simulators use a variety of mechanical and electrical devices to achieve the desired variations in absorbance. The more difficult aspect is simulating the scattering properties of whole blood. One of the most successful studies in this area (Marble *et al* 1994) used a combination of nondairy creamer mixed with solutions of red and green dye.

10.2.2.1 Bulb device. The bulb device described in section 10.2.1 above can also be used with liquids having differing optical absorbance properties corresponding to oxyhemoglobin. A commercial version of this device is currently being marketed by Nonin under the trade name *finger phantom.* This product (Nonin 1995) provides three translucent white *artificial fingers* that simulate arterial blood at nominally 80%, 90%, and 97% saturation levels. The operator gently presses the finger phantom about once every second to generate a pulse. The typical infrared percent modulation when squeezed is 0 to 5%.

10.2.2.2 Wedge device. The wedge device described in section 10.2.1 above can also be used with liquids other than blood having optical absorbance properties corresponding to those of the human finger.

10.2.2.3 Polyester resin device. Figure 10.8 shows a simple test object proposed by Munley *et al* (1989). This device consists of a piece of polyester resin that is formed in the shape of a finger. The resin is adapted to allow a core to be placed inside the artificial finger. At the end of the core, in the area exposed to the pulse oximeter LED's light path, a slotted piece of suitably colored Plexiglas is placed. As the device handle is rotated, the slot allows varying levels of LED light to reach the pulse oximeter photodiode. Speed of rotation of the crank will determine the *pulse rate* that the oximeter reads. Changing the color characteristics of Plexiglas will change the oxygen saturation reading that the pulse oximeter registers. This device was also shown to produce similar oxygen saturation readings among multiple devices of the same make and model of pulse oximeter.

10.2.2.4 Colored colloid simulator. Leuthner (1994) proposed the pulse oximeter development system shown in figure 10.9. A transparent bag is filled with a colored colloid solution. The color determines the extinction coefficients at the two wavelengths of interest. This system uses a water–gelatin mix which is heated and colored with red and black ink. To simulate different oxygen saturation levels, multiple bags with varying ratios of red and black dye need to be prepared. The bag is positioned between two acrylic disks. The disks and bag are then rotated by a stepper motor under microcontroller control. With this configuration, both the DC and AC absorbance ratio can be adjusted. Increasing the angle between the two plates increases change of absorbance over each rotation for an increase in relative AC signal. The simulated pulse shape is determined by speed of the disk rotation and the pulse rate is determined by the rotation frequency. A constant absorber material is placed on top of the disk to simulate the constant light absorbance of fingers of different people. In practice, it can vary by a factor of four. Generally a piece of white paper of varying thickness is used as the constant absorber. Two optic fibers are integrated into an artificial finger which then plugs into the finger probe of the pulse oximeter. The other ends of fibers are connected opposite each other near the rotating plates. If testing is done using different waveforms, the angular velocity of the rotation has to change and as such is controlled through the stepper motor via microcontroller control. The whole system is enclosed in a box to prevent disturbances from ambient light.

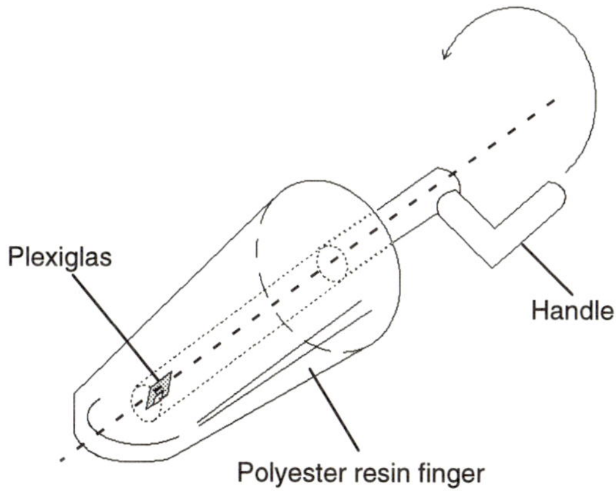

Figure 10.8 Polyester resin system proposed by Munley et al (1989).

The physical behavior of this system can be almost totally described using Beer's law, but the system cannot be used for finding the calibration table of a pulse oximeter. The main reason is that the scattering effect in whole blood is not present in this system. However this system can be used for a rough calibration table of a new instrument and to test an existing pulse oximeter for the response it gives when different colored bags are used.

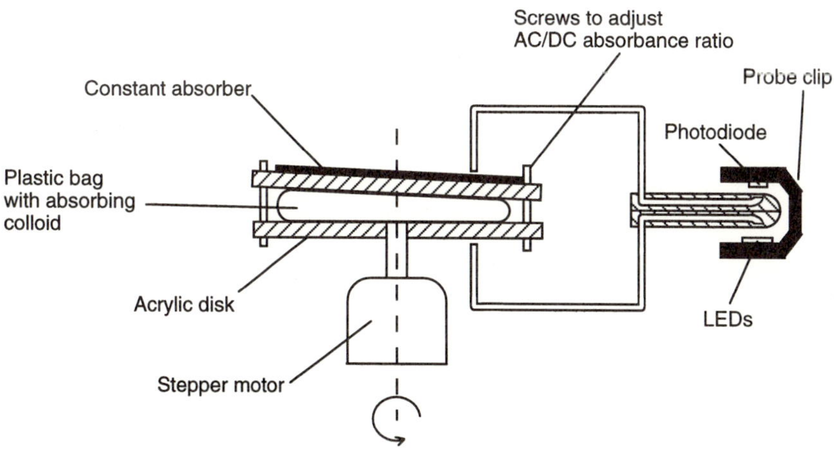

Figure 10.9 Leuthner's (1994) colored colloid disk system.

10.2.2.5 Liquid crystal retarder simulator. Zhou *et al* (1992) developed a device for generating test signals for pulse oximeters based on a voltage-controlled liquid-crystal light valve. In the first system, the pulse oximeter's LEDs are separated by an optical filter, modulated by a light valve, and recombined before detection by the probe's photodiode. The newer system does not require wavelength separation and its associated hardware as shown in figure 10.10. The transmittance characteristics are varied by taking advantage of the intrinsic wavelength dependence of a twisted-nematic liquid-crystal retarder (LCR). Polarizers are used to generate optical density variations that can be made to resemble blood perfused tissue. The intensity transmitted through the optical system can be adjusted by varying the voltage on the LCR. To simulate a pulsatile change in transmittance, the attenuation is initially made a constant DC value. A small AC voltage is then superimposed on top of the DC voltage to provide a pulsatile component. The transmittance at both the red and IR wavelengths varies depending on the voltage amplitude applied to the LCR. This allows the AC/DC ratios to be controlled by adjusting the amplitude of the voltage applied to the LCR. The polarizers are required because the angle of polarization strongly affects the range of variation of the red/IR ratio and its sensitivity to the applied voltage. Zhou *et al* are continuing work on this concept to provide the capability of simulating the shape of the plethysmographic waveform applied to the LCR.

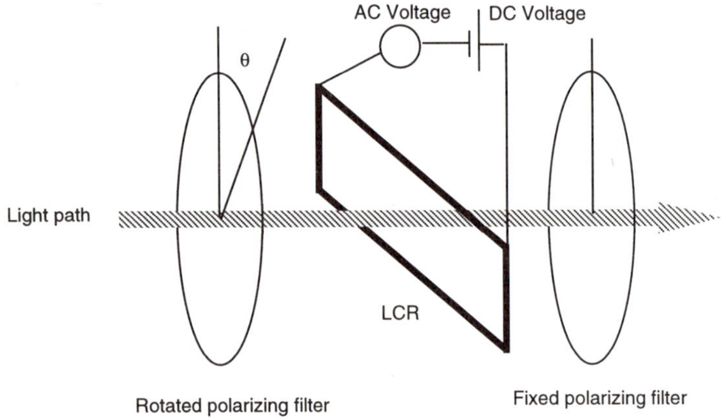

Figure 10.10 Diagram of the liquid crystal retarder (LCR) system proposed by Zhou *et al* (1992).

10.2.2.6 Aoyagi tissue model. A device based on the same general principles as the wedge system been proposed by Aoyagi *et al* (1994). Figure 10.11 shows that a static tissue model having absorption characteristics similar to a human finger is inserted into a pulse oximeter probe. A blood model having blood absorption characteristics similar to a specified oxygen saturation level is moved within the tissue model to simulate pulsatile motion and pulse rate. By altering the geometry of the blood model and/or the rate of motion of the blood model in and out of the tissue model, both the pulsatile waveform and pulse rate can be simulated.

10.2.2.7 Optoelectronic device. A number of relatively simple easy-to-use simulators have begun to appear on the market based on optoelectronic principles. Figure 10.12 shows a block diagram for one of these types of simulators. First, the user selects the parameter(s) to be simulated. The pulse oximeter probe is then attached to the device and a signal is received from the pulse oximeter probe's LEDs by the simulator. Pulse separator and timer circuitry convert the red and infrared light pulses from the pulse oximeter probe into electric signals. These signals are modulated with the appropriate level of AC/DC ratio (under computer control) and then converted back to light pulses, via the LED bar, to the probe's photodiode. Finally, the pulse oximeter responds to the converted light pulses as it would to light pulses modulated by living tissue.

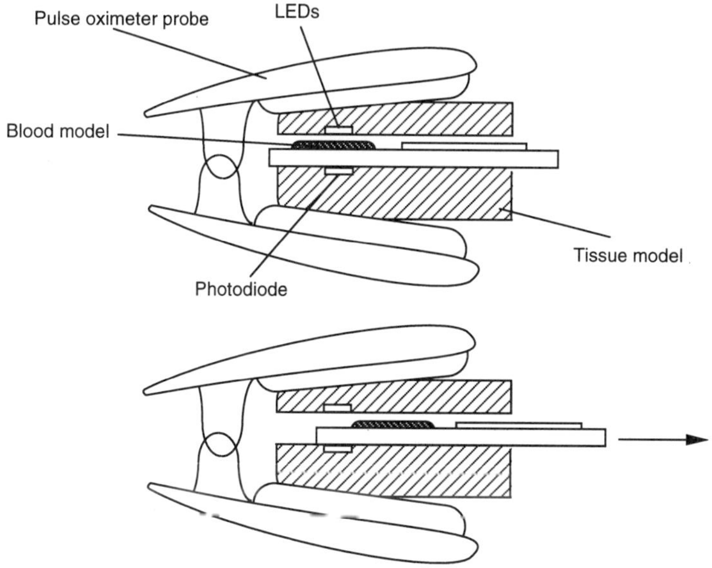

Figure 10.11 Block diagram of system as proposed by Aoyagi *et al* (1994).

These systems can test the probe and oximeter over the complete specified range of the oximeter. Also, simulation of a wide range of conditions is possible. The modulated signal can vary plethysmographic amplitude and wave shape to simulate a variety of ambient light conditions, motion artifacts, and arrhythmias. At least one system (Clinical Dynamics 1995) also includes a probe analyzer capability which independently tests LED and photodiode continuity and sensitivity. These types of simulators are primarily used by pulse oximeter manufacturers during final assembly and checkout of their products. In addition, their capability to generate automatic test sequences help document JCAHO (Joint Commission on Accreditation of Healthcare Organizations) testing requirements.

10.2.3 Electronic simulators

Electronic simulators have limited usefulness since they only simulate electronic signals to and from the probe. Usually these relatively simple devices are provided by the pulse oximeter manufacturer and only check a small number of values. These devices typically plug into the probe port on the pulse oximeter and use the drive current of the probe LEDs to generate a simulated photodiode signal back to the pulse oximeter using the device. Figure 10.13 shows an example of such a device. In remote mode, the LEDs just drive an amplifier and the output shows up on the sensor output. This is useful for simple continuity testing. In local mode, these devices are able to electronically simulate a discrete number of simulated oxygen saturation levels, pulse rates and plethysmographic waveform strengths. In addition the calibration resistor value reading capability of the pulse oximeter can be checked. These simulators are good for functional checks of the pulse oximeter's internal circuitry, but because they bypass the pulse oximeter's probe, are of limited usefulness.

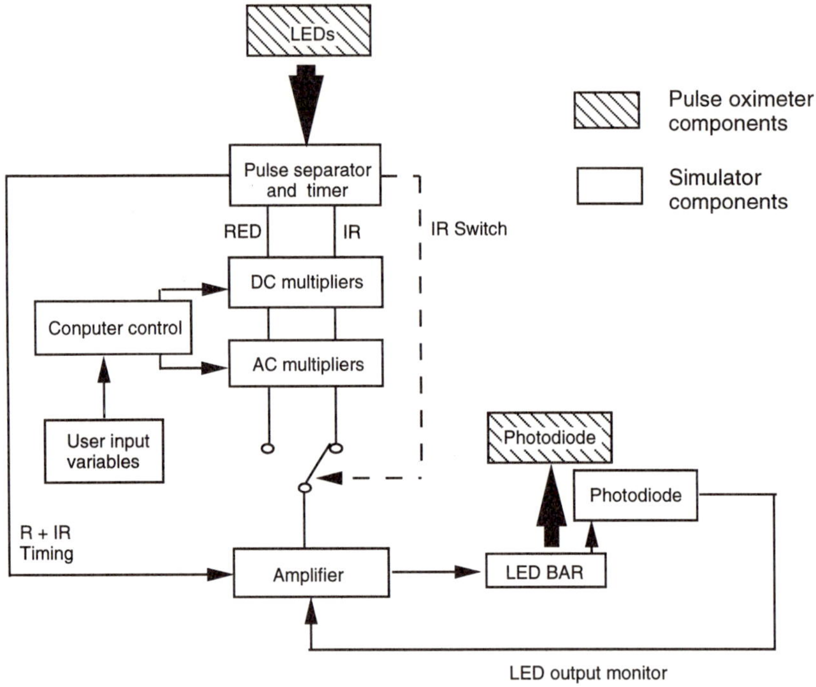

Figure 10.12 Block diagram electro-optic simulator system developed by Merrick and Haas (1994).

10.3 STANDARDS

Although the pulse oximeter has been on the market since 1977 (Santamaria and Williams 1994), surprisingly little standardization has been documented to this point. Statements like 'machines and probes are interchangeable with less than 0.5% difference', 'warm-up time factor of 0.5% to 1.0%' and 'the low perfusion light on the Ohmeda oximeter indicates the oximeter's microprocessor has low confidence level in the data' can be found in the literature. Several standards do exist, but their value from the designer's point of view is limited at best.

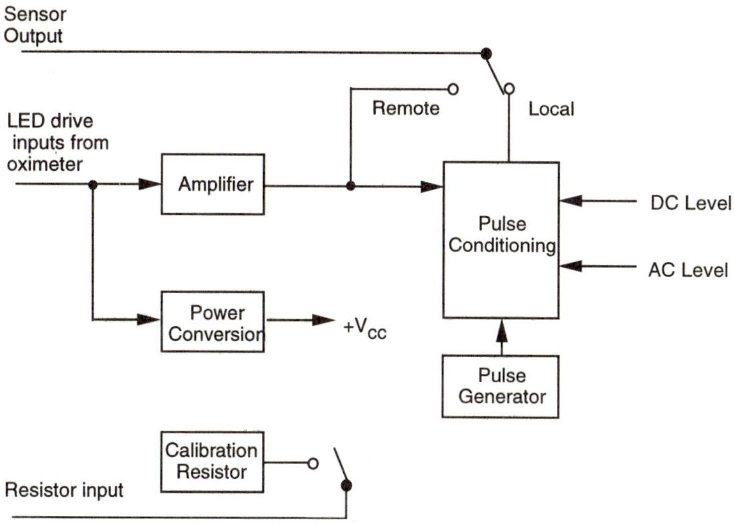

Figure 10.13 Block diagram of an electronic simulator that replaces the pulse oximeter probe (used with permission (Nellcor 1994) Pulse oximeter tester Model SRC-2).

10.3.1 ASTM F1415

The ASTM F1415 standard (ASTM 1992) contains requirements for the pulse oximeter designer in regard to marking and documenting the system, electrical safety concerns, electromagnetic interference and alarms. No specific information is provided regarding specific design requirements of the parts of the system discussed in the preceding chapters. In addition, no specific information is provided in regard to calibration or testing of these devices.

10.3.2 ISO 9919

This standard mentions a few requirements regarding calibration. These include requiring manufacturers to provide:

1. The calibration range of the pulse oximeter.
2. Whether the pulse oximeter is calibrated to display functional or fractional saturation.
3. The accuracy and range of HbO_2 saturation level displayed.
4. Whether the calibration was functional or fractional saturation.
5. Test methods for calibration need to be available from manufacturer upon request.

The ISO 9919 (International Organization for Standardization 1992) also offers this disclaimer in Annex L:

> Values derived from the pulse oximeter are not a measurement of blood or tissue oxygen tension and therefore pulse oximetry provides no direct indication of oxygen delivery to or consumption by, tissues. At present there is no widely accepted direct *in vitro* calibration

method for pulse oximeters. The only accepted *in vitro* test method for correlation of the reading from a pulse oximeter (S_pO_2) is bench-type oximetry employing more than two wavelengths of light or other methods using blood samples drawn from human subjects. Although work is progressing on the development of direct *in vitro* calibration methods, present techniques still require the use of human subjects. To include test methods in standards that require the use of human subjects, has, through past experience, been found to be unacceptable, and therefore *in vivo* test methods are not included in this International Standard.

10.3.3 Other standards

American Society of Anesthesiologists. Standards for Basic Intra-Operative Monitoring, 1986 (0696-ASA).
American Society of Anesthesiologists. Standards for Post-Anesthesia Care, 1989 (0697-ASA).
European Committee for Standardization. Drafting European norm for pulse oximeters.

REFERENCES

Ackerman S W and Weith P 1995 Knowing your pulse oximetry monitors *Med. Electron.* **26** (1) 82–6

ASTM 1992 *Standard Specification for Pulse Oximeters F1415-92* (Philadelphia PA: American Society for Testing and Materials)

Aoyagi T, Fuse M, Shindo Y and Keto M 1994 Apparatus for calibrating pulse oximeters *US patent 5,278,627*

Cheung P W, Gauglitz K F, Hunsaker S W, Prosser S J, Wagner D O and Smith R E 1993 Apparatus for the automatic calibration of signals employed in oximetry *US patent 5,259,381*

Clinical Dynamics 1995 *Technical sales brochure* (Wallingford, CT: Clinical Dynamics)

International Organization for Standardization 1992 *Pulse Oximeters for Medical Use—Requirements* ISO9919:1992(E)

Leuthner T 1994 Development system for pulse oximetry *Med. Biol. Eng. Comput.* **32** 596–8

Marble D R, Burns D H and Cheung P W 1994 Diffusion-based model of pulse oximetry: *in vitro* and *in vivo* comparison *Appl. Opt.* **33** 1279–85

Merrick E B and Haas P 1994 Simulation for pulse oximeter *US Patent 5,348,005*

Munley A J, Sik M J and Shaw A 1989 A test object for assessing pulse oximeters *Lancet* 1048–9

Nellcor 1994 *Pulse Oximeter Tester Model SRC-2* (Pleasanton, CA: Nellcor)

Nonin Medical 1995 Nonin finger phantom *Technical Note* (Plymouth, MN: Nonin Medical)

Moyle J T B 1994 *Pulse Oximetry* (London: BMG)

Payne J P and Severinghaus J W (eds) 1986 *Pulse Oximetry* (New York: Springer)

Pologe J A 1989 Functional saturation versus fractional saturation: what does the pulse oximeter read *J. Clin. Monit.* **5** 288–9

Reynolds K J, deKock J P, Tarssenko L and Moyle J T B 1991 Temperature dependence of LED and its theoretical effect on pulse oximetry *Brit. J. Anaesthesiol.* **67** 638–43

Reynolds K J, Moyle J T B, Gale L B, Sykes M K and Hahn C E W 1992 *In vitro* performance test system for pulse oximeters *Med. Biol. Eng. Comput.* **30** 629–35

Reynolds K J, Moyle J T B, Sykes M K and Hahn C E W 1993a Responses of 10 pulse oximeters to an in vitro test system *Brit. J. Anaesthesiol.* **68** 265–9

Reynolds K J, Palayiwa E, Moyle J T B, Sykes M K and Hahn C E W 1993b The effects of dyshaemoglobins on pulse oximetry *J. Clin. Monit.* **9** 81–90

Santamaria T and Williams J S 1994 Device focus: pulse oximetry *Med. Device Res. Rep.* **1** (2) 8–10

Severinghaus J W, Naifeh K H and Koh S O 1989 Errors in 14 pulse oximeters during profound hypoxia *J. Clin. Monit.* **5** 72–81

Vegfors M, Lindberg L G, Oberg P A and Lennmarken C 1993 Accuracy of pulse oximetry at various haematocrits and during haemolysis in and in vitro model *Med. Biol Eng. Comput.* **31** 135–41

Volgyesi G A 1992 Method of testing the accuracy of pulse oximeters and device therefor *US patent 5,166,517*
Wukitsch M W, Petterson M T, Tobler D R and Pologe J A 1988 Pulse oximetry: analysis of theory, technology, and practice *J Clin. Monit.* **4** 290–301
Yount J E 1989 Device and procedures for in vitro calibration of pulse oximetry monitors *US patent 4,834,532*
Zhou G X, Schmitt J M and Walker E C 1992 Electro-optical simulator for pulse oximeters *Med. Biol. Eng. Comput.* **31** 534–9

INSTRUCTIONAL OBJECTIVES

10.1 Describe how *R* curves are determined through *in vivo* testing.
10.2 Explain the role that LED temperature plays in oxygen saturation level determination.
10.3 Explain why the term S_pO_2 is necessary when referring to oxygen saturation levels.
10.4 Explain the reason why different *R* curves may be needed for a manufacturer's pulse oximeter system.
10.5 Describe how oxygen saturation level is altered through an *in vitro* test system.
10.6 Explain why pulse oximeters are less accurate for S_pO_2 saturation levels below 60%.
10.7 Describe the operation of an optoelectronic simulator system.
10.8 Describe the operation of an colored colloid simulator system.
10.9 Describe the operation of polyester resin device simulator system.
10.10 Describe the operation of a wedge simulator system.
10.11 Describe the operation of the tube and bulb simulator system.
10.12 Explain the limitations of the electronic simulators used for testing pulse oximeters.

CHAPTER 11

ACCURACY AND ERRORS

Supan Tungjitkusolmun

Continuous assessment of arterial oxygen saturation (S_aO_2) is important in clinical management of critically ill patients. Pulse oximeters have been widely used as blood oxygen monitoring devices since the early 1980s. Currently, pulse oximeters can be found in virtually every operating room, recovery room, and intensive care unit. The advantages of pulse oximetry include noninvasiveness, ease of use, portability, and patient comfort. A light source generated by two LEDs, with wavelengths at approximately 660 nm and 940 nm, and a photodiode are mounted in a probe of a pulse oximeter. Circuit control, saturation calculation, and display are managed by a microprocessor instrument as described in chapter 8. Unlike earlier techniques such as the *in vivo* eight-wavelength oximeter (chapter 3), no heating or arterialization techniques are required in pulse oximetry.

All pulse oximeters work using absorption spectrophotometry, however, considerable differences exist in the way different manufacturers obtain and process the data. These differences occur in the light-emitting diodes, sampling frequency, microprocessor algorithms, and the constants used in the calculations, or the look-up tables. Since the technique has come into wide clinical use over the past decade, it is important to examine circumstances where its reliability may be questioned. The objective of this chapter is to describe several sources of error in pulse oximetry which may cause hazardous consequences to the patients. Recognizing the limitations described in this chapter and applying appropriate corrective interventions are essential to optimize the clinical use of pulse oximeters.

11.1 EVALUATION OF PULSE OXIMETERS

The *gold standard* measurement of arterial oxygen saturation is the CO-oximeter, described in chapter 3. A comparison of the pulse oximeters' readings and CO-oximeters' readings is thus required to verify the reliability of the pulse oximetry technique. Comparisons between pulse oximeters' arterial oxygen saturation values and the CO-oximeters' readings, as well as the HP eight-wavelength ear oximeter will be discussed in this section.

11.1.1 Accuracy, bias, precision, and confidence limit

Accuracy is a measure of systemic error or bias; the greater the error, the less accurate the variable. The accuracy of a measurement is the degree to which it actually represents what it is intended to represent. The location of the mean errors reflects the accuracy of the measurement. The accuracy of pulse oximeter oxygen saturations can usually be tested by comparing with the reference techique, CO-oximetry. Parameters frequently used to represent the degree of accuracy are bias, and absolute mean errors. *Bias,* in this case, is defined as the mean of the differences between the pulse oximeter readings and the CO-oximeter readings, which can be expressed as

$$\text{bias} = \frac{\sum_{i=1}^{N} x_i}{N} = \bar{x} \tag{11.1}$$

where x_i is calculated by subtracting the ith CO-oximeter measurement from the corresponding oximeter saturation displayed by a pulse oximeter. N is the total number of measurements. Units are percent saturation.

Precision is a measure of variation of random error, or degree of reproducibility. The dispersion of points around the mean reflects the precision of the measurement. Precision is often described statistically using the standard deviation (SD) of the differences between the pulse oximeter readings and the CO-oximeter readings of repeated measurements (Nickerson *et al* 1988) as in equation (11.2). Units are percent saturation.

$$\text{precision} = \text{SD} = \sqrt{\frac{\sum_{i=1}^{N} (x_i - \bar{x})^2}{N-1}}. \tag{11.2}$$

Some researchers frequently use a *95% confidence limit,* which for a normal distribution is equal to 1.96 times SD:

$$95\% \text{ confidence limit} = 1.96 \times \text{SD} \approx 2 \times \text{SD}. \tag{11.3}$$

Example 1

The results from an experiment to compare pulse oximeter and CO-oximeter readings are shown in table 11.1. Ten measurements were made.

From table 11.1,

$$\text{bias} = \bar{x} = \frac{\sum_{i=1}^{10} x_i}{10} = \frac{15}{10} = 1.5\%$$

$$\text{precision} = \sqrt{\frac{\sum_{i=1}^{10}(x_i - \bar{x})^2}{10-1}} = \sqrt{\frac{20.5}{9}} = 1.51\%,$$

Table 11.1 Comparison of pulse oximeter and CO-oximeter readings.

Measurement (i)	CO-oximeter readings (%)	Pulse oximeter readings (%)	x_i (%)	$x_i - \bar{x}$
1	97	100	3	1.5
2	98	99	1	−0.5
3	92	91	−1	−2.5
4	96	98	2	0.5
5	97	99	2	0.5
6	90	93	3	1.5
7	89	90	1	−0.5
8	95	98	3	1.5
9	88	90	2	0.5
10	93	92	−1	−2.5

and

$$95\% \text{ confidence limit} \approx 2 \times 1.51\% = 3.02\%.$$

The bias of 1.5% means that the test pulse oximeter tends to *overestimate* the oxygen saturation level (postive bias). A 95% confidence limit of 3.02% means that the pulse oximeter will give an outcome in the range between 1.5 – 3.02% and 1.5% + 3.02%, or between −1.52% and 4.52% from the true value (the CO-oximeter reading) with a probability of 0.95.

The use of bias and precision is helpful in getting a clear picture of a pulse oximeter's performance and how this compares to other units or other studies. A unit may be very precise, so that the results are highly reproducible with a low scatter, but have a high bias so that the results are not centered on the true values. In contrast, a unit may have a very low bias, but have poor precision, with values swinging widely from side to side of the true value. In clinical practice, a 95% confidence limit of less than ±3% is considered acceptable for most cases.

Other statistical terms from the regression analysis (correlation coefficient, positive error, intercept, and slope) are also used in several studies (Yelderman and New 1983, Taylor and Whitwam 1988).

11.1.2 What do pulse oximeters really measure?

Pulse oximeters only measure a ratio of transmitted red and infrared light intensities, and relate this to a look-up table of empirical oxygen saturation values (see chapter 9). The values in the table depend on the manufacturer's purpose of estimating functional or fractional oxygen saturation, but will in reality be neither of these unless the dyshemoglobin (dysfunctional hemoglobin) levels, and the pH levels in a subject's arterial blood are exactly the same as the average values of those used in the empirical calibration to create the look-up table. Choe *et al* (1989) found that the measured oxygen saturations in two instruments (Ohmeda Biox 3700 and Radiometer Pulse Oximeter) were close to the fractional oxygen saturation (fractional SO_2). On the other hand, the other four units used in the study (Minolta/Marquest Pulsox 7, Novametrix 500, Physio-Control Lifestat, and Datex Satlite) gave results in the proximity of functional oxygen saturation (functional SO_2). The data used for calibration processes are usually obtained from healthy adults breathing hypoxic gas mixtures (see section 10.1.1).

Pulse oximeters can measure neither fractional SO_2 nor functional SO_2. However, the use of fractional SO_2 as the reference in the calibration process provides the clinician with a realistic assessment of the magnitude of the errors of physiological illness which is likely to be found for the group of patients under consideration.

11.1.3 Pulse oximeter versus CO-oximeter

Pulse oximeters are empirically calibrated by the manufacturer against a CO-oximeter. The IL (Instrumentation Laboratories, Inc.) 482 and 282 model CO-oximeters use four wavelengths of light (535.0, 585.2, 594.5, 626.6 nm) to detect the concentrations of HbO_2, Hb, COHb, and MetHb, and give the oxygen saturation as a percentage of the sum of the four species. This saturation is known as *fractional saturation* (section 4.2.2).

According to its operator's manual, the IL 482 has a precision of 0.5% (95% confidence limit of 1%) for HbO_2 measurements for samples with 0 to 10% MetHb and a pH of 7.0 to 7.4. The pH sensitivity of MetHb can cause significant changes in absorption at all four wavelengths outside these MetHb and pH ranges. Accuracy is also compromised by the presence of high lipid levels which can cause light scattering. It is not feasible to validate the value of 0.5% precision claim, since there is no quality control sample of accurately known or measured saturation that can be used to verify this. It is reasonable to accept this precision, given the high degree of reproducibility of the results.

Yelderman and New (1983) conducted a study to evaluate the accuracy of pulse oximeters over a broad range of arterial blood oxygen saturation in 1983 when the first Nellcor pulse oximeter became commercially available. A comparison of a pulse oximeter and the CO-oximeter readings was performed on five healthy, nonsmoking students ranging in age from 18 to 25. The precision of the measurements was found to be 1.83%. They concluded that pulse oximetry is a reliable technique for a measurement of arterial blood oxygen saturation in the range of 100 to 70%.

11.1.4 Pulse oximeter versus in vivo eight-wavelength ear oximeter

Hewlett-Packard ear oximetry, using eight wavelengths, was considered as a standard technique of measurement of arterial oxygen saturation before pulse oximeters were invented (see chapter 3). A comparison of the two techniques is thus necessary to see whether their results agree sufficiently for the pulse oximeter to replace the previous technique. Cahan *et al* (1990) determined that the difference between the HP ear oximeter (Hewlett-Packard 47201A ear oximeter) and the CO-oximeter (IL 282) readings was 0.9 ±4.3% (expressed as bias ± 95% confidence limit).

In a study by Cahan *et al* (1990), the difference between the individual pulse oximeters and the HP ear oximeters was found to be 2.6 ± 10.3% in the range of 99 to 70%. All five pulse oximeters studied gave higher values than the HP oximeter, and the differences between pulse oximeters and the HP readings increased as oxygen saturation fell below 85%. The greater discrepancies might be due to the longer *delay* of pulse oximeters during the progressive hypoxia.

The agreement of discrete measurements of the two methods was found to be acceptable at high oxygen saturation but unacceptable for arterial oxygen saturation levels lower than 85%. We must be careful when making an assessment of the oxygen saturation levels from two experiments in which different arterial oxygen monitoring devices were used. The continuous measurements from pulse oximeters and from the HP ear oximeters cannot be assumed to be in the same range.

11.2 ACCURACY VERSUS SATURATION

Accuracy at different levels of oxygen saturation is not the same. To make the discussion more effective, oxygen saturation is divided into three ranges: normal saturation, high saturation, and hypoxic condition (low saturation level).

11.2.1 High saturation (greater than 97.5%)

Pulse oximeters are designed to give a saturation reading of less than or equal to 100%; this limits the potential for positive errors and makes precision calculations difficult to interpret in this high range. Table 11.2 offers some outcomes of the evaluations of 20 brands of pulse oximeters. Even though precision calculations cannot be determined unbiasedly due to positive errors, the correct corresponding oxygen saturation is not critical in this range. As long as the oxygen saturation is over 97%, the patients are in favorable conditions and they require no urgent medical attention.

Table 11.2 Number of S_pO_2 readings of 100% when CO-oximeter reading was 97 to 98%. The results are expressed as the ratio of S_pO_2 readings of 100% and the number of measurements (percentage). Adapted from Webb *et al* (1991). Study 1 is from ECRI (1989). Study 2 is from Clayton *et al* (1991a).

Oximeter	Study 1	Study 2
Criticare CSI 503	–	0/17 (0%)
Engstrom EOS	–	0/15 (0%)
Spectramed Pulsat	0/17 (0%)	0/15 (0%)
Criticare CSI 504	–	0/14 (0%)
Biochem Microspan 3040	–	0/10 (0%)
Radiometer Oximeter	1/11 (9%)	1/17 (6%)
Simed S-100	1/17 (6%)	1/15 (0%)
Invivo 4500	2/9 (22%)	2/15 (13%)
Datex Satlite	1/9 (11%)	3/22 (14%)
Datascope Accusat	4/9 (44%)	3/14 (21%)
Physio-Control 1600	2/17 (12%)	4/16 (25%)
Nonin 8604D	3/9 (33%)	4/16 (25%)
Sensormedics Oxyshutle	2/7 (29%)	6/16 (38%)
Novametrix 505	3/17 (18%)	11/22 (50%)
Pulsemate Colin BX-5	–	10/16 (63%)
Minolta Pulsox 7	–	11/17 (65%)
Ohmeda Biox 3700	4/9 (44%)	11/15 (73%)
Ohmeda Biox 3740	5/16 (31%)	13/16 (81%)
Nellcor N-200	3/17 (18%)	13/18 (83%)
Kontron 7840	–	13/15 (87%)

11.2.2 Normal saturation (90 to 97.5%)

After more than a decade of development since first becoming commercially available, most models of pulse oximeters have a reliable performance in this range. In an experiment by Webb *et al* (1991), 10 of the 13 units had absolute mean errors of less than 1.0%; the standard deviation was less than 2% in eight units, and between 2 and 3% in the remaining five. Choe *et al* (1989), Taylor and Whitwam (1988), and Yelderman and New (1983) also found similar results. Pulse oximeters are well calibrated in this range since it is the most commonly found condition.

11.2.3 Low saturation (less than 80%)

Pulse oximeters have a high potential for errors at low saturations, mainly because ethically manufacturers cannot induce severe hypoxia repeatedly in volunteers for calibration purposes. Also, figure 11.1 illustrates that the absorption characteristics of 0% oxygen saturation blood are much steeper than that of 100% oxygen saturation blood at a 660 nm wavelength. At this range, when the percentage of hemoglobin saturation decreases, the slope of the absorption spectrum increases. Any slight error in the LED peak wavelength will change the readings of the pulse oximeter drastically.

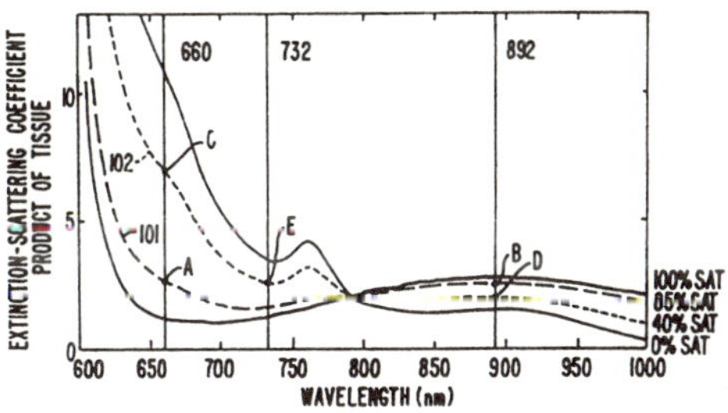

Figure 11.1 Variation in extinction coefficients over a range of wavelengths of 600 to 1000 nm at different saturation values. At 660 nm (red) wavelength, the slope of an absorption spectrum increases as oxygen saturation level decreases. From Casciani *et al* (1995).

The error associated with low saturations can also be explained by a reduction in the signal-to-noise ratio in pulse oximetry. As saturation decreases, less red light is able to penetrate through the tissues due to a high absorbance of Hb, thus the AC signal becomes weaker. To compensate for this drawback, the LED-driving current and the photodiode amplifier gain are increased to maintain the AC signal in a usable range. As the gain increases, incidental electrical and physiological noise also increase, thus resulting in a decline in the pulse oximeter's accuracy.

The accuracy of 13 pulse oximeters at low saturations was determined by ECRI (1989) in intensive care patients. Figure 11.3 shows the experimental results. All units examined were less accurate and nine out of 13 were less precise than when saturations were greater than 80%; eight out of 13 units tended to underestimate S_aO_2 by substantial amounts at low saturations.

In summary, pulse oximeters are poorly calibrated for saturations below 80%. In general, accuracy and precision are worse than for saturations above

80%, but this depends on the model and the brand. For example, Sensormedics Oxyshuttle pulse oximeter's bias only increases slightly (–0.1%), and the precisions are the same in both ranges.

Table 11.3 Accuracy of 13 pulse oximeters using finger probes on patients in the Intensive Care Unit. Adapted from Webb *et al* (1991).

Oximeter	Saturation > 80% Bias% (precision%)	Saturation < 80% Bias% (precision%)
Datascope Accusat	–0.3 (1.9)	–7.1 (3.2)
Datex Satlite	+0.0 (2.0)	+1.4 (1.5)
Invivo 4500	–0.3 (1.8)	–0.6 (4.9)
Nellcor N-200	+0.8 (1.7)	–5.5 (3.5)
Nonin 8604	+1.4 (1.8)	+8.8 (4.8)
Novametrix 505	+0.7 (1.9)	–8.1 (4.3)
Ohmeda 3700	–1.0 (2.5)	–5.3 (6.2)
Ohmeda 3740	–0.1 (2.8)	–5.5 (1.9)
Physio-Control 1600	+0.0 (1.9)	–6.0 (6.9)
Radiometer Oximeter	–1.5 (1.8)	–6.7 (3.2)
Sensormedics Oxyshuttle	–0.3 (1.8)	–0.4 (1.8)
Simed S-100	+0.1 (2.2)	+1.8 (1.6)
Spectramed Pulsat	+0.7 (1.6)	–3.4 (3.2)

11.3 ACCURACY VERSUS PERFUSION

Pulse oximeters require adequate plethysmographic (photoplethysmographic) pulsations to differentiate arterial blood absorbance from the absorbances of other substances (venous blood, tissue, and bone). A significant decrease in peripheral vascular pulsation, such as in hypothermia, vasoconstriction, hypotension, during cardiopulmonary bypass, or cardiac arrest, may result in a plethysmographic signal insufficient to be processed reliably by the oximeter. Most pulse oximeters have the ability to recognize a weak waveform which could cause an erroneous reading. They usually display a '*Low Perfusion*' or similar message to alert the user of possible problems in peripheral blood perfusion.

In a study to compare the performance of 20 pulse oximeters under the conditions of poor perfusion by Clayton *et al* (1991a), only two out of 20 oximeters had 95% confidence limits that were less than 4%. Generally the clinically acceptable range for the readings is about ±3%. Table 11.4 shows the results from the experiment.

Locally applied vasodilating drugs could be useful to enhance the plethysmographic pulsation in certain situations. The use of a pediatric warming blanket wrapped around the forearm is a simple method to increase perfusion due to a cold finger if the pulse oximeter signal is weak. Finger probes are preferable for patients with poor perfusion (see section 11.9).

11.3.1 Venous congestion

Another potential problem with pulse oximeter measurements is venous congestion, which leads to artifacts due to venous pulsation. Venous congestion is an accumulation of blood within an organ, which is the result of back pressure within its veins. Because the pulse volume amplitude of the plethysmograph is a measure of the pulsatility of the compliant vessels, some of the pulse may be attributed to venous blood of lower oxygen content mixed with the signal due to higher oxygen content in the arterial blood. Also, the decrease in venous wall compliance by congestion should decrease the pulse volume amplitude in the organs (such as the finger). The pulse oximeter is unable to distinguish between the absorption due to pulsatile veins and that caused by arteries and arterioles. Pulsatile venous flow is generated by a transmitted arterial pulse through *arteriovenous anastomoses* in the finger. Therefore, if the S_pO_2 measured by the pulse oximeter is shunted arterial blood in the vein, the S_pO_2 reading will be affected by venous blood. Pulsatile veins may lead to the pulse oximeter indicating a lower value of S_pO_2 than is the actual saturation.

Table 11.4 Accuracy of pulse oximeters, ranked according to number of readings within 3% and showing ranking for number of readings within 3% of total number of readings expressed as percentage. Each pulse oximeter was tested on 40 patients. Total = total number of measurements obtained. Adapted from Clayton *et al* (1991a).

Pulse oximeter	Total	# within ±3%	Percent ±3%/Total	Rank
Criticare CSI 503	40	40	100	1
Datex Satlite	40	38	95	2
Biochem Microspan 3040	28	26	93	3
Novametrix 505	38	35	92	4
Criticare CSI 504	39	35	90	5
Invivo 4500	38	34	89	6
Sensormedics Oxyshuttle	36	32	89	6
Physio-Control 1600	36	31	89	6
Ohmeda Biox 3740	30	26	87	9
Minolta Pulsox 7	40	34	85	10
Nellcor N-200	39	33	85	10
Simed S-100	36	30	83	12
Datascope Accusat	33	27	82	13
Radiometer Oximeter	40	32	80	14
Nonin 8604D	35	28	80	14
Spectramed Pulsat	32	25	78	16
Pulsemate Colin BX-5	39	30	77	17
Ohmeda Biox 3700	36	25	69	18
Kontron 7840	40	27	68	19
Engstrom Eos	35	20	57	20

Much of the ac display of the plethysmographic signal may be due to pulsatile cutaneous venules which have an oxygen saturation similar to the arterial saturation due to patient arteriovenous communications in the skin. However, if the large venules and veins, which carry hemoglobin with a lower oxygen saturation, are pulsating, then the technique cannot distinguish between the two. Therefore, the S_pO_2 values may be lower than the arterial oxygen saturation if venous congestion is present. Other causes of increased pulsatility in veins are arteriovenous disassociation, right atrial myxoma, and right heart block.

11.4 ACCURACY VERSUS MOTION ARTIFACTS

As with most medical devices, motion artifacts contribute a significant error to pulse oximetry. Pulse oximeters detect a pulsatile signal that normally is only a small percentage of the total plethysmographic signal. Therefore, any transient motion of the sensor relative to the skin can cause a significant artifact in the optical measurement. Furthermore, if these transient artifacts mimic a heartbeat, the instrument may be unable to differentiate between the pulsations that are due to motion artifacts and normal arterial pulsations, thereby causing erroneous readings. Practically, these artifacts can be reduced by digital signal processing and averaging the S_pO_2 values over several seconds before they are displayed. Motion artifacts, such as during shivering, seizure activity, or exercise, are usually recognized by false or erratic heart-rate displays or by distorted plethysmographic waveforms (figure 11.2).

Some manufacturers use the R wave of the patient's electrocardiogram to synchronize the optical measurements; they thereby improve the detection of noisy pulsatile signals by enhancing the signal-to-noise ratio of the measurements through the use of multiple time-averaged signals (see chapter 9).

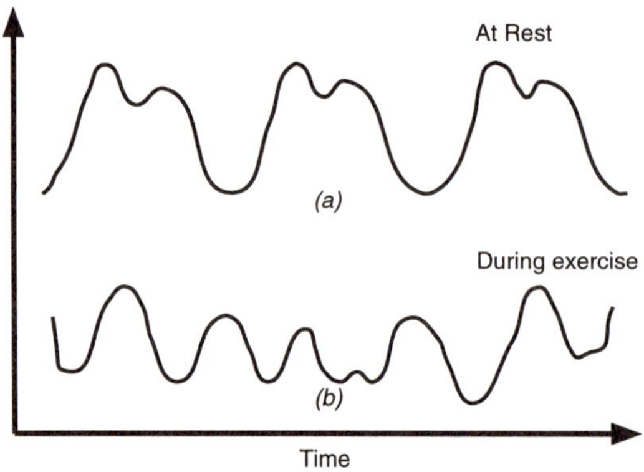

Figure 11.2 The plethysmographic waveform of a subject at rest is periodic (*a*) and during exercise is not periodic (*b*).

11.5 ACCURACY VERSUS OPTICAL INTERFERENCE

Bright external light sources are known to affect pulse oximeters and all pulse oximeters share this sensitivity. This occurs because these instruments use optical means to make their measurements. Consequently, to obtain accurate measurements, potential sources of optical interference must be controlled. Because pulse oximeters' optical components are located in the probe, proper probe application and use are key factors in reducing optical interference. Optical interference occurs when bright light from an external source (ambient light) reaches the photodiode, or when light reaches the photodiode without passing through a pulsatile arteriolar bed.

Pulse oximeters are designed to reject ambient light since the photodiodes can measure weak signals. When the intensity of ambient light is high (as from heat lamps or sunlight), the photodiode cannot sense light transmitted through tissue for S_pO_2 calculations. Protecting the photodiode from bright light obviates the problem. One solution is to cover the probe site with some opaque material, such as a surgical towel. Although this approach is generally useful, with active neonates or restless patients, the towel frequently becomes displaced and exposes the oximeter probe. One of the effective remedies to this problem is covering the probe, while it is attached to a digit, with a packaging from an alcohol swab as suggested by Siegel and Gravenstein (1987). This packaging is manufactured in a shape that makes a convenient, dark receptacle for a digit, even one on which a flexible pulse oximeter probe has been placed.

Another type of optical interference may occur when some of the light from the LEDs reaches the photodiode without passing through an arteriolar bed. Such an optical shunt results in either erratic or stable but inaccurate measurements. Figure 11.3 shows some optical interferences to pulse oximetry. Oximeter probes should be manufactured of black opaque material that does not transmit light, or enclosed in an opaque plastic housing. Although there is no substitute for continual vigilance, shielding the probes from excessive ambient light, as strongly recommended by the manufacturer, will reduce the possibility of false readings.

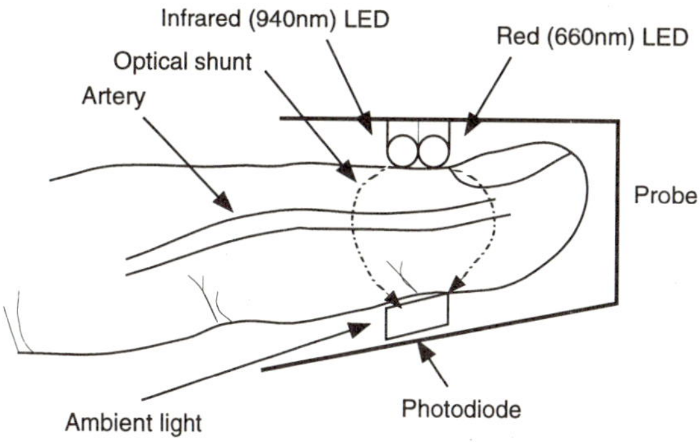

Figure 11.3 Ambient light interference and optical shunt in pulse oximetry. Optical shunt occurs when the light from the LEDs reaches the photodiode without passing through arterial blood.

11.6 ACCURACY VERSUS INTRAVENOUS DYES

During medical procedures, the use of substances such as dyes may be necessary. This section investigates the effects of dyes on pulse oximeter readings.

Several intravenous-administered dyes appeared to be associated with abrupt decreases in pulse oximetry S_pO_2 readings (Scheller *et al* 1986). Fifteen white subjects were studied, five with each of the three dyes, indigo carmine (InCa), indocyanine green (InGr), and methylene blue (MeBl). In all subjects, baseline readings were 97% or greater in both the toe and finger locations. Table 11.5 summarizes subject characteristics, the time from injection to the first noticeable decrease in S_pO_2 readings (latency), the lowest S_pO_2 reading (nadir), and the time required to return to baseline (duration), for each of the three dyes. Of the three dyes, InCa produced the fewest and smallest changes in S_pO_2 readings. Decreases from baseline were observed in three of the five subjects given indigo carmine, but only in the toe location. The magnitude of the measured oxygen saturation decreases were small following InCa, and the lowest S_pO_2 reading observed in any subject was 92%. By contrast, oxygen saturation reading decreases were observed in all subjects in both sensing locations following the administration of MeBl, with a median lowest S_pO_2 reading of 65%. The lowest S_pO_2 reading observed in any subject following MeBl was 1%. In subjects given

MeBl, measured oxygen saturations remained below baseline for between approximately 1 and 2 min in both the finger and toe. S_pO_2 reading decreases following the administration of InGr were intermediate between those observed with MeBl and InCa. Figure 11.4 shows the absorbance spectra for the three dyes as determined by spectrophotometry.

Table 11.5 Subject characteristics and S_pO_2 reading responses to IV Dyes. Adapted from Scheller *et al* (1986). Latency = the time from injection to the first noticeable decrease in S_pO_2 readings. Nadir = the lowest S_pO_2 reading. Duration = the time required to return to baseline reading. NC = no observed change.

Dye	Weight (kg)	Height (cm)	Latency (s) Finger/Toe	Duration (s) Finger/Toe	Nadir (O_2 saturation, %) Finger/Toe
MeBl	75	178	80/65	70/90	91/98
	68	175	35/30	105/80	58/65
	79	183	40/40	65/50	76/59
	93	180	40/35	50/50	80/69
	46	163	35/30	115/80	1/32
InGr	83	188	35/45	10/40	96/96
	67	175	45/40	35/25	95/93
	70	178	45/35	45/70	93/84
	86	191	50/45	70/30	93/92
	70	175	NC/65	NC/60	99/88
InCa	83	188	NC/NC	NC/NC	NC/NC
	67	178	NC/40	NC/40	NC/93
	46	163	NC/25	NC/30	NC/92
	86	175	NC/NC	NC/NC	NC/NC
	65	173	NC/20	NC/20	NC/94

All the three dyes absorb light in the region of the 660 nm wavelength at which the red LED of a pulse oximeter emitted light. Methylene blue has an extremely high absorbance in this region. This explains why methylene blue interferes to a greater degree with S_pO_2 readings than the other dyes (from Beer's law). Likewise, the absorbance of indocyanine green is slightly greater than indigo carmine at this wavelength, which is consistent with the observation that S_pO_2 readings were affected to a greater degree in those subjects given indocyanine green than in those given indigo carmine.

Absorbances of all three dyes are negligible in the region of 940 nm and thus have insignificant effects on the IR light intensities detected by photodiodes. The variable responses of the individual subject's S_pO_2 readings following dye injection may have been related to differences in cardiac output or blood volume. For example, following methylene blue, the largest S_pO_2 reading decrease and longest duration of decrease was seen in the smallest subject (body surface area = 1.34 m^2). The measurement of cardiac output by the transcutaneous detection of various intravenous dyes has been studied in both adults and children and found to correlate well with dye dilution methods that use continuous arterial blood sampling (Scheller *et al* 1986).

Saito *et al* (1995) observed that after intra-arterial injection of the blue dye *patent blue* in an anemic patient, the reduction in the pulse oximeter readings sustained for more than 20 min.

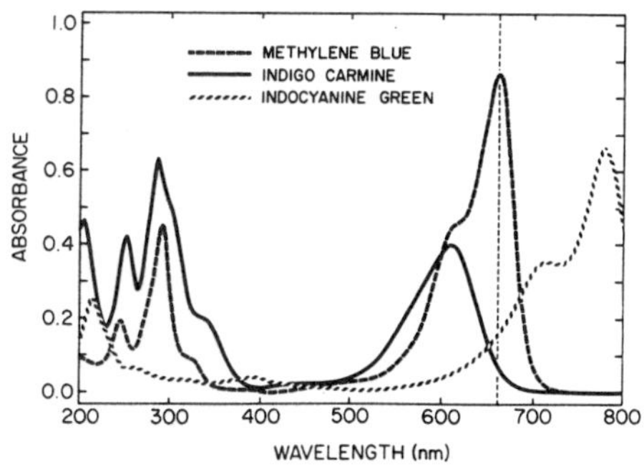

Figure 11.4 Absorbances of dyes. MeBl has the highest absorbance in the region of the 660 nm wavelength. From Scheller (1986).

Clinicians should be aware of the potential influences of intravenously administered dyes on S_pO_2 monitor readings so that operating room time is not wasted and more invasive analysis not undertaken, e.g., arterial blood gases, should falsely low S_pO_2 readings be temporarily induced by administration of these dyes (Scheller *et al* 1986).

11.7 EFFECT OF DYSHEMOGLOBINS AND FETAL HEMOGLOBIN

Dyshemoglobins are abnormal hemoglobins which cannot transport oxygen to the tissues. The presence of dyshemoglobins may cause inaccuracy in pulse oximetry. This section will discuss the two most commonly found in adults, carboxyhemoglobin and methemoglobin, as well as fetal hemoglobin.

11.7.1 Carboxyhemoglobin (COHb)

Seidler *et al* (1993) observed limitations of S_pO_2 readings in patients treated after inhalation of CO. Serial measurements of COHb concentration (IL 482 CO-oximeter) were done hourly in 6 patients until the results became normal, and arterial blood pressure, heart rate, and S_pO_2 were also monitored (by M1020 module, Hewlett-Packard). Figure 11.5 shows mean COHb values with corresponding S_pO_2 levels.

For all 18 measurements, the mean S_pO_2 reading was above 91%, which would be readily accepted as sufficient oxygenation. Decrease in COHb concentrations led to a slight increase of S_pO_2, as would be expected by the formula (Tremper and Baker 1989)

$$S_pO_2 = \frac{(c_{HbO_2} + 0.9c_{COHb})}{c_{\text{total hemoglobins}}} \times 100\%. \qquad (11.4)$$

As the level of COHb concentration in the blood reduces, the concentration of HbO_2 will rise while the concentration of total hemoglobins remain the same. Therefore, the magnitude of the numerator (c_{HbO_2} + $0.9c_{COHb}$) of equation (11.4) will increase which results in a larger $S_pO_2$value.

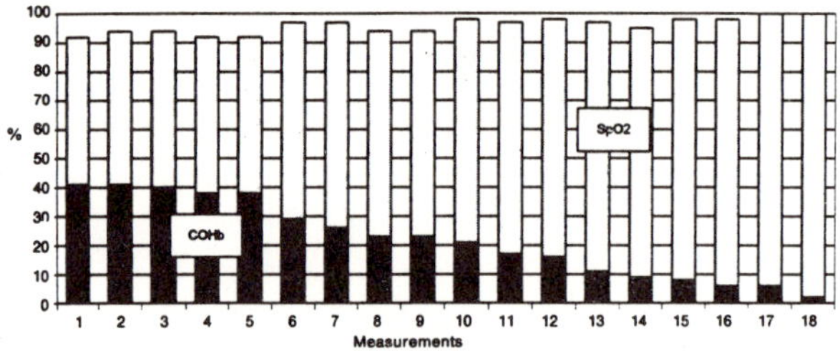

Figure 11.5 Mean measured arterial blood oxygen saturation (S_pO_2) with corresponding COHb values for 18 measurements in 6 patients. From Seidler (1993).

The increasing availability of pulse oximetry in intensive care units may lead to a false interpretation of oxygen transport capacity in cases of CO poisoning, especially if S_pO_2 is between 91% and 98%. Physicians should be aware that the diagnosis of CO poisoning still depends on a high degree of clinical suspicion and direct measurement of CO (Seidler *et al* 1993). The normal level of COHb in the arterial blood is less than 2%. Smokers or smoke-inhalation victims may have COHb levels greater than 10%. A high level of COHb overestimates the S_aO_2 values.

11.7.2 Methemoglobin (MetHb)

Methemoglobin is hemoglobin with iron oxidized from the normal (or reduced) ferrous (Fe^{2+}) state to the ferric (Fe^{3+}) state as described earlier in chapter 4. Methemoglobin is incapable of transporting oxygen.

Methemoglobinemia (high level of MetHb present in the blood) may be induced by a large number of drugs including local anesthetics (prilocaine, benzocaine), nitrates (nitroglycerin), nitrites, phenacetin, pyridium, primiquine, and sulfonamides. There are several case reports of potentially serious methemoglobin levels (greater than 30%) induced by topical anesthetics used in the airway. There are also case reports describing pulse oximeter readings during methemoglobinemia. However, the MetHb levels in these were too low (6% or less) to accurately characterize pulse oximeter behaviour.

At 660 nm the extinction coefficient of MetHb is similar to that of Hb and much greater than that of HbO_2 (figure 4.2). At 940 nm MetHb has a greater extinction coefficient than either Hb or HbO_2. MetHb thus adds to the pulse additional absorbance at both wavelengths. In contrast, COHb adds significant absorbance only at the shorter wavelength, where COHb has an extinction coefficient comparable to that of HbO_2. S_pO_2 is computed from the ratio *R* of the pulse-added absorbances at the two wavelengths. The presence of MetHb increases both the numerator and denominator of this ratio, which tends to drive *R* toward unity.

The arterial oxygen saturation can be expressed as

$$S_pO_2 = HbO_2\,\% = \frac{c_{HbO_2}}{c_{\text{total hemoglobins}}} \times 100\% \qquad (11.5)$$

while the functional hemoglobin saturation (measured arterial saturation)

$$S_pO_2 = \frac{c_{HbO_2}}{c_{Hb} + c_{HbO_2}} \times 100\% \qquad (11.6)$$

$$= \frac{c_{HbO_2}}{c_{\text{total hemoglobins}} - c_{MetHb} - c_{COHb}} \times 100\%. \qquad (11.7)$$

Theoretically, from equations (11.5) and (11.7), we can see that in the presence of MetHb, pulse oximeters overestimate the value of oxygen saturation in arterial blood, i.e., S_pO_2 is greater than S_aO_2.

11.7.3 Fetal hemoglobin

One of the concerns clinicians often have related to the interpretation of pulse oximeter readings in newborn infants is the fetal hemoglobin (HbF) present in the blood because pulse oximeters are calibrated empirically by inducing hypoxia in healthy adults. At birth, newborns have approximately 60 to 95% of the total hemoglobin in the form of fetal hemoglobin while the remainder is adult hemoglobin (HbA). In infants older than nine months, HbF levels higher than 2% often indicate an anemia such as sickle-cell anemia.

Mendelson and Kent (1989), and Zijlstra *et al* (1991) demonstrated that there is no significance difference in absorption spectra of adult and fetal hemolyzed blood in the 650 to 1000 nm wavelength region, which is used in pulse oximetry. On the other hand, adult and fetal hemoglobin absorption characteristics differ in the range of wavelengths below 650 nm.

The theoretical S_pO_2 readings for the adult and fetal hemoglobin can be determined by substituting the extinction coefficients given in table 11.6 into equation (4.19), which is

$$S_aO_2 = \frac{\varepsilon_{Hb}(\lambda_R) - \varepsilon_{Hb}(\lambda_{IR})R}{\varepsilon_{Hb}(\lambda_R) - \varepsilon_{HbO_2}(\lambda_R) + [\varepsilon_{HbO_2}(\lambda_{IR}) - \varepsilon_{Hb}(\lambda_{IR})]R} \times 100\%. \qquad (11.8)$$

Table 11.6 Extinction coefficients of adult and fetal blood expressed in ($L \cdot mmol^{-1} \cdot cm^{-1}$) (from Mendelson and Kent 1991).

	Hb		HbO_2	
λ	Adult	Fetal	Adult	Fetal
660 nm	0.86	0.90	0.12	0.16
940 nm	0.20	0.20	0.29	0.30

Figure 11.6 shows the results of the theoretical simulation. Mendelson and Kent (1989) suggested that a maximum error of approximately 3% in pulse oximeter oxygen saturation readings could be expected when measurements from adult and fetal blood are compared.

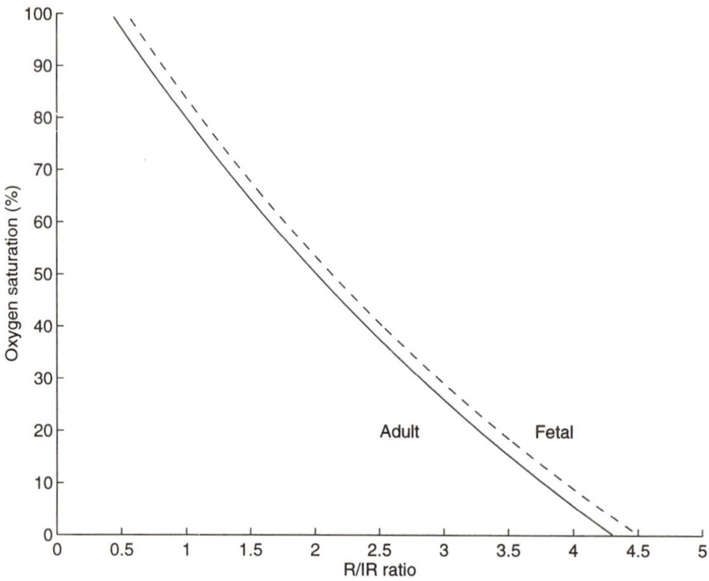

Figure 11.6 The calibration curves derived from a theoretical simulation show that pulse oximeters will read about 3% high for fetal hemoglobin. The R/IR ratio is *R* in equation (11.6).

11.7.4 Bilirubin

Bilirubin is an orange or yellow colored compound which is a breakdown product of heme. High levels of bilirubin can affect absorbance at lower wavelengths used by the CO-oximeters. A bilirubin concentration of 20 mg/dl will cause up to 1% error in the measurement of four main hemoglobin species. The absorption spectrum of bilirubin has a peak at 460 nm and much smaller peaks at 560 and 600 nm. Veyckemans *et al* (1989) showed that there was no significant error detected from the influence of high bilirubin plasma levels. The presence of bilirubin in the arterial blood will not induce any significant errors in pulse oximetry measurements.

11.8 EFFECT OF TEMPERATURE

11.8.1 Ambient temperature

An exposure of the body to cold temperatures can cause changes in peripheral perfusion which may cause inaccuracy. The temperature dependence of LEDs in pulse oximeter probes is unlikely to affect the pulse oximetric values. Reynolds *et al* (1991) showed that there was a 5.5 nm increase in the peak wavelength for a 660 nm LED, and a 7.8 nm increase in the peak wavelength for a 950 nm LED as temperature increased from 0 to 50 °C (see chapter 10).

Table 11.7 Extinction coefficients at 0 °C and 50 °C expressed in (L mmol^{-1} cm^{-1}). Adapted from Reynolds *et al* (1991).

	Hb		HbO_2	
λ	0 °C	50 °C	0 °C	50 °C
660 nm	0.856	0.811	0.123	0.117
950 nm	0.153	0.139	0.274	0.265

Table 11.7 lists the extinction coefficients of Hb and HbO_2 at different wavelengths and temperatures. Substituting these values into the relationship between S_pO_2 and R given in equation (11.6) which is derived from Beer's law, theoretical calibration curves can be obtained as in figure 11.7. Thus the effect of shifts in wavelength of the LEDs on pulse oximeter accuracy is neglibible as the temperature increases from 0 °C to 50 °C.

The reduced amplitude of the ac signals occurring during cold exposure causes the pulse oximeter to be more sensitive to motion artifacts, for example those caused by shivering or coughing. These artifacts may cause the pulse oximeter to give an erroneous value of S_pO_2. Reynold *et al* (1991) concluded that inaccuracies in pulse oximeter readings at extreme temperatures are far more likely to be caused by reductions in peripheral perfusion, rather than a result of the temperature dependence of the LEDs in the pulse oximeter probe.

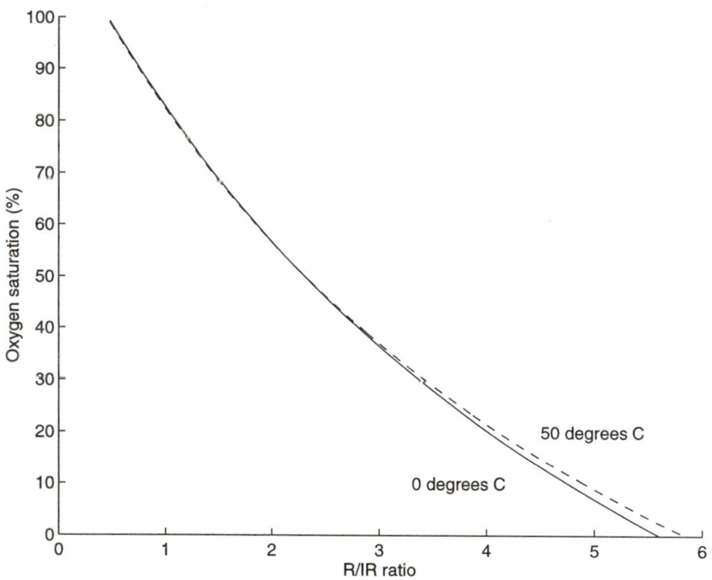

Figure 11.7 The calibration curves from a theoretical model show a shift from 0 °C to 50 °C.

11.8.2 Patient temperature

Errors in pulse oximetry readings do not increase significantly with a decrease in patient temperature. Palve and Vuori (1989) found that, in a recovery room study of the Nellcor N-100 and Ohmeda Biox 3700 pulse oximeters, they were reliable on patients with low cardiac output and hypothermia after open heart surgery, with standard deviations ranging from 1.8% to 3.9% for finger probes and from 0.9% to 2.1% for ear probes which are comparable to the outcomes of normal cases.

11.9 ACCURACY VERSUS MEDICAL CONDITIONS

Although pulse oximeters are designed to help detect pathophysiological oxygenation conditions of patients that might lead to a life threatening situation, some medical conditions cause pulse oximeters to be unreliable. Fortunately, pulse oximetry works well in the majority of the cases. The following are some frequent encounters where the accuracy of pulse oximeters is often questioned.

11.9.1 Cardiac arrhythmia

The heart rate derived from a pulse oximeter should match that from an ECG signal for the patient with a healthy heart. If the two differ, either of the monitors may be in error because of poor signal quality, or the electrical activity of the heart may appear to be normal while it produces beats with inadequate stroke volume output due to inadequate filling or contraction (Webb *et al* 1991).

Wong *et al* (1989) conducted an experiment to test the accuracy of pulse oximeters for 163 patients with cardiac arrhythmias. They found that for the group of 24 patients with a pulse oximeter to ECG pulse rate discrepancy of greater than 3 beats/min, S_pO_2 measurements were as accurate as those for the group of 139 patients with pulse rate agreement, as long as the S_pO_2 reading was stable on the pulse oximeter and there was reasonable signal strength.

11.9.2 Myxoma

Fearley and Manners (1993) described a case of inaccurate oximetry in a patient with a right ventricular myxoma. The ventilation/perfusion scan of the patient was normal but cardiac angiography revealed a rounded mass in the right ventricular outflow tract. Before cardiopulmonary bypass, pulse oximetry using an ear probe gave a consistent hemoglobin saturation of 75%, but repeated arterial blood gas analysis showed an arterial oxygen saturation exceeding 95%. The ear probe gave readings of 96 to 98% on volunteers in the operating room. Postoperative oximetry was consistently more than 96%. It was concluded that a ventricular contribution to the central venous pressure due to the dilated tricuspid ring might lead to the inaccuracy in pulse oximetry. The pulsatile venous pressure presumably induced an alternating current in the oximeter giving rise to a saturation not related to the arterial oxygen saturation. Thus pulse oximeter readings must be interpreted carefully in the clinical context of the patient being monitored.

11.10 ACCURACY VERSUS PROBE POSITION

Severinghaus *et al* (1989) found that ear and forehead probes generally had a much faster response to changing S_pO_2 values than finger probes. It was suggested that finger probes require a greater transit time for blood to reach the finger compared to ear. Kagle *et al* (1987) found the Ohmeda 3700 finger probe to be on average 24 s behind the Ohmeda 3700 ear probe in its response to rapid desaturation. West *et al* (1987) found that measurement accuracy was related to response delay times, with longer delays associated with lower accuracy. The ear probe with the shortest delay had some accuracy problems at low saturations, and the slowest responding finger probe was claimed to be totally inadequate as a monitor of rapid changes in saturation due to its delayed and highly damped response.

Forehead probes have been tested at stable low saturations on volunteers by Cheung and Stommel (1989) using a commercially available unit and Mendelson *et al* (1988) using a custom-built reflectance probe. Both groups found good correlation between the forehead measured values and CO-oximetry measurements for saturations down to 65%. Severinghaus *et al* (1989) found the accuracy of seven forehead probes to be comparable to that of finger probes during rapidly induced desaturation in volunteers.

Table 11.8 Accuracy of pulse oximeters ranked according to percentage of readings within 3% of the CO-oximeter readings out of the total number of readings. From Clayton *et al* (1991b).

Pulse oximeter	Total	# within ±3%	Percent ±3%/Total	Rank
Criticare CSI 503 finger	40	40	100	1
Datex Satlite finger	40	38	95	2
Criticare CSI 503 ear	17	16	94	3
Novametrix 505 finger	38	35	92	4
Criticare CSI 504 finger	39	35	90	5
Datex Satlite ear	35	31	89	6
Physio-Control 1600 ear	36	32	89	6
Invivo 4500 finger	38	34	89	6
Radiometer Oximeter ear	36	32	89	6
Sensormedics Oxyshuttle finger	36	32	89	6
Ohmeda Biox 3740 finger	28	26	87	11
Criticare CSI 504 ear	14	12	86	12
Physio-Control 1600 finger	36	31	86	12
Sensormedics Oxyshuttle ear	35	30	86	12
Radiometer Oximeter finger	40	32	80	15
Ohmeda Biox 3700 ear	40	30	75	16
Ohmeda Biox 3740 ear	34	25	74	17
Ohmeda Biox 3700 finger	36	25	69	18
Datex Satlite forehead	37	22	59	19
Novamatrix 505 nose	34	19	56	20
Invivo 4500 nose	26	8	31	21

Under poor perfusion conditions, pulse oximeters might either fail to provide a reading or give a '*Low signal quality*' warning. Clayton *et al* (1991b) studied the performance of probes under conditions of poor peripheral perfusion in patients who have undergone cardiopulmonary bypass in the immediate postoperative period. The results are shown in table 11.8. Finger probes were to found to have better performances than the ear, nose, and forehead probes and the authors recommended using them during poor perfusion situations. It was also noted that ear probes generally had the faster response as reported by other studies (Severinghaus *et al* 1989, West *et al* 1987, Kagle *et al* 1987). The delay of finger probes should be taken into account when planning critical management algorithms.

11.11 ELECTROMAGNETIC INTERFERENCE

Electromagnetic interference (EMI) includes several different sources of interference from the electromagnetic spectrum. It may be generated by many sources, mostly man made but also results from atmospheric events and cosmic noise. Even nuclear explosions produce an enormous electromagnetic pulse interference. All electronic devices are affected by EMI, but the consequences are more serious when affecting medical devices such as pacemakers and pulse oximeters. Frequent sources of interference are electrostatically charged operators, communications transmissions, other medical devices, and other electrical and electronic equipment.

Pulse oximeters contain a microprocessor and many other electronic circuits that are very sensitive to EMI. The requirement in their design for a high degree of electromagnetic compatibility (EMC) is now required by statute, such as the Food and Drug Administration (FDA) in the United States. The Center for Devices and Radiological Health (CDRH) is developing a comprehensive strategy of EMC requirements for medical devices.

A performance degradation in pulse oximetry due to radiated interference was reported by Silberberg (1996). A pulse oximeter displayed a hemoglobin saturation level of 100% and a pulse rate of 60 for a patient who had deceased earlier that day. This anomalous performance was because a telemetry transceiver had been placed too close to the pulse oximeter. Thus, EMI can contribute a large error to pulse oximetry. Care should be taken to make sure that there is no significance presence of EMI in the environment.

11.11.1 Interference from magnetic resonance imaging (MRI)

The radio frequency transmissions from the magnet and rapidly switching magnetic field gradients are two majors sources of artifact generated in medical devices during magnetic resonance imaging (MRI).

The magnetic resonance scanner places unusual demands on the equipment and practices of patients' safety. As sedation or anesthesia is necessary for successful MRI of some patients (particularly infants and young children), reliable patient monitoring is essential. The strong magnetic field, radio frequency (RF) radiation, and reduced patient access complicate traditional methods of patient monitoring. Conventional ECG monitoring, for instance, is subject to artifactual changes during MRI. Moreover, infants have smaller oxygen reserves which, coupled with their higher metabolic rate, can lead to rapid decreases in blood oxygenation S_pO_2. Pulse oximetry is ideal to use during MRI. It is flexible as to the choice of monitoring site, and suffers few problems from induced electromagnetic noise.

The difficulties in using pulse oximetry in MRI stem largely from the design of the monitor unit. Pulse oximeters adapted to the MRI environment have a compact nonmetallic housing and are battery operated. Extended fiber optic leads are also used to keep the electronics outside the bore of the MRI magnet as described in chapter 7. Because there are no electric cables extending through the magnetic resonance imager bore, there is no possibility of RF burns to the patient or RF-induced noise in the signal conveyed to the processor and display unit. However, the fiber optic leads tend to be relatively delicate and easily broken. Once damaged, the cost of repair is very high. Furthermore, fiber optic systems normally require different probes for patients of different size, particularly separate adult and pediatric probes. Individual probes are very expensive.

Blakeley *et al* (1994) proposed a design system for a MRI-compatible pulse oximeter which is shown in figure 11.8. The radio frequency signals can be eliminated by using notch filters and a low-pass filter. This system can prevent radio frequency burns in patients. The proposed system also worked with existing pulse oximeters. No modifications of pulse oximeters are needed.

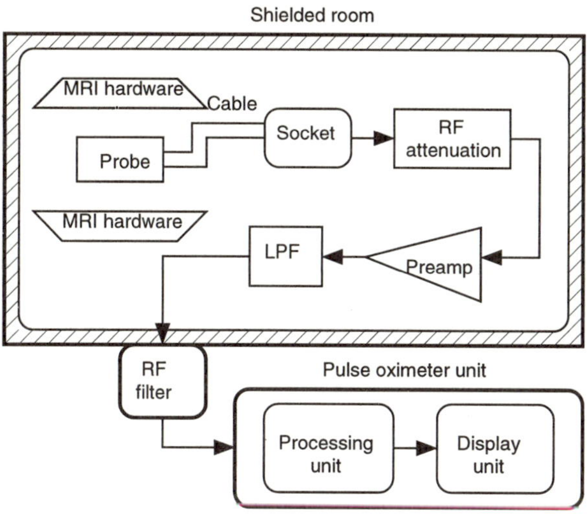

Figure 11.8 MRI compatible pulse oximetry (adapted from Blakeley *et al* 1994).

11.12 OTHER EFFECTS ON ACCURACY

Besides the major sources of errors described in previous sections, users frequently encounter other circumstances where the accuracy of a pulse oximeter is questioned. The followings are some factors which have some effects on the performance of a pulse oximeter, although no large error is expected.

11.12.1 Exercise

The presence of cardiorespiratory abnormalities during physical stress may not be noticeable under resting conditions. These abnormalities can be investigated by exercise stress testing which requires the pulse oximetry technique (see section 13.6.2). Powers *et al* (1989) and Williams *et al* (1986) found that pulse oximeters using ear probes underestimated arterial saturation by 10 to 15% during heavy exercise. It was suggested that this is caused by reduced ear perfusion in exercise. Smyth *et al* (1986) in contrast found up to 15% overestimation by a pulse oximeter using an ear probe during exercise under hypoxic conditions.

In a more recent study by Norton *et al* (1992), 10 subjects were used to perform strenuous exercise on a bicycle ergometer. Blood oxygen saturations were measured using the Ohmeda Biox pulse oximeter 3700R with the ear probe, and the blood gas analyzer (Ciba-Corning, model 278). The results of oxygen saturation levels obtained indicated that relatively large underestimations of S_aO_2 can occur when a pulse oximeter is used, and these errors increase as the severity of exercise increases. Powers *et al* (1989) found similar results. Further studies are still needed to investigate the performance of pulse oximeters during exercise. Estimations of arterial blood oxygen saturation during severe exercise using the pulse oximetry technique should be viewed with caution, as potentially large errors may occur.

11.12.2 Dried blood

Trauma patients may have significant quantities of dried blood remaining on their hands upon arrival in the emergency department. There is often insufficient time to clean the patient's hand thoroughly before the application of the pulse oximeter probe (Rosewarne and Reynolds 1991). In a study by Rosewarne and Reynolds (1991), the finger probes of six commercially available pulse oximeters were applied to the fingers of a healthy male Caucasian volunteer. Two of the fingers had previously been coated in whole blood which was allowed to dry. Rosewarne and Reynolds (1991) found that there was no significant difference in saturation range among those fingers with or without dried blood. The variation in readings between brands of pulse oximeter was of the same order as between fingers.

In emergency situations, the presence of dried blood is unlikely to cause a decline in pulse oximeter accuracy and performance as long as adequate perfusion is maintained.

11.12.3 Pigments

In theory, skin pigmentation and other surface light absorbers such as nail polish, should not cause errors in S_pO_2 readings since the pigments absorb a constant fraction of the incident light, and the pulse oximeters use only pulsatile absorption data. The absorbances of light by the pigments are nonpulsatile and, just as for tissue absorption, are cancelled out of the saturation calculation.

However, Cote *et al* (1988) found that black, blue, and green nail polishes caused a significant lowering of S_pO_2 readings of the Nellcor N-100, while red and purple nail polish did not. Cecil *et al* (1988) also showed apparently greater inaccuracy in pulse oximeter readings for black patients. This is probably caused by the fact that N-100 increases its light output in response to low detected light levels, and the higher LED current caused a shift in the output spectrum (see chapter 5). The shifting of the peak wavelength of LEDs affects the measured transmitted red and infrared light intensities, and thus alters the oxygen saturation reading. For the nail polish problem, the solution is to mount the probe side-to-side on the finger (White and Boyle 1989). This technique may also help to avoid the saturation underestimation problem caused by only partial placement of the LEDs over the finger because of very long fingernails.

REFERENCES

Blakeley D G, Gauss R C and Flugan D C 1994 MRI compatible pulse oximetry *US patent: 5,323,776*

Cahan C, Decker M J, Hoekje P L and Strohl K P 1990 Agreement between noninvasive oximetric values for oxygen saturation *Chest* **97** 814–9

Casciani J R, Mannheimer P D, Nierlich S L and Ruskewicz S J 1995 Pulse oximeter sensor optimized for low saturation *US patent 5,421,329*

Cecil W T, Thorpe K J, Fibuch E E and Tuohy G F 1988 A clinical evaluation of the accuracy of the Nellcor N-100 and the Ohmeda 3700 pulse oximeters *J. Clin. Monit.* **4** 31–6

Cheung E Y and Stommel K A 1989 Quantitative evaluation of a combined pulse oximetry and end-tidal CO_2 monitor *Biomed. Instrum. Technol.* **23** 216–21

Choe H, Tashiro C, Fukumitsu K, Masahuru Y and Yoshiya I 1989 Comparison of recorded values from six pulse oximeters *Crit. Care Med.* **17** 678–81

Clayton D G, Webb R K, Ralston A C, Duthie D and Runciman W B 1991a A comparison of the performance of 20 pulse oximeters under conditions of poor perfusion *Anaesthesia* **46** 3–10

Clayton D G, Webb R K, Ralston A C, Duthie D and Runciman W B 1991b Pulse oximeter probe: A comparison between finger, nose, ear and forehead probes under conditions of poor perfusion *Anaesthesia* **46** 260–5

Cote C J, Goldstein E A, Fuchsman W H and Hoaglin D C 1988 The effect of nail polish on pulse oximetry *Anesth. Analg.* **67** 683–6

ECRI 1989 Pulse oximeters *Health Devices* **18** 185–230

Fearley S J and Manners J M 1993 Pulse oximetry artefact in a patient with a right atrial myxoma *Anaesthesia* **48** 87–8

Kagle D M, Alexander C M, Berko R S, Giuffre M and Gross J B 1987 Evaluation of the Ohmeda 3700 pulse oximeter: steady-state and transient response characteristics *Anesthesiology* **66** 376–80

Mendelson Y, Kent J C, Yocum B L and Birle M J 1988 Design and evaluation of a new reflectance pulse oximeter sensor *Med. Instrum.* **22** 167–73

Mendelson Y and Kent J C 1989 Variations in optical absorption spectra of adult and fetal hemoglobins and its effect on pulse oximetry *IEEE Trans. Biomed. Eng.* **36** 844–8

Nickerson B G, Sarkisian C and Tremper K 1988 Bias and precision of pulse oximeters and arterial oximeters *Chest* **93** 515–7

Norton L H, Squires B, Craig N P, McLeay G, McGrath P and Norton K I 1992 Accuracy of pulse oximetry during exercise stress testing *Int. J. Sports Med.* **13** 523–7

Palve H and Vuori A 1989 Pulse oximetry during low cardiac output and hypothermia states immediately after open heart surgery *Crit. Care Med.* **17** 66–9

Powers S K, Dodd S, Freeman J, Ayers G D, Samson H and McKnight T 1989 Accuracy of pulse oximetry to estimate HbO_2 fraction of total Hb during exercise *J. Appl. Physiol.* **67** 300–4

Reynolds K J, de Kock J P, Tarassenko L and Moyle J T B 1991 Temperature dependence of LED and its theoretical effect on pulse oximetry *Br. J. Anaesth.* **67** 638–43

Rosewarne F A and Reynolds K J 1991 Dried blood does not affect pulse oximetry *Anaesthesia* **46** 886–70

Saito S, Fukura H, Shimada H and Fujita T 1995 Prolonged interference of blue dye "patent blue" with pulse oximetry readings *Acta Anaesthesiol. Scand.* **39** 268–9

Scheller M S, Unger R J and Kelner M J 1986 Effects of intravenously administered dyes on pulse oximetry readings *Anesthesiology* **65** 550–2

Seidler D, Hirschl M M and Roeggla G 1993 Limitations of pulse oximetry *Lancet* **341** 1600–1

Severinghaus J W, Naifeh K H and Koh S O 1989 Errors in 14 pulse oximeters during profound hypoxia *J. Clin. Monit.* **5** 72–81

Siegel M N and Gravenstein N 1987 Preventing ambient light from affecting pulse oximetry *Anesthesiology* **67** 280

Silberberg J L 1996 Electronic medical devices and EMI *Compliance Eng.* **XIII** (2) D14–21

Smyth R J, D'urzo A D, Slutsky A S, Galko B M and Rebuck A S 1986 Ear oximetry during combined hypoxia and exercise *J. Appl. Physiol.* **60** 716–9

Taylor M B and Whitwam J G 1988 The accuracy of pulse oximeters: a comparative clinical evaluation of five pulse oximeters *Anaesthesia* **43** 229–32

Tremper K K and Barker S J 1989 Pulse oximetry *Anesthesiology* **70** 98–108

Veyckemans F, Baele P, Guillaume J E, Willems E, Robert A and Clerbaux T 1989 Hyperbilirubinemia does not interfere with hemoglobin saturation measured by pulse oximetry *Anesthesiology* **70** 118–22

Webb R K, Ralston A C and Runciman W B 1991 Potential errors in pulse oximetry, II. Effects of changes in saturation and signal quality *Anaesthesia* **46** 207–12

West P, George C F and Kryger M H 1987 Dynamic *in vivo* response characteristics of three oximeters. Hewlett-Packard 47201A, Biox III, and Nellcor N-100 *Sleep* **10** 263–71

White P F and Boyle W A 1989 Nail polish and oximetry *Anesth. Analg.* **68** 546–7

Williams J, Powers S and Stuart M 1986 Hemoglobin desaturation in highly trained endurance athletes during heavy exercise *Med. Sci. Sports Exercise* **18** 168–73

Wong D H, Tremper K K, Davidson J, Zaccari J, Weidoff P, Wilbur S and Stemmer E A 1989 Pulse oximetry is accurate in patients with dysrhythmias and a pulse deficit *Anesthesiology* **70** 1024–5

Yelderman M and New W 1983 Evaluation of pulse oximetry *Anesthesiology* **59** 349–52

Zijlstra W G, Buursma A and Meeuwsen-van der Roest W P 1991 Absorption spectra of human fetal and adult oxyhemoglobin, de-oxyhemoglobin, carboxyhemoglobin, and methemoglobin *Clin. Chem.* **37** 1633–8

INSTRUCTIONAL OBJECTIVES

11.1 Explain the differences between between bias, precision, and the 95% confidence limit.
11.2 Describe the accuracy of pulse oximeters in the three ranges of oxygen saturation levels.
11.3 Using the absorption spectra shown in figure 11.1, explain why the accuracy is worse at low oxygen saturation level.
11.4 Describe the accuracy of pulse oximeters at low perfusion and how to prevent the errors.
11.5 Explain how venous congestion occurs and its results on pulse oximeter accuracy.
11.6 Describe two sources of optical intereferences and their effects on pulse oximeter accuracy.
11.7 Describe how to prevent errors from high intensity ambient light.
11.8 Describe how the absorbance of dyes affects the accuracy of pulse oximeters.
11.9 Explain the effects of MeBl on pulse oximeter readings.
11.10 Given $c_{\mathrm{HbO_2}}$ and c_{COHb}, calculate the estimated $S_{\mathrm{p}}\mathrm{O_2}$
11.11 Describe how MetHb and bilirubin affect the readings of pulse oximeters.
11.12 Describe how fetal hemoglobin affects the readings of pulse oximeters.
11.13 Explain how temperature affects pulse oximeter accuracy and describe how the theoretical calibration curve shifts from 0 °C to 50 °C.
11.14 Describe the accuracy and response time of finger probes and ear probes during rapid desaturation and low perfusion.
11.15 Explain the effect of EMI on pulse oximeter accuracy.
11.16 Describe the effect of MRI on pulse oximetry and explain the system of MRI-compatible pulse oximetry.
11.17 Describe the effect of pigments on the accuracy of pulse oximeters.

CHAPTER 12

USER INTERFACE FOR A PULSE OXIMETER

Albert Lozano-Nieto

12.1 INTRODUCTION

This chapter deals with some important aspects that need to be considered when designing any kind of product whose final goal is to be marketed rather than be used as a laboratory prototype. The product has to be built so that it will solve a need for the customer. A product that is technologically perfect can result in an economic failure if it is not sold because it does not meet the user's expectations or needs, it is not sold because it is too complicated to operate, or is removed from the market by the regulatory agencies because it does not meet the applicable regulations.

This chapter will highlight those aspects of the overall design for a pulse oximeter that may not receive enough attention when designing the hardware and software that make up the core of the system. These aspects are the design of an optimal user interface system, so that the final product will comply with all the regulations that apply to that specific product.

The chapter is organized by discussing the options available to the designers for the different stages that form a pulse oximeter, by reviewing how choices have been made in commercially available equipment, and by discussing which standards are applicable to the different parts of the pulse oximeter and their consequences for the design. The standards are a collection of rules, most of them based on common sense, used to ensure the best results in the use of pulse oximeters. In particular, we must comply with the *Standard Specifications for Pulse Oximeters, F1415-1992* from the American Society for Testing and Materials (ASTM) that compiles the current regulations for the design of pulse oximeters (ASTM 1992). This Standard references the *Safety of Medical Electrical Equipment—Part 1, General Safety Requirements, IEC 601-1* standard from the International Electrical Commission (IEC) for many general requirements concerning safety, and discusses the specific variations from the IEC 601-1 in the case of pulse oximeters (IEC 1988). A more detailed discussion about some aspects of the IEC 601-1 Standard for pulse oximeters is in the ISO 9919 Standard, *Pulse Oximeters for medical use—Requirements* (IOS 1992). Nevertheless, all the standards are subjected to revision, and undergo changes with the development of technology and other standards that affect related equipment. For example, in 1996, development began on a standard that will apply to all medical devices used during anesthesia. So, it is the responsibility of the designer to know and comply with the current applicable standards.

12.2 FRONT PANEL

The front panel of a pulse oximeter communicates between the patient and the healthcare professionals. This communication is expected to be accurate and clear. The accuracy problems are related to the core design, discussed in the previous chapters. This chapter will focus on how to make this communication as effective as possible, designing the pulse oximeter to display the necessary information in the way that is most useful to healthcare professionals.

Figure 12.1 shows how to model a pulse oximeter as a transducing system that transforms a variable from the chemical domain (arterial oxygen saturation), to a variable in the electrical domain that can be further processed, stored or displayed. While previous chapters have treated the first conversion stages, we will discuss the last conversion stages, that is, how the information is presented to the operator.

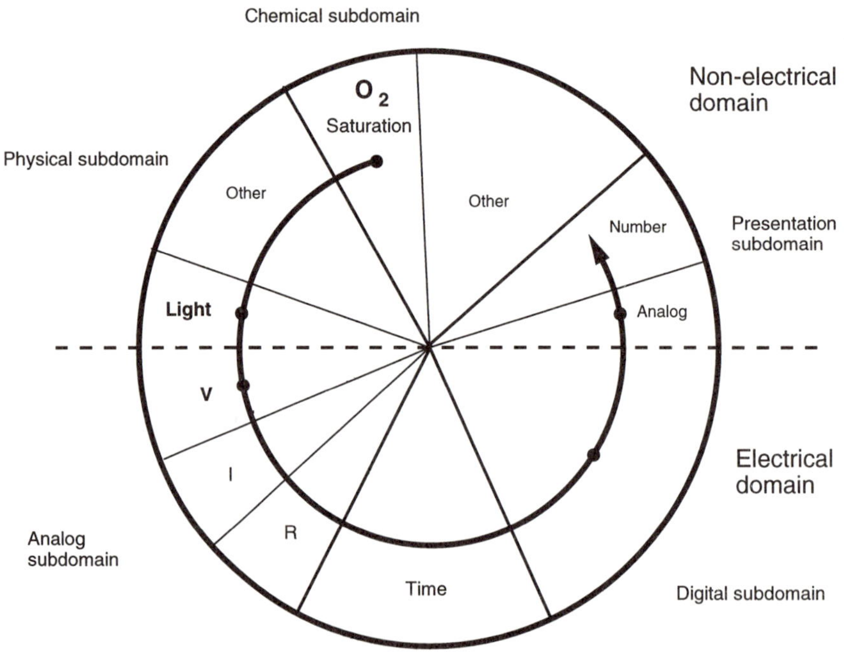

Figure 12.1 Change of domains of information in a pulse oximeter. Adapted from Malmstad *et al* (1973).

We will consider two main ways of presenting information to a human operator. These are visually and acoustically. The acoustic way is mainly used to alert the operator of a possible malfunction of the monitoring equipment or a medical problem. Other applications are to provide feedback to an input from the operator, and in some units to codify the patient pulse strength by changing the sound pitch accordingly to the strength as defined in the Standards (ASTM 1992).

Despite these acoustical outputs, the primary output of a pulse oximeter is visual. Pulse oximeters can be primarily classified based on the technique used to present visual information into two categories:

1. Graphical displays that present analog and digital information.
2. Numerical displays that only present digital information.

12.2.1 Graphical displays

It is common knowledge that 'a picture is worth thousand words', and pulse oximeters are not an exception. Graphs produce a spatial presentation to communicate quantitative information to the exterior world, making them very flexible (Gillan and Lewis 1994). The displays used in pulse oximeters are normally liquid crystal displays (LCDs), although some models from Protocol Systems Inc. (Propaq 102/104/106) also have versions with an electroluminescent display (ELD). ELD displays perform better when it is necessary to view them from long distances. They are aimed toward bedside monitoring, where the units can be plugged to a power line source, because of the higher power that these displays require. On the other hand, LCD displays are better in direct sunlight and require much less power, which increases both the display life and the battery discharge cycle (Bosman 1989). Most of the commercially available LCD units have a backlight that increases display readability but also dramatically decreases the battery operating time. For example, Criticare specifies for its 503 model, a battery use time of 20 h when the backlight is turned off, while it decreases to 10 h when the backlight is turned on.

Graphical displays present one or more real-time waveforms. Normally, the units that incorporate graphical displays are also the ones that acquire more physiological signals, so there are more choices for display. All the units with graphical displays can simultaneously present different waveforms, although for readability it is not convenient to present more than two. The most common waveforms are the plethysmographic waveform and the ECG. The model POET TE Plus from Criticare also monitors CO_2 and can display the capnographic waveform. The Propaq models from Protocol Systems, Inc., have different modular systems that can measure oxygen saturation, ECG, CO_2 consumption, and invasive and noninvasive blood pressure. The units from Medical Research Laboratories, Inc. can be used as stand-alone systems or as a part of an integrated monitoring system as previously described. The model 9500 from Magnetic Resonance Equipment Co. is a multigas monitoring system that measures oxygen saturation, CO_2, NO_2, O_2 and invasive and noninvasive blood pressure. The model BIOX 3700 from Ohmeda, shown in figure 12.2 has two separated LCD displays with different functions for each one. One displays different waveforms, while the other displays the values of oxygen saturation and pulse rate.

In addition to real-time waveforms, displays can also present the trend from a past period of time. This feature does not involve a major increase in the complexity of the electronic design because it only requires storage of the already digitized values and further processing. The length of time that is available for display depends on the amount of the memory used in the design, but also on the sampling frequency, which is normally user selectable. There is a large variation among the length of time that different models store trend display. In the Biox 3700 from Ohmeda, the length of the trend can be selected between 20 and 60 min, by pressing a key in the front panel, as shown in figure 12.2. The model N-3000 from Nellcor has three different ways of recording data for trend analysis. In the first two modes, the unit stores the average of oxygen saturation and heart rate measured over a period of 5 or 10 s, with a total duration of 12 or 24 h respectively. In the third mode, the unit stores the maximum and minimum values obtained over a period of 20 s, with a total duration of 32 h. The length of the recording also changes with presentation. The Propaq models from Protocol Systems, Inc. can display a total of 5 h on the screen and 8 h on a printer with a resolution of 2 min. The data can be presented in graphical or tabular form.

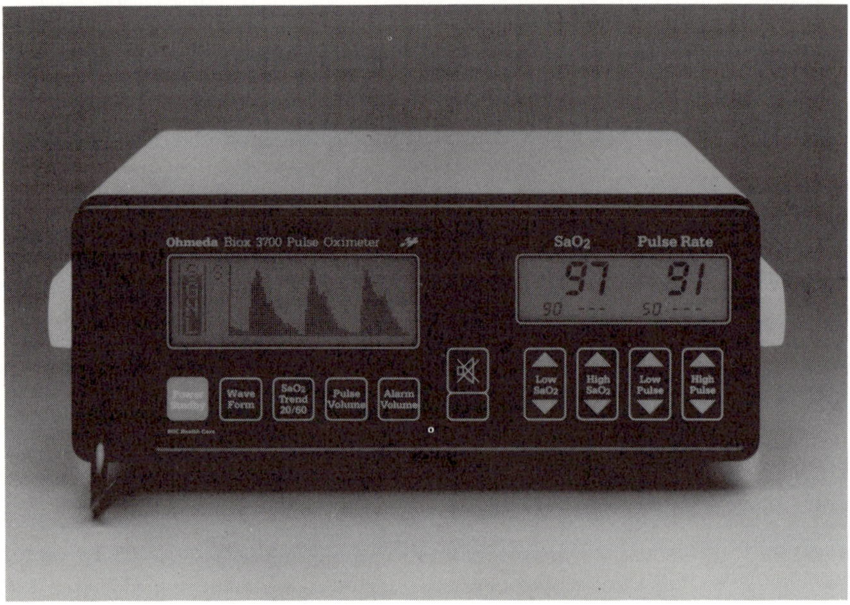

Figure 12.2 Front panel of Ohmeda Biox 3700 pulse oximeter (Courtesy of Ohmeda). The display at the left side shows real-time waveforms and pulse strength, while the display at the right side is used for alarm settings.

In some units, for example the model 504 from Criticare, the memory in which the trend data are being stored is not erasable on power-off. When trend data that contain periods of time in which the unit was turned off are displayed, the time during which the unit was turned off is shown as special characters so that the operator can be aware of this situation. This feature allows us to follow a patient during a long duration in which constant monitoring is not required. The drawback of this feature is that it can acquire the trend from the wrong patient if the previous data are not erased before starting to monitor a new patient. The trend display also marks times during which alarm set points have been exceeded or the pulse has been lost.

Trend graphs incorporate cursors that can be scrolled through the display with numeric readouts that normally show the values of the waveform and the time. This feature is particularly useful when the screen displays different waveforms because they do not incorporate a numerical vertical axis and it is not possible to distinguish magnitude by only reading the screen. It is also very important to properly label the different waveforms, as the most commonly displayed waveforms (pulse rate versus time, and oxygen saturation versus time) can present numerical values very similar to each other and confuse the system operator. It is also important that the trend display have the capability to use the dynamic range available in the screen to more clearly show small changes. For example, the model 504 from Criticare displays the trend in oxygen saturation between 75% and 100%. Although large changes in oxygen saturation can be easily recognized, it is difficult to notice small changes at a glance because most of the monitored patients will not have such a large oxygen saturation change.

A series of menus that appear on the screen normally permit the operator to select between the displayed waveforms, cursor displacement, and other function controls. The selection keys are placed under the display screen or at its sides, and the function of a particular key is automatically changed depending on the displayed screen mode.

The display of pulse strength is mandatory for those pulse oximeters that display a normalized pulse waveform (ASTM 1992). The reason for this feature is because the amplitude of the plethysmographic signal can be changed by the operator in order to achieve a good dynamic range on the screen, and it is desirable to have an indication of pulse strength regardless of the operator settings. In units with graphical displays, it is commonly done by a graphic bar whose amplitude is proportional to the pulse strength, situated on one side of the screen, as for example the unit shown in figure 12.2. The display of the pulse strength must be accompanied by acoustical signals.

Other information commonly found in graphical display units is the values at which alarms have been set, their status, low battery indication, system malfunctions, and other messages of interest to operators.

12.2.2 Numerical displays

The majority of the marketed pulse oximeters use only a numerical display made of red LED segments. In all the units examined, information on oxygen saturation and heart rate is presented. In addition to these variables, the models 507 and 5070 from Criticare Systems, Inc., that are complex monitoring units, also display the values of systolic, diastolic, and mean blood pressure. POET TE Plus from Protocol Systems, Inc. displays the values of oxygen saturation and CO_2. Because in some patients, oxygen saturation and heart rate can reach the same numerical values, it is highly desirable that the displays incorporate a fast and reliable way for the operator to associate the number on the panel with the physiological variable of interest. However, only a few units have this feature. For example, the model POET TE Plus from Criticare Systems, Inc. uses different colors for LED segments to display oxygen saturation and CO_2. The model 3500 from Magnetic Resonance Equipment Co. and the models 504 and 504S from Criticare Systems, use different size LED segments to display oxygen saturation and pulse rate, and Medical Research Laboratories, Inc. uses larger green LEDs for oxygen saturation display and smaller red LEDs for heart rate display. The pulse strength in all the units with numerical output is displayed using a LED bargraph.

12.3 FUNCTION CONTROLS

Function controls carry out communication from the healthcare professionals to the pulse oximeter to achieve the proper monitoring and care for the patient. Function controls are basically used to operate alarms (set alarm values, activate, deactivate and silence alarms) and displace the cursors along the graphical screen in those units with this feature.

It is possible to distinguish three different function controls: switches, turning knobs and keys. They do not all need to exist in the same unit.

The main function of switches is to turn the device on or off. In some units switches are replaced by keys. Since this is the most basic function in a pulse oximeter, it is important that it cannot be turned off accidentally. For this reason, some units have the main power switch or key in a lateral panel where it is unlikely to turn the power off by accident.

There are few units that incorporate turning knobs. The model N-200 shown in figure 12.3 and model N-3000 from Nellcor use turning knobs as an intuitive and quick way to increase or decrease the alarm settings. The turning knobs are placed on the front panel or on the top of the unit, where they are large and thus are easier to manipulate without affecting other controls. A function that uses turning knobs for control has to be designed so that a movement upwards, to the right or in a clockwise direction increases the control function (ASTM 1992).

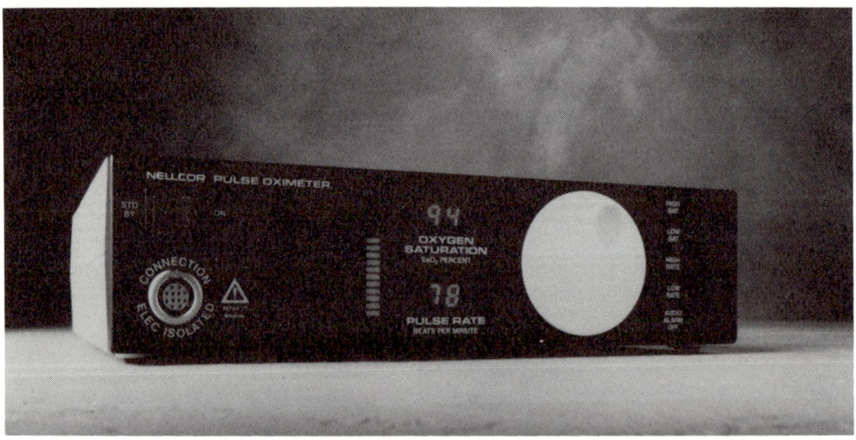

Figure 12.3 Nellcor Puritan Bennett N-200 pulse oximeter (Reprinted by permission of Nellcor Puritan Bennett, Pleasanton, California).

The majority of pulse oximeters use keys as input devices to control the instrument. We can distinguish between units that use touch panel keys and units that use push buttons. Touch panel keys have the advantage that they are cheaper to manufacture and insert during the manufacturing process and can accommodate LEDs to indicate that the function is active. They also contribute to a better seal of the unit's front panel, thus making it more suitable for use in hostile environments. Figure 12.2 shows a unit that uses these kind of keys for front panel functions. On the other hand, push buttons have a better feel and require lower pressure to activate. However, they have open spaces around them, can permit dust, humidity and other chemical agents to shorten the life of their electrical contacts.

For any kind of keys used, it is desirable that the operator has a feedback that the key has been pressed successfully, either by a visual stimulus such as turning on a LED in a touch panel key, an audio stimulus by emitting a characteristic sound, or tactile feedback from the release of the pressed key pressing on the operator's finger (Cakir *et al* 1980).

It is important to consider the number of different keys that are available in a pulse oximeter. In general, it is best to have as few keys as possible to simplify the access to the most common and critical functions, such as setting the alarm values. For example, the model N-200 from Nellcor has a very intuitive way of setting the alarm values (low oxygen saturation, high oxygen saturation, low pulse rate, and high pulse rate) that consists of pressing a single key to select the alarm, and modify the actual value by rotating the turning knob as can be seen in figure 12.3. However, this device has only five different keys, so the operator needs to press two different keys simultaneously to activate other functions. Because the key labeling only refers to the basic function, it can become difficult to remember which keys need to be pressed in order to activate the desired function, and it is therefore harder to perform. In this particular unit, the manufacturer supplies a quick reference card to be placed on the bottom of the unit. It provides a helpful reminder to the operator if the operator knows where to look.

On the other hand, the model 504US from Criticare uses the dynamic key function and labeling that has been described in previous sections. With only three touch panel keys for menu purposes, the operator enters a series of menus and submenus, changing the function of the keys according to the menu that is active. Although this way of controlling the functions has the advantage that the operator always knows the function of the set of keys, it is very easy to forget the depth of the menu entered, in which submenu a particular function of interest is located. It can also be time consuming to move between functions located in different submenus.

In the same way that the operator needs feedback to indicate that a particular key has been pressed successfully, the operator also needs some feedback that indicates that the key, or combination of pressed keys, is valid, and a control function has been executed. The most common way to produce this feedback is by turning on a visual indicator that is related to the function executed, or by emitting a characteristic sound in the case of invalid keys.

It is also important to pay attention to the layout of displays and indicators and their control keys, selecting the position of the controls in a place that is consistent with the display. Figure 12.4 shows different examples of good and poor relative positions between displays or indicators and controls, based on the idea that they have to be laid out in such a way that the relationship between controls and their indicators is obvious.

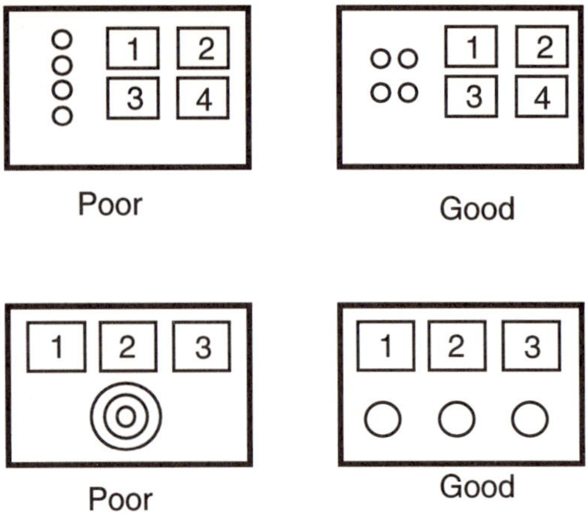

Figure 12.4 Layout of controls and indicators to ensure good operator interaction. From Salvendy (1987).

12.4 ALARM CONTROLS

The alarms communicate the patient to the healthcare professionals, alerting of a potentially dangerous situation. Because the alarms are the most critical functions in a pulse oximeter, it is absolutely necessary to be sure of their proper working condition, as well as to take extra effort to design them in such a way that they cannot be disconnected accidentally.

The design of alarms and their controls section is by far the most regulated by the standards. The most common type of pulse oximeters, the units that display the oxygen saturation and heart rate, provide alarms for the following situations:

1. High oxygen saturation.
2. Low oxygen saturation.
3. High pulse rate.
4. Low pulse rate.

Other sections in the ASTM Standard regarding the operation of alarms require that the alarm set points be operator adjustable, that the default limits on low oxygen saturation be 80% saturation or greater, and the difference between the alarm set point and the actual value of arterial oxygen saturation when the alarm is activated not exceed 2% of oxygen saturation (ASTM 1992).

In most of the units, it is possible to deactivate at least the alarm for high oxygen saturation, except in the case when the pulse oximeter is configured for neonatal monitoring. The pulse oximeter shown in figure 12.2 has deactivated the alarms for high oxygen saturation and high puse rate.

From these alarms, only the low oxygen saturation alarm is required for the pulse oximeter to be qualified as a monitoring device. Those devices without low oxygen saturation the alarm shall be marked as 'NOT FOR MONITORING' (ASTM 1992). All the marketed units examined provide these four alarm situations, except the models 8500 and 9500 from Nonin Medical, Inc. These units have been designed not for a bedside monitoring situation in a hospital where the alarms are used to attract the operator's attention, but in a one-on-one working situation where a healthcare professional is always present with the patient, using the pulse oximeter to measure the oxygen saturation, for example, during ambulance transport.

The 9500 unit, shown in figure 12.5, is the smallest available in the market. With a weight of only 36 g without batteries and an extremely small size, just slightly larger than most of the reusable finger probes, it displays heart rate and oxygen saturation. The unit 8500 is a hand-held pulse oximeter that has been designed to provide 100 h of continuous operation with batteries. Both units have been designed for evacuation situations. They both comply with the USAF vibration standards for helicopter flight use, can operate at temperatures below freezing, and the manufacturer stresses their use in helicopter evacuation.

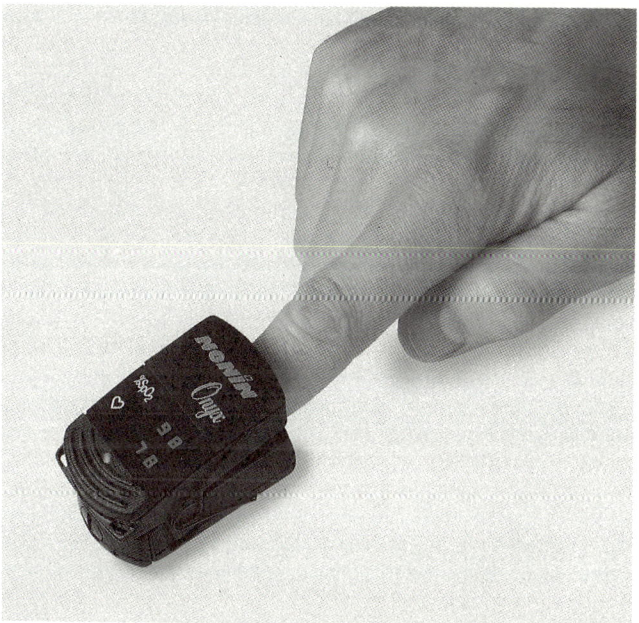

Figure 12.5 A small Nonin model 9500 pulse oximeter designed for emergency evacuation purposes (courtesy of Nonin Medical Inc.).

The visual and acoustic characteristics of the alarms are also regulated by the ASTM Standards, as shown in table 12.1. The ASTM differentiates three kinds of alarms based on their priority, assigning different colors and flashing frequency to each one.

Table 12.1 Alarm characteristics for pulse oximeters (ASTM 1992).

Alarm category	Operator response	Audible indicators	Indicator color	Flashing frequency (Hz)
High priority	Immediate	Not medium or low priority	Red	1.4 to 2.8
Medium priority	Prompt	Not high or low priority	Yellow	0.4 to 0.8
Low priority	Awareness	Not high or medium priority	Yellow	Constant

The current ASTM Standard specifies neither the frequency nor the volume of the acoustic alarm sounds. Good practice suggests that the frequency of warning sounds should be between 150 Hz and 1000 Hz. It should have at least four frequency components in order to avoid masking from environmental noise. The acoustic level recommended is 15 dB to 16 dB above the masked threshold for signals that are triggered by situations that require a rapid response, and levels between 6 dB and 10 dB above the masked threshold for all other kinds of signals, to achieve 100% detectability in controlled situations. In all cases, the level should be less than 30 dB above the masked threshold to minimize operator annoyance and disruption of communications (Salvendy 1987).

The alarms in a pulse oximeter can be disconnected or silenced. Temporary silencing should be used when the operator has been alerted of the potentially dangerous situation and has taken steps in order to solve the problem. The Standard specifies that if this feature is provided in the pulse oximeter, it should not exceed 120 s, and a visual condition of the alarm has to remain on until the condition that triggered the alarm is corrected (ASTM 1992). The reason pulse oximeters incorporate a permanent silencing alarm is to avoid nuisance noise when the device and probe are being connected to the patient. The permanent alarm silencing activation must be designed in such a way that it requires a deliberate action for deactivation by the operator to be sure that it is not done in error. It also requires a visual indication of this condition.

As most of the pulse oximeters monitor heart rate from the plethysmographic waveform, they also incorporate alarms in case the pulse is lost. This increases security for the patient by monitoring more vital signs, but it also triggers false alarms, in particular due to motion artifacts. To avoid this problem, Nellcor has developed what they call Oxismart, which, for loss of pulse, aims to distinguish between a real clinical condition and a motion artifact. This feature is incorporated in the latest models, such as the N-3000.

Motion artifacts are detected by processing the plethysmographic waveform and before validating a pulse, requiring three different steps. Only the signals that pass all the steps are used to calculate S_pO_2 (Nellcor 1995). To differentiate between a loss of pulse due to motion artifact from a loss of pulse due to a clinical condition, the system assumes that if the pulse is lost, but the patient is moving, the patient has pulse and the loss is due to a motion artifact. Figure 12.6 illustrates this fact. If the pulse oximeter fails to detect at least one pulse in 10 s, it enters into pulse search mode. The operator is aware of this situation because the PULSE SEARCH indicator lights, and the display alternates between data and dashes. In this condition, the pulse oximeter enters an evaluation period of 50 s. If the patient is moving, each time that the pulse oximeter detects a valid pulse, readings for heart rate and oxygen saturation are validated. The device returns to its normal operation after detecting an adequate sequence of validated pulses. If during the 50 s evaluation period, an adequate pulse sequence is not detected, a low-priority alarm sounds, and there is a visual indication of this condition as shown in table 12.1. On the other hand, if the pulse oximeter does not detect motion after 60 s in pulse search mode, a high-priority alarm sounds, and there is also a visual indication of this condition. With this feature, it is possible to track the oxygen saturation even in patients that produce signals of poor quality, and at the same time warning can be given of a potentially dangerous condition.

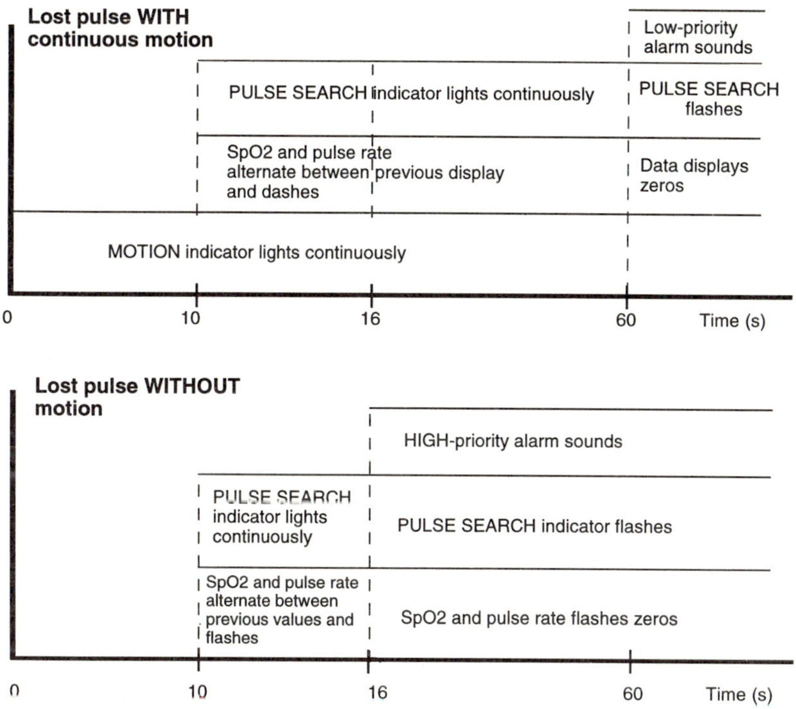

Figure 12.6 Oxismart© alarm detectors used in some Nellcor units to reduce false alarms due to motion artifacts (Nellcor 1995).

12.5 COMMUNICATION FUNCTIONS

Communication functions are not a primary function, but an added value feature for a pulse oximeter. Communication functions can be found in all types of devices, but they provide a great improvement to the units with only numerical display, because it gives them the graphical features that otherwise are missing. They are used to send data to a printer or plotter. The most common use is to print the trend for both oxygen saturation and heart rate for a patient. This feature converts the most simple units into units that act like solid state Holter monitors, with the clinical advantage associated with the knowledge of trend over time. There are few units that incorporate an internal printer, normally a thermal one, thus eliminating the need for extra connectors and cables.

The most common method of communication is using the RS-232 protocol. It is also possible to obtain analog signals proportional to the plethysmographic and pulse rate waveforms. The voltage output is normally selectable between a range of 0 to 1 V dc and 1 to 10 V dc.

12.6 CABLES AND CONNECTORS

The cables and connectors are used to transmit power and signals between the device and the surrounding accessories and power supplies. We can roughly distinguish three levels of communication:

1. Interface with the power source.
2. Interface with the lead and probe.
3. Interface with auxiliary equipment.

The power connector is used to transfer the energy required from the power source to the unit for its operation. The Standard requires that it should be designed so that it protects the patient from human errors (ASTM 1992). This means that it has to be clearly different from the connectors that will be attached to the patient. Power connectors are used to operate the units when it is turned on, and to recharge the battery when the unit is turned off.

The connector for the lead and probe is usually placed on the front panel, and it is usually mechanically incompatible between different manufacturers, unless they specify that the probe is compatible. For example, Protocol Systems advertises that their Propaq models can use probes from Nellcor. The most common types of probe connectors are DB9 and DIN. In all cases, the connectors are mechanically designed with physical alignment aids and visual indicators to be sure that the lead is inserted the correct way into the connector. It is important that the connectors be constructed robustly, because the unit can be subjected to severe mechanical stress and vibration. Because most of the units can be synchronized with the ECG signal, obtained through a separate module, it is common to have an ECG connector.

The auxiliary connectors are normally located on the side or the back panels, and they are normally used for communication functions. The most common ones are the transfer of digital data to a printer or analog data for further recording or to a graphical plotter. For these auxiliary functions almost every manufacturer uses their own set of connectors, voltage levels, and communication protocols that make them work only with their own peripheral units.

12.7 OTHER FEATURES

Other indications that need to be displayed in a pulse oximeter are those regarding the correct labeling of all inputs, outputs, control knobs, and keys. Some models of pulse oximeters are manufactured in different levels of electrical isolation. For example, Criticare manufactures the unit 504/504US in BF (body floating) and CF (cardiac floating) versions. Because they look externally very similar, if not the same, it is very important to carefully mark its application on the front panel to avoid connecting a patient that needs a CF unit to a BF unit.

For those units that can be operated using an internal rechargeable battery, or disposable batteries, it is important to have an external indication of the approximate level of charge of the batteries and the remaining operating time, to control their replacement. The units from Medical Research Laboratories, Inc. display the charge level on an indicator. Most other units display a low-battery warning signal.

12.8 COMPLIANCE REQUIREMENTS

The Electromagnetic Compliance (EMC) requirements for electrical equipment in general, and biomedical equipment in particular, are changing at a fast pace. Because most of the new regulations have long transition periods during which they are not mandatory, it is wise to design products for future compliance with those regulations. We do not describe the current applicable regulations and standards, but describe their existence and probably future evolution.

The basic idea behind the set of EMC regulations is to ensure the safety of operation of electrical equipment during normal circumstances. This means that a particular device should not cause harmful interference to other devices and this device should not be affected by interference from other devices. Figure 12.7 illustrates these effects. They can be summarized as conducted emissions, radiated emissions, and immunity from interference generated by other equipment that can be either radiated or conducted to the device in question (Gerke and Kimmel 1994a). For the interference generated in the unit, most of the problems are caused by the radiated emissions, because the use of microprocessors running at high clock frequencies is becoming more common in medical devices and these generate radiated interference.

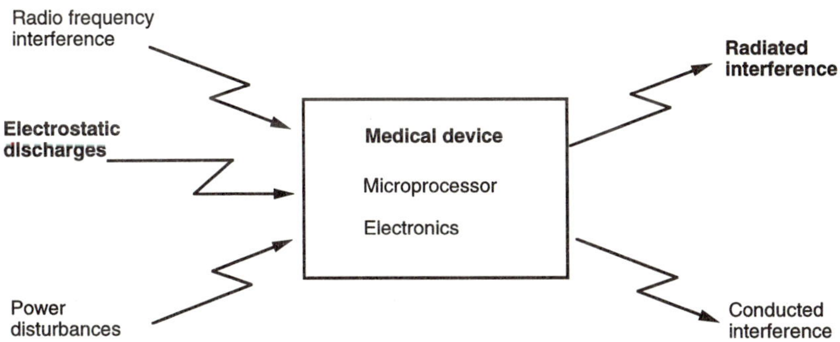

Figure 12.7 Different sources of disturbances and interferences for EMC purposes.

The ASTM Standard refers to the IEC 601-1 and IEC 801-2 Standards for electromagnetic compatibility requirements in pulse oximeters (IEC 1988, 1990). The IEC 601-1 Standard describes a general set of requirements for the safety of electrical equipment for medical use. The unit only needs to be tested against electrostatic discharges (ESD) for its accessible parts, rather than in the interior of the device. The Standard justifies this procedure based on the fact that pulse oximeters are not life-support devices, but vigilance adjuncts. Therefore, the cost to provide immunity against ESD in the interior of the system is not justified (ISO 9919, annex L). The same Standard, however, serves as a reminder to exercise common sense and provides acceptable work procedures for maintenance personnel that require them to open the device.

However, many times the manufacturers try to expand their market by exporting their products to other countries. Therefore the designers must be aware of the existence of other EMC regulations, which are generally less strict in the US and more strict in European, Asian and most other countries. As a rule of thumb, the European Economic Community (EEC) countries have more regulations and fewer exceptions to those regulations than the US, where most of the regulations are voluntary for most of the medical equipment. However, medical regulations are undergoing significant changes, and we may expect mandatory EMI regulations in the future, regarding ESD, RF fields and power disturbances, driven by the Food and Drug Administration (FDA) and the regulations in the EEC. At the present time, there are no mandatory regulations in the US, as medical devices are exempted from Federal Communications Commission (FCC) emission regulations, and they are covered only by voluntary susceptibility requirements. On the other hand, in the EEC countries, the equipment is required to be tested for emissions but not for immunity (Gerke and Kimmel 1994b). This situation is expected to change soon, and in the future we may expect mandatory regulations for RFI, ESD, and power disturbances in the US. Because of the need to be competitive in international markets, designers should consider that the best way to avoid unnecessary delays, and to lower the economic impact of changing a design, is to design for compliance from the first stages, without overdesign that implies an increment of cost with no additional value.

REFERENCES

ASTM 1992 *Standard Specification for Pulse Oximeters, F1415-1992* (Philadelphia, PA: American Society for Testing and Materials)

IOS 1992 *Pulse oximeters for medical use—Requirements, ISO 9919:1992 (E)* (Geneva: International Organization for Standardization)

IEC 1988 *Safety of Medical Electrical Equipment—Part I, General Safety Requirements IEC 601-1: 1988* (Geneva: International Electrical Commission)

IEC 1990. *Electromagnetic Compatibility for Industrial Process Measurement and Control Equipment. Part 2: Electrostatic Discharge Requirements IEC 801-2* (Geneva: International Electrical Commission)

Bosman D (ed) 1989 *Display Engineering. Conditioning, Technology, Applications* (New York: Elsevier)

Cakir A, Hart D J and Stewart T F M 1980 *Visual Display Terminals* (New York: Wiley)

Gerke D and Kimmel B 1994a Noise and interference: a different game *Electron. Design News* **39** (2) 5–14

Gerke D and Kimmel B 1994b EMI regulations. Why, where and what do they mean *Electron. Design News* **39** (2) 15–22

Gillan D J and Lewis R 1994 A compartmental model of human interaction with graphs: 1. Linear regression modeling *Human Factors* **36** 419–40

Malmstad H V, Enke C G and Crouch S R 1973 *Electronic Analog Measurements and Transducers* (Menlo Park CA: Benjamin)

Nellcor 1995 *Technology Overview: Nellcor Symphony N-3000—The next generation on Nellcor Pulse Oximetry. Reference Note: Pulse Oximetry Note Number 8* (Pleasanton, CA: Nellcor)

Salvendy G (ed) 1987 *Handbook of Human Factors* (New York: Wiley)

INSTRUCTIONAL OBJECTIVES

12.1 Describe the role of the user interface in a pulse oximeter.
12.2 Discuss the advantages and drawbacks of graphical representation of information.
12.3 Describe the most important features when designing a pulse oximeter user interface.
12.4 For pulse oximeters that only have a numerical output, describe how they can present oxygen saturation over a long period of time.
12.5 Describe and compare different types of alarms in a pulse oximeter.
12.6 Discuss how the number of keys in a pulse oximeter affect its use.
12.7 Name and describe the mandatory alarms in a pulse oximeter.
12.8 Describe the need to comply with EMC regulations.

CHAPTER 13

APPLICATIONS OF PULSE OXIMETRY

Joanna B Ruchala

Pulse oximetry is noninvasive, easy to use, readily available, and accurate. It provides information about blood oxygen saturation, heart rate, and pulse amplitude. Due to these characteristics, it has an abundance of clinical uses. Some of the main areas in which it is used are anesthesia, patient transport, childbirth, neonatal and pediatric care, sleep studies, and veterinary medicine. This chapter will discuss the causes of patient desaturation in these and other areas and how pulse oximetry is used to detect it and prevent severe hypoxemia from occurring. Some of the applications require special apparatus for pulse oximetry. Some require special calibration or specific methods of measurement.

13.1 ANESTHESIA

Air contains 20.9% oxygen which is often not sufficient during anesthesia due to problems such as airway closure, ventilation/perfusion imbalance, and CO_2 retention (Tyler *et al* 1985). Also, most anesthetics cause respiratory depression. This is when the pons and *medulla oblongata*, which control respiration, are not functioning properly. Respiratory depression reduces ventilation and can cause desaturation. Due to these problems, patients are generally preoxygenated and given a 30% oxygen mixture while under anesthesia. This, however, does not ensure prevention of desaturation. Episodes of desaturation are most often caused by human error. J B Cooper of the Department of Anesthesia at Harvard University found that human error caused 82% of incidents of desaturation during anesthesia, while equipment failure caused only 4.3% (Cooper *et al* 1984). Human error includes such things as misreading the flow meter and inadvertently allowing a lower inspired oxygen pressure than required by the patient, positioning the patient incorrectly such that the airway is obstructed, performing tracheal intubation incorrectly, administering sedatives which hinder alveolar ventilation, and encountering complications during surgical retraction. Equipment failure includes blocks in the flow meter and leaks in the anesthesia machine or breathing apparatus.

Cyanosis, a bluish tint to the skin caused by lack of oxygen, cannot be detected by a physician until the S_aO_2 is around 80% (Payne and Severinghaus 1986). Once the arterial oxygen saturation is that low, any decrease in partial pressure will cause a dramatic decrease in S_aO_2 due to the steepness of the oxygen dissociation curve (see figure 1.7). Other physiological signs of desaturation such as a drop in blood pressure or reduced heart rate also do not occur until the patient's arterial oxygen saturation is dangerously low. Blood gas analysis is very accurate, but it is invasive and slow (it takes approximately 5 min to obtain a measurement). Pulse oximetry can detect desaturation quickly and accurately and has significantly reduced the number of anesthesia-related deaths. The Datex Satlite is a pulse oximeter specially designed for anesthesia monitoring. The plethysmograph reveals circulatory depression and arrhythmia. Signal processing algorithms detect trends in pulse amplitude, S_aO_2, and pulse rate. Amplitude trends describe the course of the anesthetic (trends during a 1 h, 45 min period) and recovery (trends during a 7 h period). It can also display the CO_2, O_2, or agent waveforms.

Patients who have been under general anesthesia for surgery are often given supplemental oxygen during the procedure and in recovery. However, it is important to monitor their arterial oxygen saturation during transfer as well. Their ventilation is often poor due to residual anesthetics and muscle relaxants (Tyler *et al* 1985). Also, their alveolar–arterial oxygen tension gradient may be abnormal due to a ventilation/perfusion imbalance.

13.1.1 Problems encountered during induction to anesthesia

Desaturation is often a problem during induction to anesthesia. Moller *et al* (1991) found that during this phase, arterial oxygen saturations of 90% or less occur with a frequency of 25% of patients. Pulse oximetry can detect desaturation in real time and indicate the need for an increased oxygen mixture or adjustment of an endotracheal tube. Tracheal intubation can be a problem during anesthesia due to improper tube placement or subsequent tube movement.

Buchanan (1991) combined an endotracheal tube and a pulse oximeter probe to allow monitoring of both tube placement and arterial oxygen saturation as shown in figure 13.1. Light emitting diodes (LEDs) are attached to the leading end of the tube. Lead wires are embedded in the body of the tube, extending out of the patient's mouth. A photodiode is located outside the patient's body and placed on the anterior surface of the neck, opposite the LEDs. The photodiode's position is adjusted to detect the maximum amount of light from the LEDs and then secured with surgical tape. The LEDs and photodiode are connected to a pulse oximeter to measure arterial oxygen saturation. Measurement in this location as opposed to at extremities such as the finger or ear is more accurate and more sensitive to rapid changes in oxygen saturation. This is because blood flow in the arteries of the neck leading to the brain is preserved at the expense of blood flow to peripheral regions. If the physician prefers to keep the LEDs outside the patient's body, fearing burns to sensitive tracheal tissue, optical fiber can be used to transport the light into the trachea. If the signal to the pulse oximeter is lost during the surgical procedure, this indicates that the tracheal tube has been displaced.

Application of a laryngeal mask can also be troublesome. Haynes *et al* (1992) determined a 3% failure rate in the insertion of a laryngeal mask, and application difficulty in 18% of patients. Difficulty applying the mask sometimes occurred because the depth of the anesthesia was not great enough.

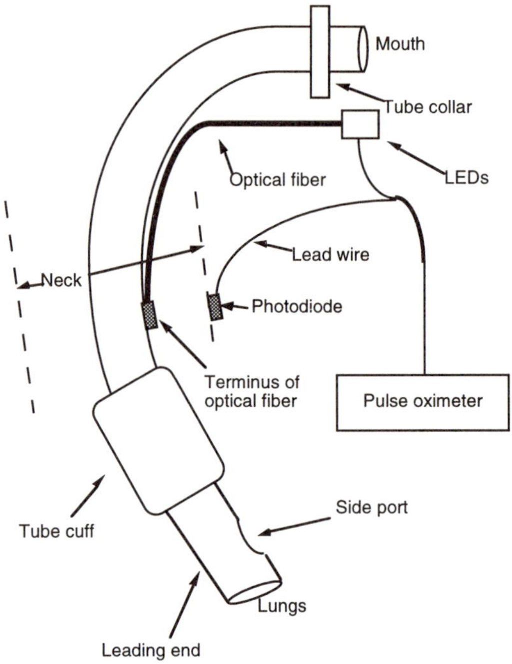

Figure 13.1 Endotracheal tube with pulse oximetry attachment. In this version, the light source is located outside the mouth, and light is transported into the trachea via optical fiber (adapted from Buchanan 1991).

13.1.2 Surgery under anesthesia

13.1.2.1 Abdominal surgery. Use of anesthesia during abdominal surgery can cause patient desaturation. During this type of surgery gas exchange in the lungs can become impaired due to a reduction in functional residual capacity (FRC). This condition can persist for several days after the operation (Knudsen 1970). Reduced FRC is thought to be caused by a reduction in the resting tone of the inspiratory muscles of the rib cage and diaphragm, which oppose the elastic recoil of the lungs (Roberts *et al* 1993). Reduced FRC can in turn cause alveolar collapse. The chance of collapse is increased by the presence of gases such as nitrous oxide which have a high solubility in blood (Roberts *et al* 1993). Alveolar collapse causes *atelectatic areas* (airless pockets) to develop in the lungs which cause desaturation (Strandberg *et al* 1986). Pulse oximetry can also be used to test the viability of internal organs in the abdomen by applying a reflectance probe covered with a sterile plastic bag directly to the organ (Moyle 1994). After the operation, postoperative pain and analgesia (sedatives) have also been found to increase desaturation (Catley *et al* 1985).

13.1.2.2 Thoracic surgery. During thoracic surgery, the anesthetic agent is often introduced to one lung. This causes a reduction in the volume of that lung, and the ventilation/perfusion of the lungs becomes unequal. The lung with the anesthetic agent has poor ventilation and good perfusion, while the other lung has good ventilation and poor perfusion (Payne and Severinghaus 1986). Therefore, while the patient has no trouble expelling CO_2, the one active lung cannot accommodate enough oxygen to sustain the patient. This results in desaturation. Pulse oximetry monitoring is necessary to determine the need for increased oxygen mixtures.

13.1.2.3 Dental surgery. In dental surgery, desaturation often occurs during particular stages such as induction to anesthesia, laryngeal mask application, prop insertion (to keep the mouth open), and dental extraction. Sometimes respiration can be detected simply by observing the reservoir bag (Bone *et al* 1987). After anesthesia, lateral positioning, oral packs, and a loose fitting face mask make it harder to detect. In a study by Lanigan (1992), 32 out of 120 patients experienced significant desaturation after dental surgery. Nitrous oxide is often used for anesthesia in dental surgery. It is 40 times more soluble in blood than nitrogen. Therefore when a patient is removed from nitrous oxide, the nitrous oxide diffuses out of the lungs faster than nitrogen diffuses into the lungs. This can cause diffusion hypoxia if the oxygen–nitrogen mixture is not high enough. At least 40% oxygen should be inspired for 10 min after nitrous oxide is stopped (Moyle 1994). Also, combining sedatives such as diazepam and midazolem with anesthesia can increase desaturation (Payne and Severinghaus 1986).

13.2 MONITORING TISSUE BLOOD SUPPLY AND ORGAN VIABILITY

A specific organ or tissue bed may not be receiving an adequate blood supply even though the patient's S_pO_2 as measured from an extremity is normal. Direct application of pulse oximetry to an organ or tissue bed can be used to determine its blood flow and viability.

13.2.1 Intestinal blood flow and bowel viability following surgery

Macdonald *et al* (1993) conducted a study to determine if pulse oximetry could be used to monitor intestinal blood flow. Oxygen saturation was measured using a Nellcor D-20 transmission probe folded around the intestine of dogs at three different sites. Blood flow was measured by an ultrasonic flow probe at the root of the superior mesenteric artery. Just prior to the flow probe, a clamp was placed for reducing the blood flow by 50% and 75%. A 15 min equilibration period was given after each reduction before measurements were taken. Blood gas analysis was used to compare with pulse oximeter measurements. The S_pO_2 reduced from 93 ±1% to 83 ±1% and then to 76 ±1%, respectively, for the reductions in blood flow. Macdonald *et al* (1993) concluded that in tissue beds that are not very metabolically active such as the ear lobe or finger tip, blood flow will not have much effect on arterial oxygen saturation. In tissue beds which are very metabolically active such as the intestine, blood flow can have a significant effect on arterial oxygen saturation. Therefore, pulse oximetry is useful for determining intestinal viability after surgery.

13.2.2 Tissue transfer and setting of limb fractures

When transferring tissue such as skin, muscle flaps, and digits, it is important to detect whether the tissue is getting an adequate blood supply. The muscle should be monitored via pulse oximetry for 24 to 48 h to determine for certain whether it will survive (Lindsey *et al* 1991). If a transmission probe is used, it is important to avoid pressure *necrosis* of the delicate muscle. In patients with limb fracture, the pulse oximeter can detect inadequate blood flow distal to the fracture. Two pulse oximeters should be used for this test, one on the injured limb and another on the healthy limb (Moyle 1994). Both pulse oximeters should obtain the same oxygen saturation measurement. Inadequate blood flow could be the sign of an entrapped artery or other complications due to incorrect setting of the fracture (David 1991). One drawback of measuring limb perfusion with a pulse oximeter is that although a signal may be obtained at the extremity of the limb, it does not ensure that muscle beds are well perfused (Clay and Dent 1991).

13.2.3 Dental pulp blood supply and viability

Pulse oximetry can also be used to diagnose dental pulp viability (Schmitt *et al* 1991). A tooth may be degenerating even though it appears normal to the naked eye or via x-ray images. Also, pulp inflammation can occasionally subside without intervention. Blood flow determines the viability of dental pulp and S_aO_2 determines the state of degeneration of a still viable tooth. Past techniques to determine dental pulp viability involved nerve stimulation. This was painful, often inaccurate, and gave no information about the state of degeneration. Nerves can sometimes function although blood flow is impaired.

To understand how monitoring oxygen saturation of dental pulp with pulse oximetry is accomplished, it is important to be familiar with the morphology of the tooth. The outer layers of the tooth consist of bone-like enamel and *dentin*. Collagen fibers connect the jaw bone to a layer of cementum at the base of the tooth, fixing the tooth in its socket. *Apical foramen*, small holes in the roots of the tooth, allow nerves and blood vessels to access the dental pulp. The blood provides oxygen, mineral salts, and nutrients to sustain the *odontoblasts* and neural tissues. Figure 13.2 shows that to measure the blood oxygen saturation, an adapted transmission probe in the shape of a U is applied over the tooth (Schmitt *et al* 1991). A black-foam insert conforms to the tooth and provides shielding from ambient light. It can be replaced and the probe reused on successive patients. The U-shaped probe is flat on top, providing a surface for the patient to bite down on to increase probe stability. The bone-like layers which surround the pulp create an optical shunt of sorts, allowing some light from the LEDs to be transmitted to the photodiode without passing through blood. Due to this extra variable, three wavelengths are needed to isolate the extinction coefficients of the blood. A wavelength in the range of 540 to 570 nm (green) is used because the extinction coefficients of enamel and dentin at this wavelength are similar to their extinction coefficients at 660 and 940 nm. Also, the extinction coefficients of oxygenated and deoxygenated blood at this wavelength greatly exceed their values at the red and IR wavelengths. The hardware of the pulse oximeter is similar to that of a two-wavelength pulse oximeter and the ratio of ratios computation of oxygen saturation is used. However, the denominator of each ratio is adjusted by subtracting off the detected DC value of the green wavelength (see equation (9.30)).

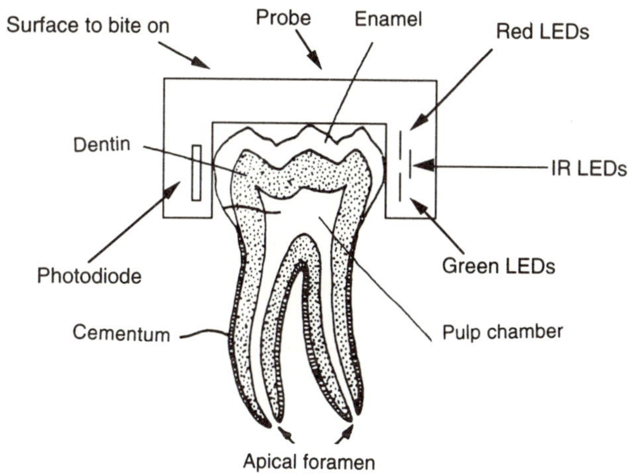

Figure 13.2 Morphology of the tooth and adapted pulse oximeter probe for determining the viability of dental pulp (adapted from Schmitt *et al* 1991).

13.3 MONITORING ON THE ROAD AND IN THE AIR

Pulse oximeters provide accurate, continuous, real-time oxygen saturation monitoring. Since they are also noninvasive, easy to use, and portable they are beneficial for monitoring in ambulances and aircraft. Both ambulances and helicopters are used for patient transport, during which vital signs need to be monitored. Altitude can cause desaturation, especially in critically ill patients. Pilots in the military are also subject to strong forces due to high acceleration, which can move blood out of the brain. These factors can cause loss of consciousness. Pulse oximeters intended for use in these types of environments are subject to special design considerations due to noise and vibration.

13.3.1 Ambulances

Pulse oximeters to be used in ambulances should be light weight and portable so the ambulatory team can apply the monitor as soon as they reach the patient. This provides immediate feedback as to the patient's condition and continuous monitoring while moving the patient into the ambulance. Once inside the ambulance the team is often very busy applying supplemental oxygen, tracheal intubation, CPR, etc. Therefore it is important that the pulse oximeter display is easy to read and the alarms are loud and distinct. During transport, the bouncing of the vehicle can cause the probe to be displaced and temporarily lose the signal. It is important for finger and ear probes to fit properly and snugly on the patient. Poorly fitting probes are often a problem when monitoring children. Vehicle motion can also create artifacts, increasing the need for signal-processing algorithms such as ECG synchronization and signal averaging.

13.3.2 Flight

13.3.2.1 Patient transport. In rural areas and in the military, helicopters rather than ambulances are used for patient transport. Patients are transported from the field to trauma centers as well as from smaller medical facilities to metropolitan hospitals (Short *et al* 1989). During flight, it is again important for the pulse oximeter to be lightweight and portable. However, since flights may take longer than ambulance rides, battery life is also an important consideration. The American Academy of Pediatrics in their 1986 air and ground transportation guideline stated that equipment battery life should be twice the expected travel time (Committee on Hospital Care). In helicopters, noise interferes a great deal with the ability to hear alarms. Therefore it is crucial that displays be readable. Visual indications of problems such as a lighted display which flashes when a patient's oxygen saturation falls below a particular level would be useful. Once again the stability of the signal is important. Rotary wing aircrafts create more vibration than either planes or ambulances (Campbell *et al* 1984).

13.3.2.2 Commercial flight regulations. Pulse oximetry monitoring during flight can also help to set commercial plane regulations. Modern planes can fly at very high altitudes and it is necessary to determine at what altitude cabin conditions become dangerous for both passengers and crew. Currently federal regulations require aircraft to maintain an equivalent cabin altitude of 2438 m or less. As altitude increases, barometric pressure decreases, and partial pressure of oxygen decreases as well. Recall that partial pressure of oxygen is related to oxygen saturation by the dissociation curve. Increasing altitude can cause hypoxia. Exposure to mild hypoxia during air travel is not generally a problem for a healthy person, though altitudes over 2438 m can cause impaired night vision (Ernest and Krill 1971) and color discrimination (Kobrick 1970). Even slight hypoxia which could affect the cognitive and decision-making skills of the crew could be dangerous. Also, passengers on board form a mixed population, some of whom could suffer from heart or lung disease. Even slight desaturation could put them at risk (Cottrell *et al* 1995).

13.3.2.3 High performance aircraft. Pilots flying high performance military aircraft, such as fighter pilots are often affected by both low partial pressure of oxygen and G-loading. *G-forces* are the forces of acceleration acting on the pilot. The pilot can lose consciousness if the partial pressure of oxygen is low and G-forces become too great. Monitoring the oxygen saturation in the head and pulse rate of the pilot during flight can determine if the pilot is in danger of losing consciousness. Once this determination has been made, control of the aircraft can be directed to an automatic pilot system and the aircraft unloaded (slowed down or taken out of a sharp turn or dive). One of the problems with monitoring a pilot during flight is that many of the methods are invasive or require equipment which can hinder the pilot's movement or ability to fly. For example, a finger probe in this situation would not be possible.

Tripp (1993) patented a design, modifying a Nellcor R-15 pulse oximeter probe such that the LEDs and photodiode are mounted on an ear plug as shown in figure 13.3. The LEDs and photodiode face outward such that light is reflected around the ear canal through the vascular tissue and detected by the photodiode. There are several advantages to his design. First, ear plugs are already worn by the pilots to protect them from the loud operating noise of the crafts. Second, placement in the ear canal reduces interference from ambient light. Third, the oxygen saturation monitored in the ear canal is closer to the oxygen saturation in the brain than the level measured at an extremity. Head movement was found not to affect the ability of the oximeter to obtain accurate measurements (Tripp 1993). The ear canal probes were constructed by drilling a 3 mm hole through the length of the plug and a second hole perpendicular to the first. The LEDs and photodiode could then be threaded into the channels and mounted on each side of the plug. Alternatively, a clay mold could be used with the LEDs and photodiode pressed into the clay on opposite sides. Silicone rubber is then poured into the mold and allowed to harden. The leads of the sensor are connected to a portable pulse oximeter. Further, the oximetry data can be input into a data bus and eventually into the aircraft computer system. In this way the aircraft can automatically unload if the S_pO_2 of the pilot falls below a specified level.

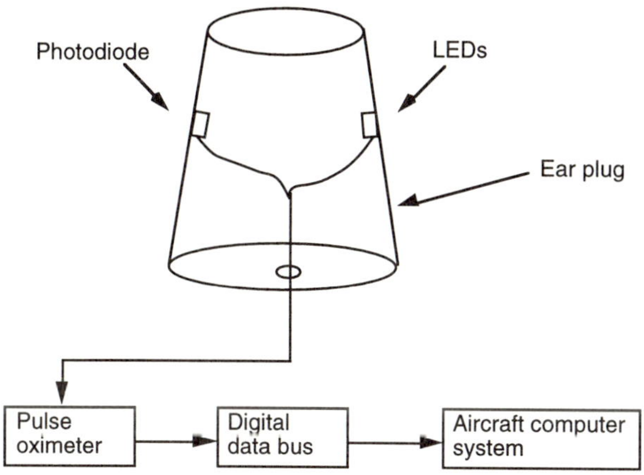

Figure 13.3 Modified pulse oximeter probe for use in the ear canal. In this version, a clay mold has been used to produce the ear plug with the LEDs and photodiode pressed into the sides (adapted from Tripp 1993).

13.4 CHILDBIRTH

Pulse oximetry is used to monitor arterial oxygen saturation of both the mother and the fetus during childbirth. Due to the inaccessibility of the fetus, special apparatus is needed for monitoring.

13.4.1 Causes of desaturation in mother and fetus

Many factors can cause desaturation and hypoxemia in a woman during labor and delivery: *hypovolemia* (diminished blood volume), hypertension (high blood pressure), anemia, maternal position, and anesthesia (Minnich *et al* 1988, Cunningham *et al* 1989). Pope and Hankins (1991) also found that desaturation frequently occurs during the administration of Demerol (a pain killing drug) and during vaginal examinations. Amniotic fluid *embolism* (AFE) can occur when amniotic fluid escapes into the mother's circulatory system. The embolism can cause the mother to develop a pulmonary shunt and thus experience arterial desaturation. If the embolism is not treated early the patient can suffer cardiorespiratory collapse, neurologic compromise, and *coagulopathy*, resulting in death (Quance 1988). A small amount of amniotic fluid can be found in the pulmonary circulation of pregnant and even nonpregnant women, which complicates the diagnosis of AFE (Clark *et al* 1986). Pulse oximetry monitoring during labor can help detect problems early (Quance 1988).

Fetal monitoring can indicate fetal distress and hypoxia. Chapter 1 noted that fetal hemoglobin has a higher affinity for oxygen than normal hemoglobin so the fetus's oxygen needs are met before those of the mother. This seems to indicate that if the mother's oxygen saturation is adequate, so is that of the fetus. However, maternal monitoring will not detect if oxygen being delivered by the mother is properly reaching the fetal blood stream. Pulse oximetry monitoring is crucial during difficult births such as breech presentation and cesarean section. These types of births put added stress on the fetus. Gardosi *et al* (1991) found that fetal oxygen saturation levels are generally lower in the breech presentation than in the vertex presentation. Fetal monitoring can also detect ăcidemia which results when a fetus experiences an increase in hydrogen ion concentration. Pulse oximeters can detect this problem because increasing pH causes the oxygen dissociation curve to shift to the right, resulting in low saturation levels. Fetal acidemia can result in acidotic and hyperoxemic infants. It is important to note that infants often experience mild hypoxemia due to the normal stress of labor (Kubli 1968). Johnson *et al* (1991) found that average S_pO_2 values of 68% ±13% occurred at cervical dilation of less than 5 cm and 58% ±17% at cervical dilation greater than or equal to 9 cm. Dildy *et al* (1994) determined even lower values of 62% ±9% and 53% ±10% respectively.

13.4.2 Special apparatus for fetal monitoring

Physicians have encountered many difficulties when attempting to monitor fetal S_aO_2 via pulse oximetry. The first problem is that the fetus is not very accessible. A device is needed to advance the probe into the uterus and position it properly on the fetus. Correct initial placement, however, does not necessarily lead to successful monitoring. During cervical dilation of early labor, the probe position can become unstable. Also, the fetal head is often covered with hair, vernix (a waxy, cheese-like substance), amniotic fluid, and maternal blood, all of which hinder the ability to obtain a stable and accurate signal. Hair not only attenuates the light from the LEDs, but also can create a shunt from the LEDs to the photodiode. During cesarean section, bleeding from the uterine incision can prevent signal detection (Johnson *et al* 1990). Other considerations when performing fetal pulse oximetry include the risk of burns to sensitive fetal skin and the risk of trauma to the fetus.

Several designs for fetal apparatus have been developed to overcome these monitoring difficulties. Two such patented designs follow.

Figure 13.4 shows a reflectance pulse oximeter probe (Chung and McNamara 1993). An abdominal examination is performed to define the position of the fetus and the state of the cervix. A cable, which is stiffer near the probe, is used to guide the probe into the correct position. The probe must be placed beyond the presenting part and the transcervical region (just beyond the cervix).

This is because cervical pressure on the presenting part creates local *edema* which lowers the pulse amplitude and makes signal detection more difficult. Also, the amplitude will vary due to cervical dilation. Figure 13.5 shows that the cable bends around the head of the fetus and conforms to the curve of the mother's pelvis. The cable contains calibration grooves and markings to aid physician placement. The probe is positioned on the temple of the fetus and therefore has less interference from hair. As labor progresses, the probe moves along with the fetus and calibration markings indicate the station of head. The Nellcor N-400 Fetal Oxygen Saturation Monitor uses this type of design. The system electronics have increased sensitivity to small signals to accommodate low amplitude fetal pulses. The probe can also detect if it becomes displaced. Within the probe body there are two small surface electrodes which measure skin impedance. If the impedance is too low, implying contact with amniotic fluid as opposed to fetal tissue, the system does not accept the data (Dildy *et al* 1993).

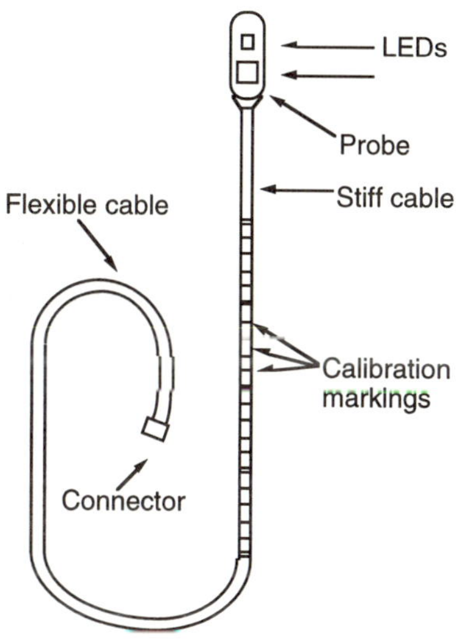

Figure 13.4 Apparatus for fetal pulse oximetry (adapted from Chung and McNamara 1993).

Figure 13.6 shows a design containing a light source located external to the mother. Light is transmitted to the fetus via an optical fiber (Joseph and Guzman 1995). This is advantageous for preventing burns due to high intensity LEDs. Wires from an external monitor and the optical fiber from the light source are threaded through a handle and a plastic tube. Figure 13.7 shows that at the end of the tube is a cylindrical base in which one monitor wire connects to a photodiode and the other connects to a reference electrode. A spiral probe containing the optical fiber extends from the base. By twisting the handle, the probe is inserted 1 to 2 mm into the scalp. The photodiode rests on top. Inserting the probe into the fetal scalp lessens interference from hair and increases the stability of the probe during labor.

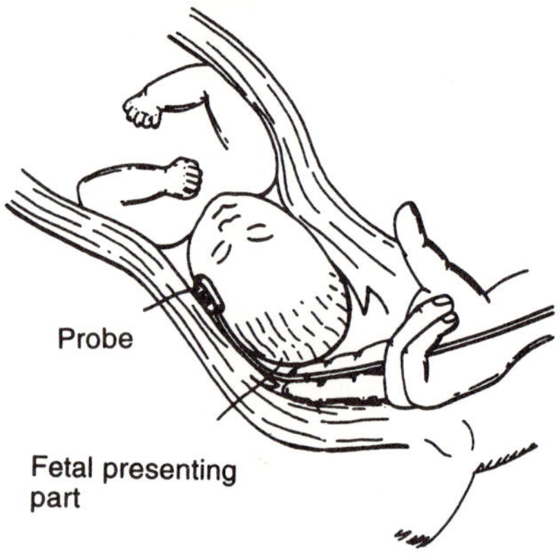

Figure 13.5 Placement of fetal probe within the uterus (Chung and McNamara 1993). The sensor rests on the infant's temple when the physician's fingers reach the saggital suture of the fetus's head.

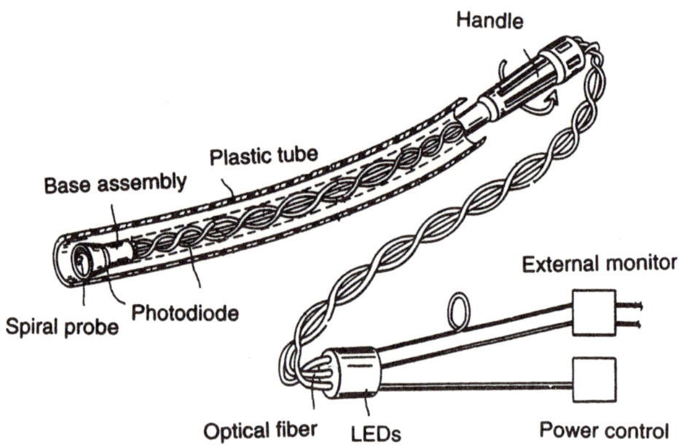

Figure 13.6 Fetal pulse oximetry apparatus with the LEDs located outside of the uterus and transmitted via optical fiber (Joseph and Guzman 1995).

13.5 NEONATAL AND PEDIATRIC CARE

A fetus generally has an S_aO_2 of about 50%. Within the first 15 min after birth, it normally rises to 90% (Oliver *et al* 1961). It is important to monitor the progress of this process and provide ventilatory aid if needed. Infants who experience problematic births are especially vulnerable. For example, infants delivered by cesarean section may be desaturated due to complications which made this type of delivery necessary. Premature infants sometimes develop *retinopathy* due to hyperoxia. High levels of retinal oxygen cause spasm of the developing vasculature, leading to *ischemia* and blindness (Moyle 1994). Pulse oximeters are often used by new parents in the home as a precaution to prevent sudden infant death syndrome.

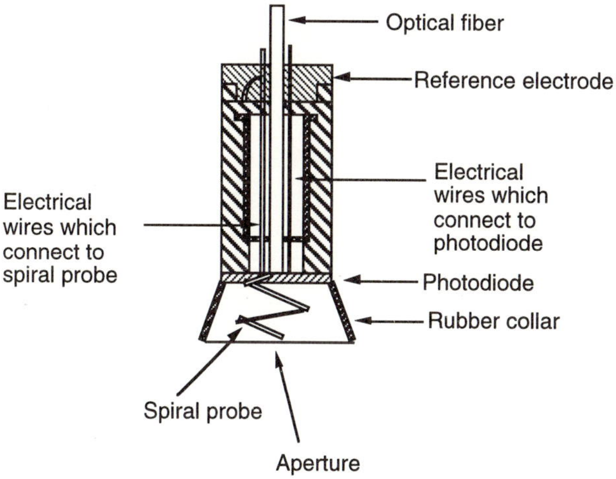

Figure 13.7 Close up cross sectional view of the sensor, showing the helical termination of the optical fiber which is inserted in the fetus's scalp (adapted from Joseph and Guzman 1995).

Determining alarm limits for pulse oximetry in neonatal care can be difficult. Figure 13.8 shows that during the weeks following birth, fetal hemoglobin is replaced by adult hemoglobin. Since the oxyhemoglobin dissociation curve of a fetus is to the left of that of the mother, the curve moves towards the right as the transition to adult hemoglobin takes place. This means that oxygen saturation levels considered safe may correspond to unsafe P_aO_2 levels and cause hypoxia. Paky and Koeck (1995) determined limits for detecting hypoxemia and hyperoxemia in neonates and found that limits to maintain an oxygen tension of 40 to 90 mmHg could only be established with less than 90% reliability. Attempting to obtain better reliability resulted in a S_pO_2 alarm limit for hypoxemia which was greater than that for hyperoxemia. This is obviously clinically unacceptable. However, with 85% reliability the range was only 92.5% to 95%. Deckardt and Steward (1984) determined that infant S_aO_2 levels between 80% and 95% are acceptable. Fanconi (1988) found detecting hypoxia in infants problematic due to inaccuracies in pulse oximeters at arterial oxygen saturations less than 65%.

Morozoff *et al* (1993) developed a system which uses a pulse oximeter as a controller to automatically adjust the air-oxygen mixture received by a neonate. The analog signal (plethysmographic waveform) measured by the pulse oximeter is input into a controller for a motorized gas blender. The blender adjusts the infant's inspired air-oxygen mixture, replacing the need for constant manual adjustment by an attending nurse. The benefits of this system are that it increases the amount of time the infant spends at normal S_aO_2 levels, reduces the need for human intervention, and reduces hospital costs by promoting early removal of oxygen therapy.

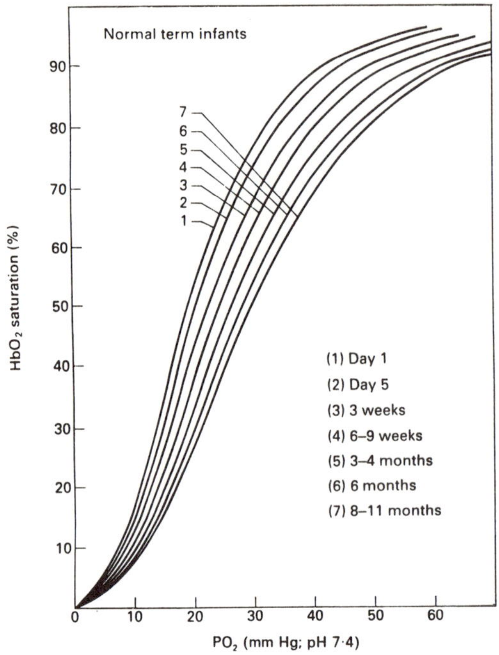

Figure 13.8 Mean oxyhemoglobin dissociation curves of infants ranging from 1 day old to 11 months. From Delivoria-Papadopoulous *et al* (1971).

The S_aO_2 controller operates according to the following algorithm. A patient's oxygen saturation is measured with a pulse oximeter. The signal is converted to a digital representation and low-pass filtered. The corner frequency of the filter is determined by the user and sets the sensitivity of the controller. The observed S_pO_2 minus the desired S_aO_2 is denoted as the error. The signs of the error's magnitude, velocity, and acceleration are input into a state machine. The state machine determines the trend of the S_aO_2 error. It analyzes the signs of the three inputs and determines if the neonate's S_aO_2 is on target, above the target, or below the target. If it is off target, the state machine goes on to determine if it is accelerating, decelerating, moving at a constant velocity, or not changing. If it is moving, it determines if the movement is toward or away from the target. Once the trend is identified by the state machine, it adjusts the F_iO_2 mixture relative to the current mixture. There is also a delay so that the system can react to the adjustment made. Alarms were added for mechanical or electrical failure as well as for S_aO_2 and F_iO_2 limits. Manual intervention can override the controller at all times.

Smaller probes are needed for both neonatal and pediatric care. Infants and children are much less willing to accept the application of a probe and remain still. Probe displacement and motion artifacts due to ill fitting probes can be a big problem. Ear probes made for adults can squeeze the softer newborn tissue too tightly. After a short time they can occlude the artery and have to be moved to regain a signal. Howell *et al* (1993) developed a modified probe design for children which uses a 5 ml syringe barrel cut in half to house the sensor. The probe is secured to the syringe and can be slipped onto the child's finger. Disposable probes with adhesive bandages are often the best for neonatal and pediatric application. The LEDs and photodiode are attached to the bandage with the proper spacing so that they are positioned correctly when the adhesive is wrapped around the infant or child's finger or toe. Meier-Stauss *et al* (1990) studied the use of pulse oximetry during the first 17 min of life and determined that signal detection occurs faster when a probe is applied to an infant's hand as opposed to its foot. They also found that saturation values from the hand were always higher than those from the foot. This observation suggests that pulse oximetry can be used to document right-to-left shunting in newborns during the first few minutes of life (Meier-Stauss *et al* 1990). This is the passage of blood from the right to the left side of the heart or from pulmonary circulation to systemic circulation.

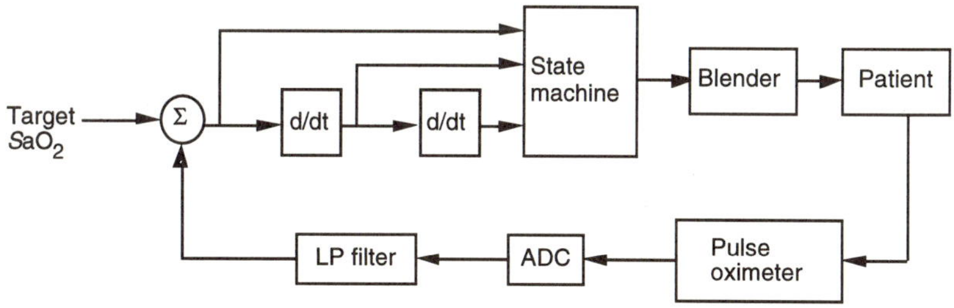

Figure 13.9 Block diagram of S_aO_2 controller. Adapted from Morozoff *et al* (1993).

13.6 SLEEP STUDIES AND PHYSICAL STRESS TESTING

Many people are able to maintain normal oxygen saturation levels while pursuing normal daily activities, but become desaturated during sleep or heavy exercise. The most common cause of desaturation during sleep is due to a disorder known as sleep apnea. Desaturation can occur during heavy exercise due to such things as poor ventilation or chronic obstructive pulmonary disease (COPD). The use of pulse oximetry during sleep and exercise aids in the diagnosis of these respiratory problems.

13.6.1 Sleep

Pulse oximetry monitoring is used during sleep to diagnose sleep disorders which cause desaturation. Sleep is composed of several stages with different characteristics. The first stage is when the person is still awake, but is drowsy and less in tune to stimuli. Two other stages which alternate throughout the night are REM (rapid eye movement) sleep and non-REM or quiet sleep. During REM sleep, rapid changes in metabolic rate do not seem to affect respiration. *Sleep apnea* is the most common sleep disorder which causes desaturation. It is defined as the cessation in breathing due to the relaxation of upper airway musculature. There are three types of sleep apnea: obstructive, central, and mixed. Obstructive sleep apnea is the most common type and is often caused by anatomical abnormalities such as a nasal obstruction, enlarged tonsils or adenoids, or an abnormal bone structure (Hauri 1992). Patients with obstructive sleep apnea often snore and are obese. They often experience bradycardia and cardiac arrhythmias and are at risk of sudden death during sleep. Central apneas are characterized by the absence of respiratory effort due to a neurological or cardiac problem. As described in chapter 1, respiratory muscles are controlled by neurons in the brainstem as well as chemoreceptors and mechanoreceptors. In patients with central sleep apnea, these neurons cease to provide control during sleep. As the muscles relax, the airway shrinks. The pressures associated with inhalation cause the airway to collapse and become completely closed off. Once breathing has stopped, the patient's oxygen saturation begins to fall. The lack of oxygen is soon detected by chemoreceptors which cause the patient to wake up, renewing control by neurons in the brainstem. The airway muscles become firm again and allow breathing to resume. However, once the patient falls asleep, the airway muscles will relax again. This cycle affects *hemodynamics*, autonomic tone, and arterial blood gas tensions (Davies and Stradling 1993).

Polysomnography is the standard for diagnosing sleep apnea. It measures and records the EEG, EMG, ECG, chest wall plethysmogram, airway flow, and arterial oxygen saturation. However, it is both expensive and of limited availability. Pulse oximetry is easy to use and widely available. Not all desaturation during sleep is indicative of sleep apnea. It could be due to hypopnea (abnormal, shallow breathing), artifact, hypoventilation, or ventilation/perfusion imbalance.

Siem *et al* (1995) used a pulse oximeter in conjunction with a polysomnograph and determined particular patterns of desaturation to be associated with sleep apnea. They divided desaturation patterns into three categories: periodic, cluster, and isolated. Periodic consisted of a minimum of four events with a fall in S_pO_2 of 2% or more with less than 2 min between events. A cluster consisted of 3 or more events with a fall in S_pO_2 of 3% or more and 2 to 10 min between events. Isolated events were separated from any other event by more than 10 min. They found that all periodic patterns were associated with sleep apnea, 65% of clusters were associated with sleep apnea, and none of the isolated events were associated with sleep apnea. Therefore, identifying patterns of desaturation with a pulse oximeter can help to identify sleep apnea.

Lynn (1995) patented a method and apparatus for specifically diagnosing moderate to severe sleep apnea using only a pulse oximeter (no polysomnograph). His method involved analyzing the slopes of the desaturation and resaturation events throughout the night, where an event was defined if the oxygen saturation fell below a specified level for a specified period of time. During an apneic event, the initial fall in arterial oxygen saturation is a function of the oxygen saturation of mixed venous blood and oxygen uptake from residual in the lungs. Then it continues to fall as a function of oxygen consumption and global oxygen stores. Oxygen stores exist first in the lungs, then arteries, tissue, and veins in that order. During apnea, oxygen depletion occurs first in the tissue, then the veins, lungs, and arteries. Therefore, desaturation of arterial blood occurs only after desaturation in other areas. The slope of the desaturation of an event must be within a certain range to be characteristic of sleep apnea. If the slope is too big (rapid desaturation) it is considered an artifact, and if the slope is to small (slow desaturation) it is considered to be due to either hypoventilation,

ventilation/perfusion imbalance, or an artifact. Lynn (1995) performed a study and found that specifically the descending slope as shown in figure 13.10 is a fall in S_pO_2 within the range of 1.1% per second and 0.3% per second. The mean was 0.8% per second. Once the desaturation is detected by the chemoreceptors, resulting in arousal, oxygen rushes into the lungs. The resaturation slope is much larger than the desaturation slope. Specifically, Lynn (1995) found that it is a rise in S_pO_2 in the range of 2.5% per second and 8.3% per second, the mean being 7.6%. The duration of an apneic event is 3 to 3.5 min.

Other parameters are considered in Lynn's diagnosis of sleep apnea. Consecutive events have similar desaturation slopes. Also, an event can increase the initial desaturation slope of a following event. This occurs because oxygen stores do not have enough time to replenish between events. The depletion of oxygen stores is not always detected by the pulse oximeter because arteries replenish their oxygen supply before tissue and veins. Desaturation slope increases occur when cyclic apneic events occur with less than 10 s between and when the depth of desaturation of the first event is larger than 15%. Thus the initial desaturation slope depends on the mixed venous saturation at the onset of sleep apnea and the amount of oxygen left in the lungs after the onset of sleep apnea. The continuing desaturation slope is a function of oxygen consumption versus stores.

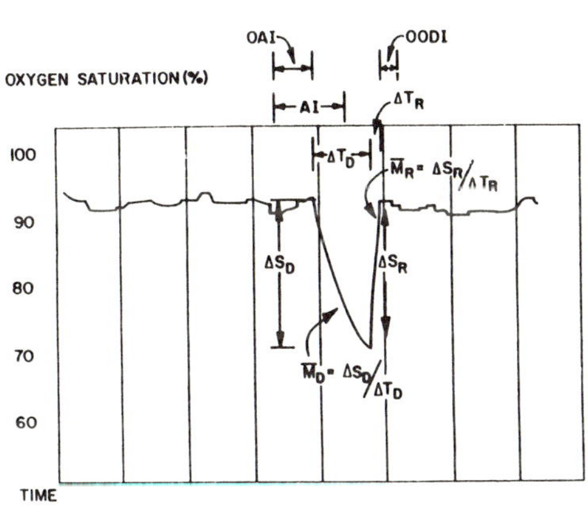

Figure 13.10 A typical apneic event. The vertical lines are 30 s apart. ΔS_D is the fall in saturation, ΔS_R is the rise in saturation, ΔT_D is the duration of the fall in saturation, ΔT_R is the duration of the rise in saturation, MD is the slope of the desaturation, MR is the slope of resaturation, AI is the apneic interval, OAI is the occult apneic interval (apnea has begun, but the arterial oxygen saturation is maintained via oxygen stores), and OODI is the occult apnea interval. This is the period following the return to baseline after a desaturation. If another apneic event occurs within this interval it will have an increased desaturation slope (Lynn 1995).

The operation of Lynn's method begins with measuring the patient's oxygen saturation with a pulse oximeter for a period of 10 min. A mean baseline measurement of arterial oxygen saturation is made during this interval. During subsequent recording, a desaturation event is defined and the duration and slope of the event is determined. The resaturation slope is then determined. Events in which the duration of the desaturation and resaturation is less than a particular value and the desaturation slope falls within a finite range are defined as phasic desaturations. The ratio of desaturation slope to resaturation slope of the phasic desaturation events is measured. From the above data, the number of apneic events which have occurred can be determined and marked. The apnea can then be treated and the diagnosis process can be repeated to confirm the success of the treatment. Figure 13.11 shows that the measurement of the slopes, computation of the ratios, and comparison of the parameters with known characteristics of sleep apnea is all done within a microprocessor. The microprocessor can be connected to a printer to obtain a hard copy of apneic event data for further analysis by a physician. A variation on the above method is to use the area under the desaturation slope and the area under the resaturation slope. A ratio of these areas can be used instead of the ratio of the slopes. When the microprocessor identifies a phasic desaturation event, it can trigger the collection and/or storage of another parameter such as sound or video. For example a microphone, either separate or as part of the probe can be used to record such things as snoring, which is characteristic of obstructive sleep apnea. Sound could be recorded throughout the night, but only stored during a suspected apneic event. In Lynn's design, sound would be stored for the duration of the event and 1 min prior to and following the event. Short, low-frequency sounds often occur prior to apnea, and high-frequency sounds due to hyperventilation often precede the recovery period.

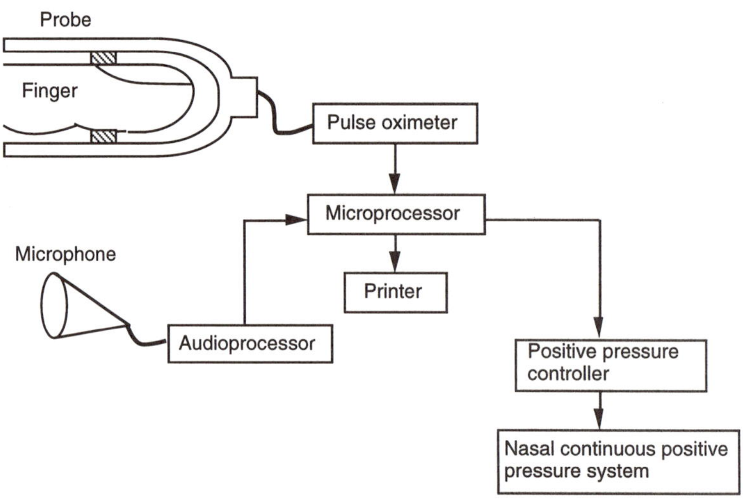

Figure 13.11 A block diagram of the apparatus used along with a pulse oximeter for sleep apnea diagnosis (adapted from Lynn 1995).

Many pulse oximeters such as the portable Protocol Propaq 106EL contain apnea delay alarms. The alarms go off when they detect more than one event of desaturation below a specified level within a specified amount of time. When patients experience recurring apnea events of 15 to 20 per hour, they are in danger and must undergo some kind of treatment. There are several methods of treating sleep apnea. A simple solution can sometimes be sleeping in a more upright position. Another way is by applying *continuous positive airway pressure*

(CPAP) via the nasal passages to support the airway and prevent the collapse of pharyngeal tissue when the muscles relax. This type of treatment requires wearing a mask while sleeping and the air flow can be uncomfortable for patients. Finally, in extreme cases uvulopalatopharyngoplasty (UPPP) or tracheostomy may be necessary. UPPP is a surgical procedure in which excess tissue or a bony abnormality is removed. A tracheostomy involves removing part of the trachea to make a new airway opening.

Other conditions which can cause desaturation during sleep although the patient maintains normal saturation levels while awake are bronchopulmonary dysplasia (BPD), chronic obstructive lung disease (COLD), cystic fibrosis, central alveolar hypoventilation syndrome (CAHS), hypopnea, airway resistance syndrome, and neuromuscular disease.

13.6.2 Exercise

Pulse oximetry can be used to evaluate pulmonary or circulatory dysfunction and performance limitations during exercise. During heavy exercise, a reduction in the partial pressure of oxygen can cause hypoxemia (Dempsey 1986). Miyachi and Tabata (1992) found that ventilation is also a major factor. Athletes tend not to desaturate as quickly as those who do not exercise as often. This is because trained athletes breathe less per unit of metabolic rate than the untrained. Monitoring the oxygen saturation of an athlete can thus determine his physical condition. Also, patients with COPD experience limited ventilation during exercise (Vas Fragoso *et al* 1993). If the condition is severe, ventilation is limited and thus the patient desaturates more quickly during exercise. Pulse oximeters tend to underestimate S_aO_2 readings during extreme exercise, possibly due to high levels of *catecholamines* and neural activity which restrict cutaneous blood flow (Norton *et al* 1992). Catecholamines are chemical compounds derived from catechol ($C_6H_6O_2$) which can affect nervous transmission and muscle tone.

13.7 MANAGEMENT OF CARDIOPULMONARY RESUSCITATION

Pulse oximeters were added as part of the emergency equipment carried by a British anesthetic resuscitation registrar to determine the effectiveness of pulse oximetry to aid in the management of CPR (Spittal 1993). The oximeters used were standard Nonin 8500 with Flex Sensor ear probes. The team found the pulse oximeter to be helpful in primary respiratory arrest, but not too useful in cardiac arrest. Its use during the cases with primary respiratory arrest helped determine if a tracheal tube was needed or if a tracheal tube already in use was not positioned properly. For example, in one patient the tube had been inadvertently placed in the esophagus. External cardiac massage often produces a distorted ECG and it is difficult to obtain reliable oxygen saturation readings. Seventeen patients the team worked on suffered from cardiac arrest and required chest compressions. During compressions, saturation readings were detected for only seven of the patients, and of the seven only three readings were thought to be reliable. The team felt that better fitting ear probes would have been useful because chest compressions cause the body to move and create motion artifacts. Also audible tones to indicate a satisfactory pulse signal and S_pO_2 level would have been useful because it is difficult to watch a display while administering CPR.

13.8 COMPUTER-CONTROLLED OXYGEN WEANING

Pulse oximetry can be used to monitor the weaning process of ventilated patients. During this process, the air/oxygen mixture is gradually adjusted, reducing the amount of oxygen until it matches that of room air. Often the amount of oxygen in the mixture has to be raised and lowered several times if the patient is not able to adjust. Strickland and Hasson (1993) developed a computer-controlled weaning system for patients with complex medical problems. The system was tested on elderly patients recovering from respiratory failure requiring ventilation. An external computer monitored the patient's oxygen saturation as measured with a pulse oximeter as well as the ventilator data. If the respiration rate, tidal volume, and oxygen saturation of the patient were normal, the computer decreased the rate of oxygen inhalation by 2 mL/kg every 2 h until a rate of 2 mL/kg was reached. If the measured values were not normal, it raised the ventilator support to the previous setting. Five minutes were allowed between measurements for the patients to stabilize. They found that the computer-controlled weaning reduced the need for blood gas sampling, shortened the weaning time, and reduced the time the patient spent with an unacceptable respiration rate and tidal volume, as compared with physician-controlled weaning.

13.9 SYSTOLIC BLOOD PRESSURE MEASUREMENT

A pulse oximeter will only obtain an oxygen saturation measurement and a plethysmographic waveform if pulsatile blood flow is detected. This characteristic was exploited by Chawla *et al* (1992) to develop a method to measure blood pressure using a pulse oximeter. An occlusive cuff and a sphygmomanometer are used along with a pulse oximeter. The cuff is occluded until the plethysmographic waveform disappears and the pressure is recorded. The disappearance of the waveform indicates that the artery has been occluded such that blood flow is too weak to be detected by the pulse oximeter. The cuff is then inflated rapidly to 200 mmHg and gradually deflated until the waveform reappears. This indicates that blood flow has increased to a detectable level in the artery. The pressure is again recorded. Specifically, the two pressure values correspond to 8.6% and 4% of the original blood flow respectively. Taking the average of the two pressure values results in a systolic blood pressure measurement which is at most 14 mmHg in error. This is within the clinically acceptable error range. This measurement technique is useful for patients with Takayasu's syndrome (pulseless disease) and critically ill patients with a weak pulse.

13.10 CEREBRAL OXYGEN SATURATION

Pulse oximetry on the *retinal fundus* allows measurement of cerebral oxygen saturation because blood supply to the retinal arteries comes from the *ophthalmic artery* which supplies cerebral tissue. Cerebral tissue is more vulnerable to permanent damage during hypoxemia. Also retinal circulation, unlike peripheral circulation is not affected during shock, hypothermia, and hemorrhage. Retinal oximetry is extremely useful in the critically ill who have weak peripheral circulation. Problems with retinal circulation and oxygen saturation are associated with diseases such as diabetic retinopathy, hypertension, sickle cell disease, and vascular occlusive diseases and can result in severe damage to retinal tissue (Delori 1988). De Kock *et al* (1993) designed special apparatus for retinal pulse oximetry monitoring. Figure 13.12 shows that a black Plexiglas cone was glued to a haptic contact lens. Holes were drilled to allow a vacuum environment and create suction. The suction kept the lens in place. Slight displacement from the center of the pupil would result in loss of a signal as the light would no longer hit the retinal arteries. An aluminum tube (8 mm in diameter and 1 mm thick) fitted inside the cone and was divided into two sections by a metallic screen. One section contained the LEDs and the other the photodiode. The LEDs and photodiode had been removed from a Nellcor finger probe. When the apparatus was tested on patients, the eye was put under local anesthesia and the pupil dilated to 6 mm. A pulsatile signal was obtained, but blinking and eye movement hindered the pulse oximeter readings.

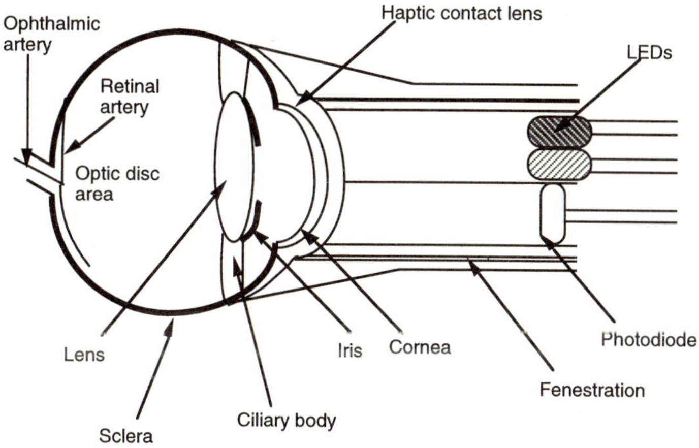

Figure 13.12. Cross section of an adapted haptic contact lens and pulse oximeter probe for use in cerebral oxygen saturation measurement. Adapted from de Kock *et al* (1993).

13.11 VETERINARY CARE

Pulse oximetry is often used in veterinary care. Pulse oximeters need to be able to monitor a wide range of heart rates to accommodate the metabolisms of different animals. Also, specialized probes are used. Common sites for probe placement include the tongue and ear for large animals, Achilles tendon, across the paw pads of dogs and cats, esophagus, nasal septum, rectum, and caudal tail. Limitations associated with the application of pulse oximetry to animals are low perfusion, motion artifacts, darkly pigmented skin, thick skin, and excessive hair (Allen 1990). Whitehair *et al* (1990) noted that oxygen saturation measurements were not obtained when a human ear probe was used on horses' nostrils, lips, and vulva. This could have been because the LEDs were not strong enough to allow sufficient light to be transmitted through the thick skin to the photodiode. Pulse oximeters are often used during anesthesia because the position of ruminant animals can cause bloating, which in turn can lead to compromised respiration, regurgitation, and death (Allen 1992). Detection of hypoxemia early can prevent the unnecessary demise of the animal. Sensor Devices Inc. and Palco both make pulse oximeters specially designed for veterinary use. The SDI Vet/Ox Plus and 4402 Pulse Oximeter can both measure pulse rates between 20 and 350 bpm with an accuracy of 2%. They also have widely variable gains to accommodate small or large pulse amplitudes.

13.12 FUTURE IMPROVEMENTS FOR PULSE OXIMETRY

Although pulse oximetry seems to be at the peak of its development, there are still improvements to be made. Many of these improvements relate to specific applications. Improvements which will increase the performance of pulse oximetry during transport are to lengthen the battery life in portable units, create even better algorithms for motion artifact reduction, and further miniaturize units. Reducing the occurrence of false alarms would be beneficial in all applications, but especially during long term monitoring when staff cannot always be in the room. In hospital environments for monitoring during surgery, recovery, and intensive care, all-in-one monitors seem to be the goal. HORNET (Hospital Operating Room Network) is a prototype for this type of monitoring (Lecky *et al* 1988). It is designed to monitor respiratory and circulatory variables such as ECG, blood pressure, oxygen saturation, and inspiratory and expiratory gas. It is also designed to handle physiological, demographic, and administrative data. It is to be used for scheduling, intraoperative monitoring, preparation of reports, permanent storage of perioperative information, and research. In addition to all-in-one monitors, the aim is to eventually equip hospitals to transmit information via radio waves to central stations.

REFERENCES

Allen J L 1990 *Proc. 1990 Am. Assoc. Zoo Veterinarians* 163

Allen J L 1992 Pulse oximetry: everyday uses in a zoological practice *Vet. Record* **130** 354–5

Bone M E, Galler D and Flynn P J 1987 Arterial oxygen desaturation during general anaesthesia for paediatric dental extractions *Anaesthesia* **42** 879–82

Buchanan D C 1991 Endotracheal tube with oximetry means *US patent 5,005,573*

Campbell M B, Lightstone A D, Smith J M, Kirpalani H and Perlman M 1984 Mechanical vibration sound levels experienced in neonatal transport *Am. J. Dis. Child.* **138** 967–70

Catley D M, Thorton C, Jordon C, Tech B, Lehane J R, Royston D and Jones J G 1985 Pronounced episodic oxygen desaturation in the post-operative period: its association with ventilatory pattern and analgesic regimen *Anesthesiology* **63** 20–8

Chawla R, Kumarvel V, Girdhar K K, Sethi A K, Indrayan A and Bhattacharya A 1992 Can pulse oximetry be used to measure systolic blood pressure? *Anesth. Analg.* **74** 196–200

Chung C and McNamara H M 1993 Fetal pulse oximetry apparatus and method of use *US patent 5,228,440*

Clark S L, Pavlov Z, Greenspoon J, Horenstein J and Phelan J P 1986 Squamous cells in the maternal pulmonary circulation *Am. J. Obstet. Gynecol.* **154** 104–6

Clay N R and Dent C M 1991 Limitations of pulse oximetry to assess limb vascularity *Br. J. Bone Joint Surg.* **71-B** 141

Committee on Hospital Care—American Academy of Pediatrics 1986 Guidelines for air and ground transportation of pediatric patients *Pediatrics* **78**: 943–50

Cooper J B et al 1984 An analysis of anesthetic mishaps from medical liability claims *Int. Anesthesia Clinics* **60**: 39

Cottrell J J, Lebovitz B L, Fennel R G and Kohn G M 1995 Inflight arterial saturation: continuous monitoring by pulse oximetry *Aviation, Space, Environ. Med.* **66** 126–30

Cunningham F G, MacDonald P C and Gant N F 1989 *Williams Obstetrics* 18th edn (Norwalk, CT: Appleton-Century-Crofts) pp 95, 96, 805

David H G 1991 Pulse oximetry in closed limb fractures *Ann. R. Coll. Surg. England* **73** 283–4

Davies R J O and Stradling J R 1993 Acute effects of obstructive sleep apnoea *Br. J. Anaesth.* **71** 725–9

de Kock J P, Tarassenko L, Glynn C J and Hill A R 1993 Reflectance pulse oximetry measurements from the retinal fundus *IEEE Trans. Biomed. Eng.* **40** 817–22

Deckardt R and Steward D J 1984 Non-invasive arterial hemoglobin oxygen saturation versus transcutaneous oxygen tension monitoring in the preterm infant *Crit. Care Med.* **12** 935–9

Delivoria-Papadopoulous M, Roncevic N and Oski F A 1971 Postnatal changes in oxygen transport of term, preterm, and sick infants: the role of red cell 2,3-diphosphoglycerate and adult haemoglobin *Pediatric Res.* **5** 235–45

Delori F C 1988 Noninvasive technique for oximetry of blood in retinal vessels *Appl. Opt.* **27** 1113–25

Dempsey J A 1986 Is the lung built for exercise? *Med. Sci. Sports Exercise* **18** 143–55

Dildy G A, Clark S L and Louks C A 1993 Preliminary experience with intrapartum fetal pulse oximetry in humans *Obstet. Gynecol.* **81** 630–4

Dildy G A, Clark S L and Louks C A 1994 Intrapartum fetal pulse oximetry: the effects of maternal hyperoxia on fetal arterial saturation *Am. J. Obstet. Gynecol.* **171** 1120–4

Ernest J T and Krill A E 1971 The effect of hypoxia on visual function *Invest. Ophthalmol.* **10** 323–8

Fanconi S 1988 Reliability of pulse oximetry in hypoxic infants *J. Pediatr.* **112** 424–7

Gardosi J O, Schram C M and Symonds E M 1991 Adaptation of pulse oximetry for fetal monitoring during labour *Lancet* **337** (8752) 1265–7

Hauri P J 1992 *The Sleep Disorders* (Kalamazoo, MI: Upjohn Company)

Haynes S R, Allsop J R and Gillies G W A 1992 Arterial oxygen saturation during induction of anaesthesia and laryngeal mask insertion: prospective evaluation of four techniques *Br. J. Anaesthesia* **68** 519–22

Howell S J, Blogg C E and Ashby M W 1993 A modified sensor for pulse oximetry in children *Anaesthesia* **48** 1083–5

Johnson N, Johnson V A, Bannister J, Lyons G, Lilford R J, Griffiths-Jones M, Tuffnell D and Onwude J L 1990 Monitoring the fetus with a pulse oximeter during caesarean section *Br. J. Obstet. Gynaecol.* **97** 653–8

Johnson N, Johnson V A, Fisher J, Jobbings B, Bannister J and Lilford R J 1991 Fetal monitoring with pulse oximetry *Br. J. Obstet. Gynaecol.* **98** 36–41

Joseph B M and Guzman F A 1995 Internal apparatus for continuous electrical and oximetric intrapartum monitoring of the fetus *US patent 5,419,322*

Knudsen J 1970 Duration of hypoxemia after uncomplicated upper abdominal and thoracoabdominal operations *Anaesthesia* **25** 372–7

Kobrick J L 1970 Effects of hypoxia and acetazolamide on color sensitivity zones in the visual field *J. Appl. Physiol.* **28** 741–7

Kubli F W 1968 Influence of labor on fetal acid-base balance *Clin. Obstet. Gynecol.* **11** 168–91

Lanigan C J 1992 Oxygen desaturation after dental anaesthesia *Br. J. Anaesthesia* **68** 142–5

Lecky J H, Matsiras P V, Garfinkel D, Aukburg S J and Carson E R 1988 PONI: A prototype respiratory and circulatory monitoring system for operating rooms *Proc. Ann. Int. Conf. IEEE Eng. Med. Biol. Soc.* **10** 1406

Lindsey L A, Watson J D D and Quaba A A 1991 Pulse oximetry in postoperative monitoring of free muscle flaps *Br. J. Plastic Surg.* **44** 27–9

Lynn L A 1995 Method and apparatus for the diagnosis of sleep apnea utilizing a single interface with a human body part *US patent 5,398,682*

Macdonald P H, Dinda P K, Beck I T and Mercer C D 1993. The use of pulse oximetry in determining intestinal blood flow *Surgery* **176** 451–8

Meier-Stauss P, Bucher H U, Hurlimann R, Konig V and Huch R 1990 Pulse oximetry used for documenting oxygen saturation and right-to-left shunting immediately after birth *Eur. J. Pediatr.* **149** 851–5

Minnich M E, Clark R B, Miller F C, et al 1988 Pulse oximetry during labor and delivery *Perinatol. Neonatol.* **12** 24

Miyachi M and Tabata I 1992 Relationship between arterial oxygen desaturation and ventilation during maximal exercise *J. Appl. Physiol.* **73** 2588–91

Moller J T, Johannessen N W, Berg H, Esperson K and Larsen L E 1991 Hypoxemia during anaesthesia: an observer study *Br. J. Anaesth.* **66** 437–44

Morozoff P E, Evans R W and Smyth J A 1993 Automatic control of blood oxygen saturation in premature infants. *Proc. 2nd IEEE Conf. on Control Applications* Vancouver, BC pp 415–9

Moyle J T B 1994 *Pulse Oximetry* (London: BMJ)
Norton L H, Squires B, Craig N P, McLeay G, McGrath P and Norton K I 1992 Accuracy of pulse oximetry during exercise stress testing *Int. J. Sports Med.* **13** 523–7
Oliver T K Jr, Demis J A and Bates G D 1961 Serial blood gas tensions and acid-base balance during the first hour of life in human infants *ACTA Paediatr.* **50** 364–360
Paky F and C M Koeck 1995 Pulse oximetry in ventilated preterm newborns: reliability of detection of hyperoxaemia and hypoxaemia, and feasibility of alarm settings *ACTA Paediatr.* **84** 613–6
Payne J P and Severinghaus J W (eds) 1986 *Pulse Oximetry* (New York: Springer)
Pope C L L and Hankins D V 1991 Pulse oximetry: application in the labor-and-delivery unit of a tertiary care center *J. Reproductive Med.* **36** 853–6
Quance D 1988 Amniotic fluid embolism: detection by pulse oximetry *Anesthesiology* **68** 951–2
Roberts C J, Parke T J and Sykes M K 1993 Effect of intraoperative inspired gas mixtures on postoperative nocturnal oxygen saturation *Br. J. Anaesth.* **71** 476–80
Schmitt J M, Webber R L and Walker E C 1991 Pulse oximeter for diagnosis of dental pulp pathology *US patent 5,040,539*
Short L, Hecker R B, Middaugh R E and Menk E J 1989 A comparison of pulse oximeters during helicopter flights *J. Emer. Med.* **7** 639–43
Siem K, Pennock B E, Koliner C M and Kaplan P D 1995 Can pulse oximetry identify episodes of sleep apnea? *Sleep Res.* **24** 497
Spittal M J 1993 Evaluation of pulse oximetry during cardiopulmonary resuscitation *Anaesthesia* **48** 701–3
Strandberg A, Tokics L, Brismar B, Lundqvist H and Hedenstierna G 1986 Atelectasis during anaesthesia and in the postoperative period *ACTA Anaesthesiol. Scand.* **30** 154–8
Strickland J H and J H Hasson 1993 A computer controlled ventilator weaning system *Chest* **103** 1220–6
Tripp L D 1993 Ear canal pulse/oxygen saturation measurement device *US patent 5,213,099*
Tyler I L, Tantisira B, Winter P M and Motoyama E K 1985 Continuous monitoring of arterial saturation with pulse oximetry during transfer to the recovery room *Anesth. Analg.* **64** 1108–12
Vas Fragoso C A, Clark T and Kotch A 1993 The tidal volume response to incremental exercise in COPD *Chest* **103** 1438–41
Whitehair K J, Watney G C, Leith D E and Debowes R M 1990 Pulse oximetry in horses *Vet. Surg.* **19** 243–8

INSTRUCTIONAL OBJECTIVES

13.1 Describe possible causes of desaturation during induction to anesthesia.
13.2 Describe applications of pulse oximetry to determine organ/tissue viability.
13.3 Discuss the special problems encountered when using pulse oximetry in moving environments such as ambulances and helicopters.
13.4 Describe the affects of altitude and G-forces on oxygen saturation levels.
13.5 Explain why it is important to monitor oxygen saturation levels of both the mother and the fetus during labor and delivery.
13.6 Explain difficulties of fetal pulse oximetry monitoring and describe apparatus which overcome these difficulties.
13.7 Explain the need for pulse oximetry monitoring in neonatal care and discuss problems with obtaining alarm limits.
13.8 Explain why monitoring oxygen saturation during heavy exercise is useful for diagnosing pulmonary and circulatory dysfunction and discuss the problems with obtaining accurate measurements.
13.9 Describe a method for diagnosing sleep apnea using pulse oximetry monitoring.
13.10 Explain why pulse oximetry on the retinal fundus is useful and describe how it is administered.
13.11 Describe how systolic blood pressure can be measured using pulse oximetry.
13.12 Discuss the special difficulties of pulse oximetry monitoring in the field of veterinary care.
13.13 Describe future goals for improvement of pulse oximetry monitoring.

GLOSSARY

95% confidence limit: 1.96 times the standard deviation; a 4% confidence limit means that 95% of the S_pO_2 readings should be within 4% when compared to the readings from the CO-oximeter

absorbance: negative logarithm of the transmittance in a light absorbing medium; a measured value equal to the product of the extinction coefficient, optical path length and concentration of the light absorbers

absorption spectrum: extinction coefficients of a certain absorber versus wavelength

absorptivity: see extinction coefficient

accuracy: a statistical term used to represent the correctness of data

acid-base imbalance: abnormal pH of the blood

ADC: analog to digital converter

address/data bus: the bus is the link or path between one unit and the other on the processor board; address information required for data retrieval or storage is passed via the address bus; buses can be unidirectional or bidirectional; data information is passed on data buses

airway resistance: resistance to air flowing through passageways of the lungs; work used to overcome airway resistance during inhalation and expiration is lost as heat

alveolar ducts: respiratory tract between the alveoli and the bronchioles

alveolar sacs: a group of alveoli clustered together

alveolus: small sacs at the end of the respiratory tract where gas exchange occurs

anoxia: total lack of oxygen, e.g., cardiac arrest

apical foramen: small holes in the roots of the tooth; they provide an opening for blood vessels to reach dental pulp

ARDS: adult respiratory distress syndrome

arterial oxygen saturation: oxygen saturation of arterial blood, which delivers oxygen to the tissue; can be either functional or fractional oxygen saturation of the arterial; usually measured in percent

arterial pulsation: pressure and volume change in the arteries and arterioles due to pump function of the heart

arterialization: when venous blood near the skin is brought to nearly arterial oxygen levels by some external influence such as heating

arteriovenous anastomoses: a thick-walled blood vessel that connects an arteriole directly with a venule, thus bypassing the capillaries

artificial finger: a man-made device that simulates the absorbance properties of a human finger.

asthma: sudden dyspnea with wheezing caused by spasms of the bronchial or swelling of its mucous membrane

atelectic areas: shrunken, airless portions of the lung

ATP (adenosine triphosphate): a compound in cells composed of adenine, ribose, and three phosphate groups found in cells; the phosphate bonds store energy needed by the cell

atrium: either of two upper muscular chambers of the heart

baseline component: the signal which doesn't vary with time

beam angle: the angular measure of radiated power measured on an axis from half-power point to half-power point

Beer's law: describes the exponential light attenuation in an absorbing medium

BF equipment: body floating; equipment with parts in direct contact with the patient are isolated

bias: the fixed DC voltage applied between the base and emitter of a transistor, to keep the device on; the average of the differences between the pulse oximeter readings and the CO-oximeter readings

bilirubin: the orange or yellow compounds which are the breakdown products of hemoglobin

blue dye 'patent blue': a commonly used dye to distinguish the area of arterial perfusion
body surface area: the surface area of a body which can be estimated from charts related to height and weight
bronchioles: respiratory tract between the alveolar duct and the bronchus
bronchitis: inflammation of bronchial tube mucus membrane
bronchus: one of two respiratory tracts between the trachea and the bronchioles, which provide a pathway into the lungs
calibrating resistor: the resistance associated with the probe; this is useful in disposable probes, as this is helpful in determining the wavelengths associated with that probe, so the suitable calibration curves may be used
calibration: the process of fine tuning the device to measure accurately; due to constant use or change in one of the measuring parameters it is necessary to retune the device
cardiac index: the cardiac output normalized by the body surface area of an individual
cardiac output: the rate of blood flow through the heart
cardiopulmonary bypass: a method to maintain the circulation of the body while the heart is deliberately stopped during heart surgery
cardiopulmonary resuscitation: providing assistance in order to restore respiration and cardiac contraction
catecholamines: amino compounds derived from catechol ($C_6H_6O_2$); they have sympathomimetic activity and affect nervous transmission and muscular tone
Central Processing Unit (CPU): the heart of any digital control system; this unit is used to generate control signals for the various operating systems on the processor board
CF equipment: cardiac floating; equipment suitable to be used in direct cardiac applications
chemoreceptors: function in sensing the chemical concentrations of CO_2, O_2, and H^+ of the blood
circulatory system: sends blood around the body to deliver oxygen and transport waste products
coagulopathy: a disease which affects blood clotting
comparator: a device used to determine whether two numbers or bits of information are equal
compliance: set of rules and standards that apply to a specific product; see lung compliance
control bus: information relating to control of various units on the processor board are passed on the control bus; these buses originate at the CPU and are bidirectional
CPAP: continuous positive airway pressure; applying air pressure through the nasal cavities to create a pneumatic splint, keeping the airway from collapsing
current noise (I_n): the random variation of input bias current in an op amp; it produces noise when it reacts with the feedback resistance of the transimpedance amplifier
cuvette: in spectrophotometers, the container that holds the blood sample; it is designed so that it affects the transmission of light as little as possible
cyanosis: bluish discoloration of skin and lips due to severe hypoxemia and excessive reduced hemoglobin
DCT: discrete Fourier transform
decoder: a digital device used to decode the input signal into a set of output signals
defibrillation: application of high energy pulses to the heart when the heart loses synchronization of the heart muscle fibers; the fibers contract irregularly, usually at a rapid rate
delay: response time of a pulse oximeter
demodulate: used to demultiplex the R and IR signals from the output of the photodiode
dentin: hard tissue surrounding dental pulp; it forms most of the tooth and is covered by enamel
desaturation: a process which lowers the oxygen level in the blood
diastole: a rhythmically recurrent expansion especially the dilation of the cavities of the heart during which they fill with blood
diffusion: transport of particles across membranes
dissociation: the process of a molecule being separated into ions, atoms, molecules or free radicals
drift: baseline wandering due to the changing characteristics of the components
driven right leg circuit: during ECG measurements the electrical signals are given a separate ground path thus maintaining patient safety
duty cycle: the ratio of the on time to the total operation time; it is the fraction of the time the device remains on; for a Nellcor system, this is usually 25%; it is 33% for an Ohmeda system
dysfunctional hemoglobins: do not support oxygen transport to the tissue
dyshemoglobins: see dysfunctional hemoglobins
ECG: electrocardiogram
edema: swelling of tissue
eigenvalue spread: ratio of largest eigenvalue to smallest eigenvalue

eigenvalue: a scalar λ which satisfies $\mathbf{A}v = \lambda v$ where $\mathbf{A}$ is a square matrix and v are associated eigenvectors
electrocautery: destroying tissue by electrical heating of a wire
electroluminescence: the emission of light by electrons falling from the higher-energy conduction band to the lower-energy valence band; the electrons emit light energy in the form of photons of light
embolism: obstruction of a blood vessel by a clot or foreign object
EMC: electromagnetic compatibility
EMI: electromagnetic interference
emission spectrum: the frequency response of an LED, displayed on a wavelength scale
emphysema: accumulation of air in tissue; usually refers to destruction of the walls of the respiratory bronchioles
Erasable Programmable Read Only Memory (EPROM): ROMs that can be written into only through specialized means; as ROMs are read only devices, we cannot reprogram them through conventional ways; ultraviolet rays are used to erase the locations, and new data can be burned into them
error: the difference between the pulse oximeter reading and the CO-oximeter reading
ESD: electrostatic discharge
extinction coefficient: numeric measure of opaqueness; the greater the value, the greater the opaqueness
feedback: the technique of returning to a machine or system part of its output so that the machine or system exercises self-correction or control of the process
FFT: fast Fourier transform
fibrosis: formation of scar tissue in the connective tissue of the lungs; results from pneumonia or similar type of infection
filter: a filter is basically a voltage dividing network arranged to possess frequency-discriminating properties
finger phantom: see artificial finger
F_iO_2: fraction of inspired oxygen (normal atmospheric fraction is 0.21)
flip flop: a storage device which can be used to retain one bit of information
fractional oxygen saturation: ratio of oxygenated hemoglobin over total hemoglobin; usually measured in percent
fractional S_aO_2: see fractional oxygen saturation and arterial oxygen saturation
functional hemoglobins: capable of carrying oxygen molecules (Hb and HbO_2)
functional oxygen saturation: ratio of oxygenated hemoglobin over functional hemoglobin; usually measured in percent
functional S_aO_2: see functional oxygen saturation and arterial oxygen saturation
G forces: forces created by large accelerations
Hb: see reduced hemoglobin
HbO_2: see oxygenated hemoglobin
hemodynamics: relating to blood circulation
hemoglobin: a molecule in red blood cells for transport of oxygen molecules
hemolyzed: describes blood in which the red blood cells have been destroyed and the hemoglobins released into the plasma; this process is commonly done on blood for *in vitro* oximetry measurements to reduce the effects of scattering
heparinized blood: blood that has been treated to prevent clotting
hypercapnia: excess carbon dioxide in the blood
hyperoxia: excess oxygen in the system due to high P_iO_2
hypotension: condition in which the arterial blood pressure is abnormally low
hypothermia: reduction of body temperature below the normal range in the absence of protective reflex actions, such as shivering; sometimes body temperature is lowered for therapeutic purposes such as during surgery to reduce the patient's requirement for O_2
hypovolemia: diminished blood volume
hypoxemia: deficient oxygenation of blood
hypoxia: deficient oxygenation of tissue
hypoxic hypoxemia: hypoxemia caused by a drop in oxygen tension as a consequence of decreased lung function
illumination: the respective luminous or radiant flux density incident on a photodetector
***in vitro*:** outside the body
***in vivo*:** within the body
instrumentation amplifier: a very high precision differential amplifier
interrupt: an event which invokes higher priority algorithms to attend to emergencies
interstitial fluid: fluid in between the cells, other than blood cells

intubation: placing a ventilation tube into trachea of the patient to assist mechanical ventilation
ischemia: lack of blood in an area of the body due to blood vessel constriction or mechanical obstruction
isosbestic point: wavelength at which the extinction coefficients of oxyhemoglobin and reduced hemoglobin are equal (805 nm)
Kreb's cycle: a series of chemical reactions during which molecules are oxidized and energy is released
larynx: structure located between the pharynx and trachea which contains the vocal cords
latch: a digital unit used to latch data/address
linear extrapolation: predict values beyond measured values
linear regression: a straight line fit through the data points
linearly independent: a set of vectors $\{x_i\}_{i=1}^{n}$ is linearly independent if $\sum_{i=1}^{n} a_i \boldsymbol{x}_i = \mathbf{0}$ is true only if scalar $a_i = 0$ for all i
lookup table: used by hardware instead of an equation to determine the oxygen saturation by the ratio of absorbances based on empirical data
lung compliance: a measure of the elasticity of the lungs; work used to overcome compliance during inhalation is restored during expiration
mechanoreceptors: sensory receptors that are stimulated by mechanical changes such as pressure
medulla oblongata: portion of the brain stem which connects the pons and the spinal cord
memory: the storage element in a digital system; it can be in the form of read only memory or random access memory
monochromatic light: light consisting of only one wavelength
motion artifact: errors introduced into the signal due to motion
MRI: magnetic resonance imaging
multiple scattering: the effect when scattering occurs more than once
myoglobin: a respiratory pigment found in the muscles which store oxygen
myxoma: benign gelatinous tumor of connective tissue
necrosis: death of tissue
odontoblasts: special cells within the dental pulp which form the dentin in the tooth
oxidation: a chemical reaction by which the molecule or atom loses an electron
oximeter: an instrument that uses optical measurements to determine the oxygen saturation of the blood
oxyhemoglobin: hemoglobin combined with an oxygen molecule which will be released freely to tissue
***p–i–n* photodiode:** a p–n junction photodiode that has a large intrinsic layer providing lower capacitance and faster response than the conventional photodiode
P_aO_2: partial pressure of oxygen dissolved in arterial blood
partial pressure: the pressure of one gas in a mixture of gases
patient isolation: to avoid the lack of ohmic continuity or physical separation; this can be provided by a transformer
pattern generator: a unit used to generate timing patterns used for synchronous detection gating, LED control, synchronizing the power supply, calibration patterns, and diagnostic timing
peak wavelength: the wavelength at which the radiated power (or light output) of an LED is a maximum
perfusion: the passage of a fluid through the vessels of an organ
pH: negative log of the concentration of hydrogen ion (H^+) concentration relative to a standard solution; a pH of 7 is neutral, below 7 is acidic and above 7 is alkaline
pharynx: respiratory tract that connects the nasal cavity to the larynx
photocell: a device whose resistance changes as a function of light intensity
photoconductors: see photocell
photodetector: a generic term for any device which is able to convert an optical signal input to an electrical signal output
photodiode: a *p–n* junction diode which converts incident light to an electrical signal; in pulse oximeters, this optical sensor is located in the probe and is configured to produce a current linearly proportional to incident light
photoplethysmograph: a plethysmograph that uses a photodetector
photoplethysmographic signal: time varying signal of transmitted light intensity in living tissue due to arterial pulsation
photoresistors: see photocell
P_iO_2: partial pressure of inspired oxygen

plethysmograph: an instrument that detects variations in size of a part due to blood contained in the part
plethysmographic signal: time varying signal in living tissue due to arterial pulsation
pneumonia: inflammation of the lungs that can occur from a variety of sources
polarization filter: an optical filter that only transmits light that is in that state of polarization; used in pairs to vary optical transmission
polypeptide: compound composed of amino acids molecules
polysomnography: monitors EEG, EMG, ECG, chest wall plethysmogram, airway flow, and oxygen saturation; it is a gold standard for sleep apnea diagnosis
precision: a measure of variation of random error or degree of reproducibility; it is usually represented by the standard deviation of the differences between the pulse oximeter and CO-oximeter readings
programmable gain amplifier: amplifiers whose gains can be adjusted and varied depending on the circuit requirement (for example, due to change in ambient light level, the dc offset can vary, thereby increasing the risk of sending the amplifier into saturation)
pulmonary: related to the lungs
pulsatile component: the signal which varies with time
pulse capability: the maximum allowable pulse current of an LED as a function of duty cycle and frequency
pulse oximeter: an oximeter that takes advantage of the pulsatile nature of the blood in the arteries
QRS: peak of the ECG waveform which corresponds to ventricular depolarization
R wave: the peak in the QRS complex of the heart beat
Random Access Memory (RAM): a memory location capable of being read from and written into. This unit is used to store and retrieve data in a digital system
Read Only Memory (ROM): a memory location capable of only being written into; this used to store calibration related information
rectifier: a device that offers a much higher resistance to current in one direction than the other; rectifiers are used to obtain a unidirectional current (DC) from an alternating current
reduced hemoglobin: functional hemoglobin unbound to oxygen
reduction: a chemical reaction by which the molecule or atom gains an electron
respiration: the process of gas exchange
respiratory quotient: ratio of volume of CO_2 produced per volume of O_2 consumed
retinal fundus: posterior portion of the interior of the eye
retinopathy of prematurity: disorder in the retina of neonates due to supplemental oxygen among many other factors
RFI: radio frequency interference
sample-and-hold: a circuit used to hold a value for a given period of time; this is very useful during analog-to-digital conversion; it consists of a high gain FET and a large capacitor
S_aO_2: see arterial oxygen saturation
scattering: light is refracted by a small object causing a deviation of the light beam from its initial direction of propagation
sensitivity: the ratio of the electrical output signal to the intensity of incident light
signal-to-noise ratio: indicates the quality of the signal
simulator: a device that behaves in a like manner to the original device
sleep apnea: cessation of breathing during sleep for episodes of 15 s or greater; there are three types: obstructive, central, and mixed
spectral bandwidth: the half-power bandwidth of the light emitted from an LED, measured in nanometers
spectral response: the relationship of output signal of a photodetector to the incident light at a particular wavelength
spectrophotometry: the process of measuring the absorbance of light at different wavelengths to determine the concentration of the substance in a solution
S_pO_2: arterial oxygen saturation as measured by the pulse oximeter; usually measured in percent
spurious pulses: erroneous pulses introduced from disturbing sources
stemmed trigger: an emitter coupled multivibrator which is used as a voltage discriminator; this is also useful as a squaring circuit as it can be used to convert the sine waves into square waves
switching time: the time required for an LED to switch from its ON state to its OFF state, or vice versa
synchronization: correlation of specific events to improve accuracy
synchronous detector: the circuit component used to synchronously demultiplex the signal from the photodiode into its R and IR components
systemic: related to the body in its entirety

systole: the period of contraction of the heart, especially that of the ventricles

thermal resistance: causes the increase in junction temperature above ambient per unit of power dissipation for the given LED's package and mounting configuration

thoracic cavity: the body cavity between the neck and the diaphragm

thresholds: signal levels established to make appropriate decisions in algorithms

transcutaneous: (transdermal) through the skin, e.g., administration of medicine via a patch

transimpedance amplifier: an amplifier used in pulse oximetry to convert the current produced by the photodiode to a voltage for further processing in the system

transmittance: ratio of transmitted light to incident light intensity in an absorbing medium; the greater the value, the less light is passing through the medium

trauma: wound or injury

UART: (universal asynchronous receiver–transmitter) a device which can be programmed to do asynchronous communication

UPPP: uvulopatatopharyngoplasty; surgical procedure to remove excess tissue in the upper airway

vasoconstriction: a decrease in the diameter of blood vessels

vasodilation: an increase in the diameter of blood vessels which results in an increase in blood flow

ventilation: passage of air into and out of the respiratory tract

ventricle: either of two lower muscular chambers of the heart

wait state generator: a combination of one clock storage devices, used to store data for one clock cycle; it is useful to slow down the microprocessor when the I/O devices are communicating at a very slow rate

watchdog timer: a fail safe timer used to turn off the pulse oximeter, if the microprocessor system fails

INDEX